APPLIED ALGEBRA
CODES, CIPHERS, AND DISCRETE ALGORITHMS
SECOND EDITION

D0168944

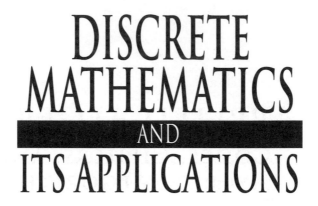

DISCRETE MATHEMATICS
AND
ITS APPLICATIONS

Series Editor
Kenneth H. Rosen, Ph.D.

Continued Titles

DISCRETE MATHEMATICS AND ITS APPLICATIONS

Series Editor KENNETH H. ROSEN

APPLIED ALGEBRA

CODES, CIPHERS, AND DISCRETE ALGORITHMS

SECOND EDITION

DAREL W. HARDY
COLORADO STATE UNIVERSITY
FORT COLLINS, U.S.A.

FRED RICHMAN
FLORIDA ATLANTIC UNIVERSITY
BOCA RATON, U.S.A.

CAROL L. WALKER
NEW MEXICO STATE UNIVERSITY
LAS CRUCES, U.S.A.

CRC Press
Taylor & Francis Group
Boca Raton London New York

CRC Press is an imprint of the
Taylor & Francis Group, an **informa** business

A CHAPMAN & HALL BOOK

Chapman & Hall/CRC
Taylor & Francis Group
6000 Broken Sound Parkway NW, Suite 300
Boca Raton, FL 33487-2742

© 2009 by Taylor & Francis Group, LLC
Chapman & Hall/CRC is an imprint of Taylor & Francis Group, an Informa business

Library of Congress Cataloging-in-Publication Data

Hardy, Darel W.
    Applied algebra : codes, ciphers, and discrete algorithms / Darel W. Hardy, Carol L. Walker. -- 2nd ed. / Fred Richman.
        p. cm. -- (Discrete mathematics, its applications)
    Includes bibliographical references and index.
    ISBN 978-1-4200-7142-9 (hardcover : alk. paper)
    1. Coding theory. 2. Computer security--Mathematics. I. Walker, Carol L. II. Richman, Fred. III. Title. IV. Series.

QA268.H365 2009
003'.54--dc22   2009000533

Visit the Taylor & Francis Web site at
http://www.taylorandfrancis.com

and the CRC Press Web site at
http://www.crcpress.com

Contents

Preface

Applied Algebra: Codes, Ciphers, and Discrete Algorithms, Second Edition deals with the mathematics of data communication and storage. It includes hints for using *Scientific Notebook*®, *Maple*®, or *MuPAD*® to do complicated calculations and to make the mathematical ideas more accessible. Two central topics are data security (how to make data visible only to friendly eyes) and data integrity (how to minimize data corruption).

Cryptography is the study of data security: How can a bank be sure that a message to transfer $1,000,000 was sent by an authorized person? Or imagine a political crisis in a remote region of the world. It is vital that sensitive issues be discussed with government leaders back home. The crisis could get out of control if these discussions were intercepted and read by some third party.

The messages are bounced off of satellites so the signals can be captured by anyone with a simple satellite dish. How can the messages be transformed so that a third party cannot read them, yet they can easily be read by friends back home?

Issues of data integrity are handled by error-control codes. The first pictures transmitted from the back side of the Moon in the late 1960s were in black and white, and of poor quality. Lost data caused vertical black streaks in the pictures. The loss of data was due to interference from solar radiation. More recent pictures from much greater distances using the Voyager series of planetary probes were beautiful, high-resolution color images with no apparent lost data. This was mostly the result of software that detects and corrects errors caused by interference.

This book discusses mathematically interesting methods for solving these problems—methods that are practical and widely used. The material was designed for a course in applied algebra for juniors and seniors majoring in mathematics and computer science.

The primary mathematical tools come from number theory and the theory of finite fields. All mathematics that will be used is developed as needed, but students who have had a prior course in abstract algebra or linear algebra have found such background to be useful.

Supercomputers perform billions of operations per second, and must store and retrieve vast amounts of data. The probability of a single read/write error is small, but doing billions of read/writes can make the probability of at least one error relatively large. Many computer codes (such as those required to do modern cryptography) will not tolerate even a single error. These fast computers must be designed so that errors—even multiple errors—can be recognized and corrected before causing trouble.

These examples reflect advances in hardware, but mostly advances in mathematics. Desktop computers can detect single errors and larger computers can correct multiple errors. The error-correction capabilities of the Voyager project resulted in thousands of flawless pictures being sent back to Earth to be analyzed.

Improvements in computer hardware since the 1950s have been incredible. In pushing technology to its limits, we are restricted by the physical size of atomic particles and the speed of light. Scientists are now considering the use of clean rooms in orbit to eliminate the few stray particles that contaminate Earth-bound labs.

In spite of these dramatic changes, the increase in speed due to improvements in mathematical algorithms has been even more spectacular. For many problems, the net effect since 1950 on computing speed due to improved algorithms has been greater than that due to improved hardware. (We will see an example of a problem in cryptography that would take more than 10^{10} years on the fastest theoretical computer that we could imagine using naive methods, but is computable in a few nanoseconds on a PC using more sophisticated algorithms.) This trend is likely to continue, because mathematics itself recognizes no physical bounds.

We will look at several algorithms that arise in the study of cryptography and error-control codes. Many of these algorithms feature common-sense approaches to relatively simple problems such as computing large powers. Other algorithms are based on interesting mathematical ideas.

Those who become hooked on applied algebra will eventually need to learn abstract algebra, and lots of it. This book attempts to show the power of algebra in a relatively simple setting. Instead of a general study of finite groups, we consider only finite groups of permutations. Just enough of the theory of finite fields is developed to allow us to construct the fields used for error-control codes and for the new Advanced Encryption Standard. Almost everything we do will be with integers, or polynomials over the integers, or remainders modulo an integer or a polynomial. Once in a while we look at rational numbers.

A floating-point number is different from a rational number or a real number. Each floating-point number corresponds to infinitely many rational numbers (and to infinitely many irrational numbers). Computer algebra systems such as *Maple* or *MuPAD* deal primarily with integers and rational numbers—not floating-point numbers.

Numerical analysis packages such as *MATLAB*® and *IMSL* use floating-point arithmetic. They trade precision for speed. Computer algebra systems are generally much slower than numerical analysis routines using floating-point arithmetic. When high precision is important—and it is essential for many problems in algebra—we have to go with computer algebra systems.

Interactive Version Using *Scientific Notebook*®

This book includes an interactive version, on CD-Rom, of *Applied Algebra: Codes, Ciphers, and Discrete Algorithms, Second Edition* and the software *Scientific Notebook*, a mathematical word processor and easy-to-use computer algebra system. This software is used as the browser for reading the interactive version of the book, and provides the text editor and computing engine for interactive examples and self-tests.

The interactive version contains all of the material from the print version. In addition, the interactive version

- Adds links that make it easy to find topics and navigate page-by-page, chapter-by-chapter, or by keywords

- Adds interactive examples

- Adds computing hints

- Adds self tests

We believe you will find it convenient to have the interactive version of *Applied Algebra: Codes, Ciphers, and Discrete Algorithms, Second Edition* and the software *Scientific Notebook* installed on your computer. After your license for *Scientific Notebook* expires, you can still use it as a browser for reading the book. Only the interactive features—such as the interactive examples and self tests—will be lost.

Computing hints are provided for using *Scientific Notebook*, *Maple*, and *MuPAD* in order to understand better the ideas developed in this book. By now, all of us tend to use a calculator for routine numerical calculations—even for balancing a checkbook. Want to compute $\sum_{i=0}^{\infty} r^i$? Need to find $543! + 2^{100}$? How about the first 37 terms of the Taylor series for $f(x) = x \sin x$ expanded about $x = \pi/4$? These are all child's play using *Scientific Notebook* (an interface to *MuPAD*) or using a computer algebra system such as *Maple* or *MuPAD* directly. With these systems you can concentrate on the mathematics and not be distracted by the computations.

Computer algebra packages (*Axiom, Derive, MuPAD, Maple, Mathematica, Reduce*, etc.) are becoming tools of the trade. In the future you might well need to know how to use such a package. You may even have such a package already installed on your own personal computer. These packages have many limitations, and it is important that you have a good idea of what they will and will not do. By reading this book and experimenting with the computer algebra hints, you should acquire a good feel for the capabilities and limitations of these packages.

You will find this software useful for your other courses as well. Entering text and mathematics in *Scientific Notebook* is so straightforward there is practically no learning curve. And, with the built-in computer algebra system, you can use the intuitive interface to solve equations right in your documents without having to master a complex syntax.

With *Scientific Notebook*, you can compute symbolically or numerically, integrate, differentiate, and solve algebraic and differential equations. You can also create 2D and 3D plots in many styles and coordinate systems, and animate the plots. *Scientific Notebook* provides a ready laboratory in which you can experiment with mathematics to develop new insights and solve interesting problems, as well as a vehicle for producing clear, well-written homework.

Acknowledgments

Our thanks go to the students who enrolled in the course Information Integrity and Security at Colorado State University. Their questions and insights led to many improvements in the original manuscript. They also wrote much of the computer code that appears on the websites. We would also like to thank our acquiring editor Bob Stern, who convinced us to sign with CRC/Taylor Francis and gave us several helpful suggestions, and our production coordinator Marsha Pronin, editorial assistant Samantha K. White, cover designer Kevin Craig, and project editor Michele A. Dimont, who skillfully led us through the task of converting our manuscript into a printed text. We thank Shashi Kumar of International Typesetting and Composition for solving technical problems with the manuscript, and David Walker, whose sharp eyes helped us create a clean manuscript. We thank the Scientific WorkPlace® team, whose product helps make technical writing fun, with special thanks to George Pearson, a TEXspert who assisted us with the final production of this manuscript.

Darel W. Hardy
Fort Collins, Colorado

Fred Richman
Boca Raton, Florida

Carol L. Walker
Las Cruces, New Mexico

Chapter 1

Integers and Computer Algebra

Number theory is the study of the integers: $\ldots, -3, -2, -1, 0, 1, 2, 3, \ldots$. Number theorists investigate how the integers behave under addition and multiplication. Often they deal with just the nonnegative integers, $0, 1, 2, 3, \ldots$, or with the positive integers $1, 2, 3, 4, \ldots$.

> *The theory of numbers is especially entitled to a separate history on account of the great interest which has been taken in it continuously through the centuries from the time of Pythagoras, an interest shared on the one extreme by nearly every noted mathematician and on the other extreme by numerous amateurs attracted by no other part of mathematics.*
>
> Leonard Eugene Dickson

1.1 Integers

As simple as the integers may seem, many mathematicians have devoted their lives to studying them. Problems in number theory are often easy to state but difficult to solve.

In the early 1600's, Pierre de Fermat said that if n is an integer greater than 2, then the equation $x^n + y^n = z^n$ has no solution in positive integers x, y, and z. He gave no proof. Hundreds of mathematicians, both amateur and professional, tried to prove or disprove this statement. In 1995, some 350 years after Fermat made the claim, Andrew Wiles[1] of Princeton University gave

---

[1]You can find information about Wiles, Fermat, and other mathematicians from *The MacTutor History of Mathematics* archive site at http://www-groups.dcs.st-andrews.ac.uk/~history/.

a complicated proof in his paper, "Modular elliptic curves and Fermat's Last Theorem," in the *Annals of Mathematics*.

Definition 1.1 *We say that an integer n **divides** an integer m, and write n|m, if there is an integer a such that na = m. We also say that m is a **multiple** of n or that n is a **divisor** of m.*

*An integer p > 1 is a **prime** if its only positive divisors are 1 and p. The first five primes are 2, 3, 5, 7, and 11.*

*An integer n > 1 that is not a prime is called a **composite**. The first five composites are 4, 6, 8, 9, and 10.*

Don't confuse the symbol $n|m$, which means that n divides m, with the fraction n/m. This is especially easy to do when you write them by hand!

Many problems in number theory deal with divisors and primes.

Problem 1.2 *Determine whether a given large number is prime or composite.*

This turns out to be relatively easy to do. We will show how you can do this on a small computer for numbers with hundreds of digits.

Problem 1.3 *Find the prime divisors of a given composite number.*

This seems to be hard. Oddly enough, we can recognize that a number is composite without being able to find any of its factors. Many of today's cryptographic systems rely for their security on this inability to factor large numbers.

Here are a few elementary properties of divisors.

Theorem 1.4 *Let a, b, c, x, and y be integers.*

    i. If a|b and b|a, then a = ±b.

   ii. If a|b and b|c, then a|c.

  iii. If c|a and c|b, then c|(ax + by).

One problem in this section is to prove this theorem. (See problem 6.)

Problems 1.1

1. Show that if $n > 0$ is composite, then n has a divisor d with $1 < d^2 \leq n$.

2. Show that 101 is prime by showing that 101 has no prime divisors d such that $1 < d^2 \leq 101$.

3. Let $a = 15$ and $b = 24$. Find integers x and y such that $ax + by$ divides both a and b.

4. Find the prime power factorization of 10!.

5. Find the prime power factorization of $2^9 + 5^{12}$.

6. Prove Theorem 1.4.

7. For each of the following claims about arbitrary integers a, b, c, and d, either show that it is true or show that it is false.

   (a) If $a|b$ and $b|c$, then $ab|c$.

   (b) If $a|b$ and $a|c$, then $a|(b+c)$.

   (c) If $a|b$ and $a|c$, then $b|c$.

   (d) If $a|b$, then $a^2|b^2$.

   (e) If $a|b$ and $c|d$, then $(a+c)|(b+d)$.

   (f) If $a|b$ and $c|d$, then $ac|bd$.

   (g) If $a|b$ and $a|c$, then $a|bc$.

8. Is it true that if a number ends in 2, like 10132, then it must be divisible by 2? Why or why not? Prove that the product of two consecutive integers is divisible by 2.

9. Is it true that if a number ends in 3, then it must be divisible by 3? Why or why not?

10. For which digits d is it true that if a number ends in d, then it must be divisible by d?

11. Prove that there are infinitely many primes by showing that if $p_1, p_2, \ldots, p_k$ are primes, then the integer $p_1 p_2 \cdots p_k + 1$ must have a prime factor distinct from each prime $p_1, p_2, \ldots, p_k$.

12. Prove that if n is odd, then $n^2 - 1$ is divisible by 8.

13. Show that every even number between 4 and 100 is the sum of two primes.

14. List all the prime numbers between 60 and 120.

15. Identify each of the following as prime or composite, and factor the composites into primes.
    a. $2! + 1$ b. $3! + 1$ c. $4! + 1$ d. $5! + 1$
    e. $6! + 1$ f. $7! + 1$ g. $8! + 1$ h. $9! + 1$

16. Why are 2 and 3 the only consecutive numbers that are both prime?

17. Why are 3, 5, and 7 the only three consecutive odd numbers that are prime?

18. Is $n^2 + n + 17$ a prime for all $n > 1$?

19. Can $n^2 + 1$ be a prime if n is odd? What if n is even?

20. If $2^n + 1$ is prime, then must n be prime?

21. If $2^n - 1$ is prime, then must n be prime?

22. If there are least four composites between two consecutive primes, then there are at least five composites between these two primes. Why?

1.2 Computer Algebra vs. Numerical Analysis

Numerical analysis includes the study of **round-off** and **truncation** errors when using floating-point arithmetic. A **floating-point number** is a number written in the form $\pm m \times 10^e$ where m is a decimal to a fixed number of digits of a number between 1 and 10, and e is an integer in some fixed range. Examples are

$$4.683940958 \times 10^{22}$$

and

$$-2.435623410 \times 10^{-17}.$$

Here we have used ten digit numbers for the number m, which is called the **mantissa**. Ranges for e might be something like $-37 < e < 38$ or $-200 < e < 200$. Addition and multiplication of floating-point numbers are very fast on computers that have special hardware, a *floating-point accelerator*, to do floating point operations.

Sums and products of floating-point numbers are not exact because of the fixed number of digits in m. Every time you add or multiply, you introduce a small error. One goal of numerical analysis is to determine the accuracy of the final answer.

Example 1.5 In computing the product of $4.683940958 \times 10^{22}$ and $7.948735673 \times 10^{13}$ on a computer that supports a 10-digit mantissa, the exact answer is

$$3.7231408583080394734 \times 10^{36}$$

but the mantissa would have to be rounded to ten digits, so that the result would be

$$3.723140858 \times 10^{36}$$

Floating-point numbers cannot represent all integers and rational numbers exactly. The integer 12345678901 cannot be represented exactly by a floating-point number with a ten-digit mantissa. The rational number $1/3$ cannot be represented exactly by any floating-point number. Computer-algebra systems represent integers and rationals exactly, and computer-algebra evaluations yield exact results:

$$738475937594759 \times 5838593589383 = 4311660\,8751543\overline{6}0261498843697$$

$$\frac{75837594375}{385793759} + \frac{2783479}{374853795738548} = \frac{28\,428\,010\,112\,222\,955\,601\,975\,061}{144\,616\,254\,933\,392\,614\,121\,932}$$

$$
\begin{aligned}
100! = \ & 93\,326\,215\,443\,944\,152\,681\,699\,238 \\
& 856\,266\,700\,490\,715\,968\,264\,381\,621 \\
& 468\,592\,963\,895\,217\,599\,993\,229\,915 \\
& 608\,941\,463\,976\,156\,518\,286\,253\,697 \\
& 920\,827\,223\,758\,251\,185\,210\,916\,864 \\
& 000\,000\,000\,000\,000\,000\,000\,000
\end{aligned}
$$

Some irrational numbers, like $\sqrt{2}$, are represented exactly in computer algebra systems. A typical computer algebra system will compute $\left(\sqrt{2}\right)^2$ as 2. Floating-point evaluations yield approximate results.

$$738475937594759 \times 5838593589383 = 4.\,311\,660\,875 \times 10^{27}$$
$$\frac{75837594375}{385793759} + \frac{2783479}{374853795738548} = 196.\,575\,482\,6$$
$$100! = 9.\,332\,621\,544 \times 10^{157}$$

Exact arithmetic is usually slower than floating-point arithmetic. Why do we need exact arithmetic? In the real world, we can rarely measure anything exactly.

In fact, exact arithmetic with very large integers is used in ATM machines, credit card transactions, and in cryptography. We will see several examples of how large integer arithmetic can be used to make internet transactions secure.

Problems 1.2

1. The two numbers 3.14 and 22/7 both claim to be the best approximation to π. Which is the better approximation, and why?

2. List the numbers $\sqrt{10}$, π, and 3.16 in increasing order. Justify your answer.

3. Use a computer algebra system to find the floating-point representation of π with a ten-digit mantissa.

4. Find at least two more numbers with the same floating-point representation as computed in Problem 3.

5. Define $x^{a/b} = \sqrt[b]{x^a}$. Explain why $\frac{1}{2}$ is not always the same as $\frac{2}{4}$.

6. Evaluate the following using floating-point arithmetic with a ten-digit mantissa.

   (a) $\frac{1}{3} + \frac{2}{3}$

   (b) $10 + 1.1 \times 10^{-10}$

7. Show that
$$\frac{1}{3} = 0.333333333333\ldots = 0.\bar{3}$$
   where the overbar indicates that 3 repeats forever.

8. Show that
$$\frac{1}{2} = 0.4999999999999\ldots = 0.4\bar{9}$$

9. Rewrite the number $x = 3.4\overline{89}$ as a rational number.

10. Find an exact repeating decimal representation for $1/61$.

11. The rational number a/b evaluates numerically to $0.469\,387\,755$. If a and b are both two-digit integers, what are they?

1.3   Sums and Products

We have a compact notation for the sum of a list of numbers:
$$\sum_{i=1}^{n} a_i = a_1 + a_2 + \cdots + a_n$$

The letter i on the left-hand side of this equation is called an **index**. It could be replaced by any other symbol without changing the meaning of the sum:
$$\sum_{i=1}^{n} a_i = \sum_{j=1}^{n} a_j$$

Notation 1.6 (Summation) *If n and m are integers such that $n \leq m$, then*
$$\sum_{i=n}^{m} a_i = a_n + a_{n+1} + \cdots + a_m$$

For example,
$$\sum_{i=-3}^{2} (5 + i) = (5 - 3) + (5 - 2) + (5 - 1) + (5 - 0) + (5 + 1) + (5 + 2) = 27$$

and
$$\sum_{k=2}^{4} k^2 = 2^2 + 3^2 + 4^2 = 29$$

Theorem 1.7 *The following equations hold for the summation notation*·

   i. $\sum_{i=n}^{m} ka_i = k \sum_{i=n}^{m} a_i$

   ii. $\sum_{i=n}^{m} (a_i + b_i) = \sum_{i=n}^{m} a_i + \sum_{i=n}^{m} b_i$

   iii. $\sum_{i=n}^{m} \sum_{j=r}^{s} a_i b_j = \left(\sum_{i=n}^{m} a_i\right) \left(\sum_{j=r}^{s} b_j\right)$

Proof. The first formula is the distributive law:

$$\sum_{i=n}^{m} ka_i = ka_n + ka_{n+1} + \cdots + ka_m$$

$$= k\left(a_n + a_{n+1} + \cdots + a_m\right)$$

$$= k \sum_{i=n}^{m} a_i$$

The other two formulas are left as problems. ∎

   There is a compact notation for products just like for sums.

Notation 1.8 (Product) *If n and m are integers such that $n \leq m$, then*

$$\prod_{i=n}^{m} a_i = a_n a_{n+1} \cdots a_m$$

Example 1.9 $\prod_{j=1}^{4} j = 24$ $\prod_{j=1}^{10} j^2 = 13\,168\,189\,440\,000$

               $\prod_{k=1}^{10} k = 3628\,800$ $10! = 3628\,800$

Problems 1.3

Use a calculator or computer algebra system to evaluate the sums and products in Problems 1–10. Justify as many as possible by hand.

1. $\sum_{j=1}^{10} 2$
               2. $\sum_{j=-8}^{5} j(j+1)$

3. $\sum_{i=0}^{10} \sum_{j=0}^{10} (i+1)(j+1)$
      4. $\sum_{k=1}^{n} \frac{1}{k(k+1)}$

5. $\sum_{k=2}^{n} k^3$
               6. $3 + 3 \cdot 5^2 + 3 \cdot 5^4 + \cdots + 3 \cdot 5^{100}$

7. $\displaystyle\sum_{k=n}^{m} 1$ 8. $\displaystyle\sum_{k=1}^{n} k$

9. $\displaystyle\sum_{k=0}^{5} \left(1 + k^3\right)$ 10. $\displaystyle 3\sum_{k=1}^{m} \left(5k^2 - 3k + 4\right)$

For problems 11–14, write the answer as a product of powers of primes. Then use a calculator or computer algebra system to evaluate the product.

11. $\displaystyle\prod_{i=1}^{10} i^2$ 12. $\displaystyle\prod_{i=1}^{10} i$

13. $\displaystyle\prod_{n=1}^{5} \left(n^2 + n\right)$ 14. $\displaystyle\prod_{n=1}^{10} \left(n^2 + n\right)$

15. Verify Part ii of Theorem 1.7.

16. Verify Part iii of Theorem 1.7.

1.4   Mathematical Induction

Equations and inequalities that hold for each positive integer n are discovered in many different ways. The most common way to verify that they are correct is **mathematical induction**.

Principle of Mathematical Induction: Let $P(n)$ be a statement that depends on the positive integer n. If $P(1)$ is true, and if $P(k + 1)$ is true whenever $P(k)$ is true, then $P(n)$ is true for each positive integer n.

Mathematical induction is like climbing a (very tall) ladder. If you can get started (stand on the first rung), and if you can always climb up one additional step, then you can climb as high as you like.

Mathematical induction is sometimes pictured as knocking down a string of dominos. The first domino falls, and each falling domino knocks down the next domino, so all the dominos fall (see Figure 1.1).

We will give several examples to show how to use mathematical induction.

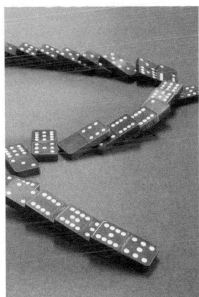

Figure 1.1   Falling dominos

Example 1.10 We will show by mathematical induction that the equation

$$1 + 2 + 3 + \cdots + n = \frac{n(n+1)}{2}$$

holds for every positive integer n. We use the compact notation $\sum_{i=1}^{n} i$ instead of $1 + 2 + 3 + \cdots + n$ so our lines won't get too long.

For $n = 1$, the equation is $1 = 2/2$, which is true. For the induction step, suppose that the equation is true for $n = k$, that is

$$\sum_{i=1}^{k} i = \frac{k(k+1)}{2}$$

Then

$$\sum_{i=1}^{k+1} i = \left(\sum_{i=1}^{k} i\right) + k + 1 = \frac{1}{2}k(k+1) + k + 1$$

$$= \left(\frac{1}{2}k + 1\right)(k+1) = \frac{k+2}{2}(k+1)$$

$$= \frac{(k+1)((k+1)+1)}{2}$$

so the equation is true for $n = k + 1$. Thus, by induction, the equation

$$\sum_{i=1}^{n} i = \frac{n(n+1)}{2}$$

is true for every positive integer n.

The right-hand side, $n\left(n+1\right)/2$, of the above equation is in **closed-form**: it is a formula that involves a finite number of standard operations. An expression using three dots (an **ellipsis**) to indicate missing terms is not in closed-form, nor is an expression using the summation symbol Σ.

Example 1.11 We will show by induction that $n < 2^n$ for each positive integer n. The statement is true for $n = 1$ because $1 < 2 = 2^1$. For the induction step, suppose that the statement is true for $n = k$, that is, $k < 2^k$. Then

$$k + 1 < 2^k + 1 < 2^k + 2^k = 2 \cdot 2^k = 2^{k+1}$$

so $k + 1 < 2^{k+1}$, that is, the statement is true for $n = k + 1$.

By induction, $n < 2^n$ for every positive integer n.

Notation 1.12 (Binomial coefficient) *The symbol $\binom{n}{k}$ is the **binomial coefficient**. It is the number of ways you can choose k things out of n things.*

The name *binomial coefficient* derives from the fact that it is the coefficient of $x^{n-k}y^k$ when you expand the n-th power of the binomial $x + y$

$$\left(x + y\right)^n = \left(x + y\right)\left(x + y\right)\cdots\left(x + y\right).$$

That's because if you take the y from any k of those n factors on the right, and the x from the remaining $n - k$ factors, you get a term of the form $x^{n-k}y^k$.

Notation 1.13 (Factorial) *The symbol $k!$ stands for the product $1 \cdot 2 \cdot 3 \cdots \cdot k$. We set $0!$ equal to 1. The symbol $k!$ is read k **factorial**.*

We will show on the next page that

$$\binom{n}{k} = \frac{n\left(n-1\right)\left(n-2\right)\cdots\left(n-k+1\right)}{k!}$$

for k a nonnegative integer. Note that the right-hand side is 0 if $k > n \geq 0$ because 0 will be a factor of the numerator. That corresponds to the fact that there is no way to choose more than n things out of n things.

Because of the formula displayed above, binomial coefficients can be computed by the following algorithm (see Algorithm 1.1).

Algorithm 1.1   Binomial coefficient algorithm

Input: n, k
Output: $\binom{n}{k}$
Function Binomial(n, k)
    Set $t = n$
    Set $p = 1$
    For b **from** 1 **to** k **do**
        Set $p = pt/b$
        Set $t = t - 1$
    End For
    Binomial $= p$
End Function

Proposition 1.14 *The binomial coefficients satisfy the identity*

$$\binom{n+1}{k} = \binom{n}{k} + \binom{n}{k-1}$$

for $n \geq 0$ and $k \geq 1$.

Proof. Call a subset of size k a k-subset. Each k-subset of $\{1, 2, 3, \ldots, n, n+1\}$ either includes the number $n + 1$ or does not include the number $n + 1$. We will count all the k-subsets by counting those two kinds of k-subsets separately. That will give us the two terms on the right-hand side of the equation.

The k-subsets of $\{1, 2, 3, \ldots, n, n+1\}$ that exclude $n + 1$ are just the k-subsets of $\{1, 2, 3, \ldots, n\}$, so there are $\binom{n}{k}$ of them. A k-subset of $\{1, 2, 3, \ldots, n, n+1\}$ that includes $n + 1$ is a $(k-1)$-subset of $\{1, 2, 3, \ldots, n\}$ together with $n + 1$. There are $\binom{n}{k-1}$ of those. So

$$\binom{n+1}{k} = \binom{n}{k} + \binom{n}{k-1}$$

as claimed. ∎

This formula is the basis of **Pascal's triangle** and suggests a method for computing binomial coefficients. Here is Pascal's triangle:

$$\binom{0}{0}$$
$$\binom{1}{0} \qquad \binom{1}{1}$$
$$\binom{2}{0} \qquad \binom{2}{1} \qquad \binom{2}{2}$$
$$\binom{3}{0} \qquad \binom{3}{1} \qquad \binom{3}{2} \qquad \binom{3}{3}$$
$$\vdots \qquad\qquad \vdots \qquad\qquad \vdots$$

Each entry can be computed by adding the two entries above it. Thus $\binom{6}{4} = 15$ because

$$\binom{5}{3} \qquad\qquad \binom{5}{4} \quad 10 \qquad\qquad\qquad 5$$

$$\searrow \ + \ \swarrow \qquad\qquad \searrow \ + \ \swarrow$$

$$\binom{6}{4} \qquad\qquad\qquad 15$$

What about the formula

$$\binom{n}{k} = \frac{n\,(n-1)\,(n-2)\cdots(n-k+1)}{k!}$$

We will show this by induction on n. Let $P\,(n)$ be the statement that, for any nonnegative integer k, this equation holds. The number of 1-subsets of the set $\{1\}$, and is $1 = \frac{1}{1!}$, so $P\,(1)$ is true.

Now suppose that $P\,(n)$ is true. We want to conclude that $P\,(n+1)$ is true. Now

$$\binom{n+1}{k} = \binom{n}{k} + \binom{n}{k-1}$$

but the right-hand side, because $P\,(n)$ is true, is equal to

$$\frac{n\,(n-1)\,(n-2)\cdots(n-k+1)}{k!} + \frac{n\,(n-1)\,(n-2)\cdots(n-k+2)}{(k-1)!}$$

You are asked to prove in Problem 10 that this sum is equal to

$$\frac{(n+1)n\,(n-1)\cdots(n-k)}{k!}$$

When you do that, the formula will have been verified by induction.

Problems 1.4

Use a computer algebra system to find a closed-form solution for each of the sums in problems 1–8. Simplify the answer if possible. Then use mathematical induction to verify these simplified answers.

1. $\displaystyle\sum_{i=1}^{n} i^2 = 1^2 + 2^2 + 3^2 + \cdots + n^2$

2. $\displaystyle\sum_{k=1}^{n} (-1)^k\, k = (-1) + (-1)^2\, 2 + (-1)^3\, 3 + \cdots + (-1)^n\, n$

3. $\displaystyle\sum_{k=1}^{n} k \cdot 2^k = 1 \cdot 2 + 2 \cdot 2^2 + 3 \cdot 2^3 + \cdots + n \cdot 2^n$

4. $\displaystyle\sum_{k=0}^{n} \binom{r+k}{r} = \binom{r}{0} + \binom{r+1}{1} + \cdots + \binom{r+n}{n}$

5. $\displaystyle\sum_{r=0}^{n}\binom{n}{r}x^r = \binom{n}{0} + \binom{n}{1}x + \cdots + \binom{n}{n}x^n$

6. $\displaystyle\sum_{k=1}^{n}k = 1 + 2 + 3 + \cdots + (n-1) + n$

7. $\displaystyle\sum_{k=1}^{n}\frac{1}{k(k+1)} = \frac{1}{1\cdot 2} + \frac{1}{2\cdot 3} + \frac{1}{3\cdot 4} + \cdots + \frac{1}{(n-1)\cdot n}$

8. $\displaystyle\sum_{k=1}^{n}k^3 = 1 + 2^3 + 3^3 + \cdots + (n-1)^3 + n^3$

9. Use the identity
$$\binom{n+1}{r+1} = \binom{n}{r} + \binom{n}{r+1}$$
and mathematical induction on n to verify that
$$\sum_{r=0}^{n}\binom{n}{r} = 2^n$$

10. Verify that for any real number n and positive integer k,
$$\frac{n(n-1)(n-2)\cdots(n-k+1)}{k!} + \frac{n(n-1)(n-2)\cdots(n-k+2)}{(k-1)!}$$
$$= \frac{(n+1)n(n-1)\cdots(n-k)}{k!}$$

11. Use the formula $\binom{n}{k} = \frac{n(n-1)(n-2)\cdots(n-k+1)}{k!}$ to extend the definition of the binomial coefficient $\binom{n}{k}$ to any real number n and positive integer k. In particular, what is $\binom{-\sqrt{3}}{2}$? What is $\binom{1/5}{3}$?

12. Use the Pascal's triangle identity
$$\binom{n}{k-1} + \binom{n}{k} = \binom{n+1}{k}$$
(see Problems 9, 10, and 11) and the formula $\binom{n}{0} = 1$ to extend the definition of the binomial coefficient $\binom{n}{k}$ to any real number n and any integer k. In particular, what is $\binom{\sqrt{3}}{-1}$? What is $\binom{\sqrt{3}}{-2}$?

13. Use the identity
$$\binom{n}{k-1} + \binom{n}{k} = \binom{n+1}{k}$$
and mathematical induction on n to prove that
$$(x+y)^n = \sum_{k=0}^{n}\binom{n}{k}x^{n-k}y^k$$

14. Use the binomial coefficient algorithm to calculate $\binom{100}{95}$. Verify that all of the intermediate values in the calculation are integers. Explain why you would expect this.

15. Show that $16! = 14!5!2!$ by using a computer algebra system to compute each side of the equation. Verify this result by hand, doing as little rewriting of the right-hand side as possible.

16. Use a computer algebra system to compute $1000!$; then count the number of trailing zeros. Give a direct argument that shows that your answer is correct.

17. The Fibonacci numbers F_n are defined by $F_0 = 0$, $F_1 = 1$, and $F_{k+1} = F_k + F_{k-1}$ for $n \geq 1$. The first few Fibonacci numbers are $0, 1, 1, 2, 3, 5, 8, 13$. Prove that

$$F_n = \frac{1}{\sqrt{5}}\left(\frac{1+\sqrt{5}}{2}\right)^n - \frac{1}{\sqrt{5}}\left(\frac{1-\sqrt{5}}{2}\right)^n$$

by using induction on n to prove that, for $n \geq 1$, both of the equations

$$F_n = \frac{1}{\sqrt{5}}\left(\frac{1+\sqrt{5}}{2}\right)^n - \frac{1}{\sqrt{5}}\left(\frac{1-\sqrt{5}}{2}\right)^n$$

$$F_{n-1} = \frac{1}{\sqrt{5}}\left(\frac{1+\sqrt{5}}{2}\right)^{n-1} - \frac{1}{\sqrt{5}}\left(\frac{1-\sqrt{5}}{2}\right)^{n-1}$$

hold.

18. Draw a picture to illustrate the identity

$$1 + 2 + 3 + \cdots + n = \frac{n(n+1)}{2}$$

19. Use calculus and the sum formula

$$\sum_{k=0}^{n} x^k = \frac{x^{n+1} - 1}{x - 1}$$

to derive a closed-form formula for

$$\sum_{k=1}^{n} k \cdot 2^k = 1 \cdot 2 + 2 \cdot 2^2 + 3 \cdot 2^3 + \cdots + n \cdot 2^n$$

20. Use mathematical induction to prove that

$$\sum_{k=0}^{n} k \binom{n}{k} = n2^{n-1}$$

Chapter 2

Codes

A **code** is a systematic way to represent words or symbols by other words or symbols. We will look at codes that aid in the transfer of information. Some codes represent information more compactly (hex versus decimal notation for numbers), some make it possible for the visually impaired to read (Braille), some help machines to read (bar codes), and some allow transmission errors to be corrected (the BCH codes).

This information may be transferred between people (Morse code, referee signals, flag codes), between people and machines (ASCII code), or between machines (bar code, binary code).

2.1 Binary and Hexadecimal Codes

If intelligent beings existed elsewhere in the universe, they probably would have a notion of the integers like our own. Unless they had ten fingers, it's unlikely that they would use decimal notation for integers. However, they almost certainly would understand a binary notation. Binary notation uses only two digits, 0 and 1, in contrast with the ten decimal digits, $0, 1, 2, 3, 4, 5, 6, 7, 8, 9$. A binary digit is often called a **bit**. Table 2.1 gives the correspondence between binary and decimal notation.

We will sometimes write $(1101)_2$ instead of 1101 to make clear that we are using binary notation rather than decimal. In general, if b_0, b_1, ..., b_n is a sequence of 0's and 1's, then

$$(b_n b_{n-1} \ldots b_2 b_1 b_0)_2 = \sum_{i=0}^{n} b_i 2^i.$$

For example,

$$(1101)_2 = 1 \cdot 2^3 + 1 \cdot 2^2 + 0 \cdot 2^1 + 1 \cdot 2^0$$
$$= 8 + 4 + 0 + 1 = 13$$

so it is fairly easy to convert from binary to decimal notation. When doing the conversion on a computer, the work is often arranged to avoid direct computation of powers:

$$(1101)_2 = 1 \cdot 2^3 + 1 \cdot 2^2 + 0 \cdot 2^1 + 1 \cdot 2^0$$
$$= \left(1 \cdot 2^2 + 1 \cdot 2 + 0\right) \cdot 2 + 1$$
$$= ((1 \cdot 2 + 1) \cdot 2 + 0) \cdot 2 + 1$$

Decimal	Binary	Decimal	Binary
0	0	8	1000
1	1	9	1001
2	10	10	1010
3	11	11	1011
4	100	12	1100
5	101	13	1101
6	110	14	1110
7	111	15	1111

Table 2.1   Binary code

In converting from decimal to binary, we use the **floor** and **ceiling** functions.

Definition 2.1 *The **floor** of a real number x, which we write as $\lfloor x \rfloor$, is the largest integer that is less than or equal to x. The **ceiling** of a real number x, which we write as $\lceil x \rceil$, is the smallest integer that is greater than or equal to x.*

Example 2.2 $\lfloor 2.649853 \rfloor = 2$ $\lfloor \pi \rfloor = 3$ $\lfloor -5/2 \rfloor = -3$ $\lfloor -5 \rfloor = -5$
$\lceil 2.649853 \rceil = 3$ $\lceil \pi \rceil = 4$ $\lceil -5/2 \rceil = -2$ $\lceil -5 \rceil = -5$

Sometimes $\lfloor x \rfloor$ is called the **greatest integer function** and written as $[x]$ instead of $\lfloor x \rfloor$.

Algorithm 2.1 computes the binary representation of a positive integer:

Algorithm 2.1   Binary representations

Input: A positive integer n
Output: The binary representation $(b_k b_{k-1} \ldots b_1 b_0)_2$ of n
    Set $i = 0$
    While $n > 0$ **do**
        Set $b_i = n - \lfloor n/2 \rfloor 2$
        Set $n = \lfloor n/2 \rfloor$
        Set $i = i + 1$
    End While
    Set $k = i - 1$
    Return k, b_0, b_1, $\ldots$, b_k

Example 2.3 To find the binary representation of 23, we do the calculations

$$b_0 = 23 - \lfloor 23/2 \rfloor 2 = 23 - 22 = 1$$
$$b_1 = 11 - \lfloor 11/2 \rfloor 2 = 11 - 10 = 1$$
$$b_2 = 5 - \lfloor 5/2 \rfloor 2 = 5 - 4 = 1$$
$$b_3 = 2 - \lfloor 2/2 \rfloor 2 = 2 - 2 = 0$$
$$b_4 = 1 - \lfloor 1/2 \rfloor 2 = 1 - 0 = 1$$

which yield the result

$$23 = (b_4 b_3 b_2 b_1 b_0)_2 = (10111)_2$$

The **hexadecimal code** is closely related to binary code. Its digits are the sixteen symbols

$$0\ 1\ 2\ 3\ 4\ 5\ 6\ 7\ 8\ 9\ A\ B\ C\ D\ E\ F$$

The correspondence between binary and hexadecimal is shown in Table 2.2.

Decimal	Hexadecimal	Binary	Decimal	Hexadecimal	Binary
0	0	0	8	8	1000
1	1	1	9	9	1001
2	2	10	10	A	1010
3	3	11	11	B	1011
4	4	100	12	C	1100
5	5	101	13	D	1101
6	6	110	14	E	1110
7	7	111	15	F	1111

Table 2.2   Hexadecimal numbers

The letters A, B, C, D, E, F stand for the numbers 10, 11, 12, 13, 14, 15. The hexadecimal representation of the number 26 is $(1A)_{16}$. In general, the hexadecimal representation $(h_n h_{n-1} \ldots h_2 h_1 h_0)_{16}$ stands for the number

$$(h_n h_{n-1} \ldots h_2 h_1 h_0)_{16} = \sum_{i=0}^{n} h_i 16^i$$

where the h_i on the right are integers between 0 and 15.

It's easy to convert from binary to hexadecimal. Given a binary number $(b_n b_{n-1} \ldots b_2 b_1 b_0)_2$, start at the right and break up the symbols into groups of four. The leftmost group may be smaller than four. Then use the preceding table to replace each group with one of the symbols 0, 1, 2, ..., E, F, ignoring the leading zeros in each group. Do you see how to convert from hexadecimal to binary?

Example 2.4 We convert $(101100101110100101101111)_2$ to hexadecimal as follows:

$$(101100101110100101101111)_2$$

$$\underbrace{1}\ \underbrace{6}\ \underbrace{5}\ \underbrace{D}\ \underbrace{1}\ \underbrace{6}\ \underbrace{F}$$

$$= ([0001][0110][0101][1101][0001][0110][1111])_2$$

$$= (165D16F)_{16}$$

Note that leading zeros do not alter the value of a binary number. Thus $(0110)_2 = (110)_2 = (6)_{16}$ and $(1)_2 = (0001)_2 = (1)_{16}$.

Conversion in the other direction is just as easy.

Example 2.5 To convert $(5F90A)_{16}$ to binary, write

$$\overset{5\qquad F\qquad 9\qquad 0\qquad A}{(5F90A)_{16} = ([0101][1111][1001][0000][1010])_2}$$

$$= (1011111100100001010)_2$$

Note that leading zeros were added to form groups of four; then the leading zero in the new binary number was deleted.

Arithmetic is easy, but tedious, in binary. The addition and multiplication tables are given by

+	0	1		×	0	1
0	0	1		0	0	0
1	1	10		1	0	1

Table 2.3a   Binary addition and multiplication

The following is a typical multiplication problem.

Example 2.6 To compute $(1011)_2 \times (101)_2$, write

$$
\begin{array}{cccccc}
 & & 1 & 0 & 1 & 1 \\
 & \times & 1 & 0 & 1 \\
\hline
 & & 1 & 0 & 1 & 1 \\
1 & 0_1 & 1 & 1 \\
\hline
1 & 1 & 0 & 1 & 1 & 1 \\
\end{array}
$$

and conclude that $(1011)_2 \times (101)_2 = (110111)_2$.

You can also note that $(101)_2 = (100)_2 + (1)_2$. Multiplication by $(100)_2$ can be accomplished by adjoining two zeros to the right. Thus, using associative and distributive and commutative laws of arithmetic,

$$\begin{aligned}(1011)_2 \times (101)_2 &= (1011)_2 \times (100)_2 + (1011)_2 \times (1)_2 \\ &= (101100)_2 + (1011)_2 \\ &= (100000)_2 + (1000)_2 + (1000)_2 + (100)_2 + (10)_2 + (1)_2 \\ &= (100000)_2 + (10000)_2 + (100)_2 + (10)_2 + (1)_2 \\ &= (110111)_2\end{aligned}$$

Long division is also straightforward.

Example 2.7 To expand $\frac{(101\,101)_2}{(1101)_2}$, use long division

$$\begin{array}{rrrr|rrrrrr} & & & & & & & & 1 & 1 \\ \cline{5-10} 1 & 1 & 0 & 1 & 1 & 0 & 1 & 1 & 0 & 1 \\ & & & & & 1 & 1 & 0 & 1 & \\ \cline{5-10} & & & & & 1 & 0 & 0 & 1 & 1 \\ & & & & & & 1 & 1 & 0 & 1 \\ \cline{6-10} & & & & & & & 1 & 1 & 0 \end{array}$$

and conclude that

$$\frac{(101101)_2}{(1101)_2} = (11)_2 + \frac{(110)_2}{(1101)_2}$$

so that the quotient is $(11)_2$ with a remainder of $(110)_2$.

Example 2.8 To compute $(3C)_{16} \times (2AB)_{16}$, write

$$\begin{aligned}(30 + C) \times (200 + A0 + B) &= (30 \times 200) + (30 \times A0) + (30 \times B) \\ &\quad + (C \times 200) + (C \times A0) + (C \times B) \\ &= 6000 + 1E00 + 210 + 1800 + 780 + 84 \\ &= 6000 + (1E00 + 1800) + (210 + 780) + 84 \\ &= 6000 + (3600) + (990) + 84 \\ &= 9600 + A14 \\ &= (A014)_{16}\end{aligned}$$

Here is the multiplication table for hexadecimal arithmetic:

×	1	2	3	4	5	6	7	8	9	A	B	C	D	E	F
1	1	2	3	4	5	6	7	8	9	A	B	C	D	E	F
2	2	4	6	8	A	C	E	10	12	14	16	18	1A	1C	1E
3	3	6	9	C	F	12	15	18	1B	1E	21	24	27	2A	2D
4	4	8	C	10	14	18	1C	20	24	28	2C	30	34	38	3C
5	5	A	F	14	19	1E	23	28	2D	32	37	3C	41	46	4B
6	6	C	12	18	1E	24	2A	30	36	3C	42	48	4E	54	5A
7	7	E	15	1C	23	2A	31	38	3F	46	4D	54	5B	62	69
8	8	10	18	20	28	30	38	40	48	50	58	60	68	70	78
9	9	12	1B	24	2D	36	3F	48	51	5A	63	6C	75	7E	87
A	A	14	1E	28	32	3C	46	50	5A	64	6E	78	82	8C	96
B	B	16	21	2C	37	42	4D	58	63	6E	79	84	8F	9A	A5
C	C	18	24	30	3C	48	54	60	6C	78	84	90	9C	A8	B4
D	D	1A	27	34	41	4E	5B	68	75	82	8F	9C	A9	B6	C3
E	E	1C	2A	38	46	54	62	70	7E	8C	9A	A8	B6	C4	D2
F	F	1E	2D	3C	4B	5A	69	78	87	96	A5	B4	C3	D2	E1

Table 2.3b　　　Hexadecimal multiplication

Example 2.9 Alternatively, convert to decimal using

$$(3C)_{16} = (3 \cdot 16 + 12) = 60$$
$$(2AB)_{16} = \left(2 \cdot 16^2 + 10 \cdot 16 + 11\right) = 683$$

Calculate the decimal product

$$60 \times 683 = 40\,980$$

Then convert the result to hexadecimal using

$$
\begin{aligned}
40980 \bmod 16 &= 4 & (40980 - 4)/16 &= 2561 \\
2561 \bmod 16 &= 1 & (2561 - 1)/16 &= 160 \\
160 \bmod 16 &= 0 & (160 - 0)/16 &= 10 \\
10 \bmod 16 &= A
\end{aligned}
$$

which implies

$$(3C)_{16} \times (2AB)_{16} = (A014)_{16}$$

Problems 2.1

Use the methods in this section to perform the following conversions from one number system to another.

1. Convert $(5AB92)_{16}$ to binary.

2. Convert $(43D69)_{16}$ to binary.

3. Convert $(5AB92)_{16}$ to decimal.

4. Convert $(43D69)_{16}$ to decimal.

5. Convert $(101010111000011101010011010101000)_2$ to hexadecimal.

6. Convert $(1110101010110111010100011010101010)_2$ to hexadecimal.

7. Convert $(101010111000011101010011010101000)_2$ to decimal.

8. Convert $(1110101010110111010100011010101010)_2$ to decimal.

9. Convert $(50927341)_{10}$ to binary.

10. Convert $(385941059)_{10}$ to binary.

11. Convert $(50927341)_{10}$ to hexadecimal.

12. Convert $(385941059)_{10}$ to hexadecimal.

13. Compute $(2B)_{16} \times (C1F)_{16}$ and express the result in hexadecimal.

14. Compute $(123)_{16} \times (ABC)_{16}$ and express the result in hexadecimal.

15. Compute $(101101)_2 \times (1101)_2$ and express the result in binary.

16. Compute $(11011011)_2 \times (1001)_2$ and express the result in binary.

17. Compute $\dfrac{(C1F)_{16}}{(2B)_{16}}$ using hexadecimal long division.

18. Compute $\dfrac{(B3C)_{16}}{(2A)_{16}}$ using hexadecimal long division.

19. Compute $\dfrac{(101101)_2}{(1101)_2}$ using binary long division.

20. Compute $\dfrac{(11011011)_2}{(1001)_2}$ using binary long division.

2.2   ASCII Code

Digital computers work with binary numbers. Hexadecimal numbers are mostly for human consumption. People seem to be able to work best with alphabets of 10 to 30 characters and have a lot of trouble with two-letter alphabets. (Can you imagine reading a 1000-page novel that used the binary alphabet 0 and 1?)

There are 26 letters in the standard alphabet. However, we use lower case letters as well as upper case letters, the digits (0 1 2 3 4 5 6 7 8 9) plus numerous punctuation marks (! ? ; : , . ” ’), as well as special mathematical symbols ($+ - = * /$) and other special-purpose symbols (@ # $ % &). The American Standard Code for Information Interchange (ASCII, pronounced `ask'--ee`) is widely used for representing these symbols (see Table 2.4). In addition to the printable characters, ASCII also includes characters for line feeds, form feeds, tabs, carriage returns, and so on.

	0	1	2	3	4	5	6	7
00	NUL	SOH	STX	ETX	EOT	ENQ	ACK	BEL
08	BS	HT	LF	VT	FF	CR	SO	SI
10	DLE	DC1	DC2	DC3	DC4	NAK	SYN	ETB
18	CAN	EM	SUB	ESC	FS	GS	RS	US
20		!	”	#	$	%	&	’
28	(	)	*	+	,	-	.	/
30	0	1	2	3	4	5	6	7
38	8	9	:	;	<	=	>	?
40	@	A	B	C	D	E	F	G
48	H	I	J	K	L	M	N	O
50	P	Q	R	S	T	U	V	W
58	X	Y	Z	[	\	]	^	_
60	‘	a	b	c	d	e	f	g
68	h	i	j	k	l	m	n	o
70	p	q	r	s	t	u	v	w
78	x	y	z	{	|	}	~	DEL

Table 2.4   ASCII code

You will probably never need many of the special characters in the range 00–1F. They are used to communicate with a printer or with another computer. Here are a few of the more useful ones: BEL is a bell, a sound to alert the user that something unusual is going on. BS is a backspace, HT is a horizontal tab, NL is a new line, NP is a new page, and ESC is the Escape key. The characters in the range 00–1F are known as **control characters**. They can be entered on a keyboard that has a CTRL key by holding down the CTRL key while pressing a letter. In particular, CTRL+A yields 01, CTRL+B yields 02, ..., CTRL+Z yields 1A. In fact, on some keyboards CTRL+M has exactly the same effect as pressing the ENTER key. The symbols

#	$	%	&	’	(	)	*	+

appear in nearly the same order as they appear on the top row of a standard keyboard.[1]

The HyperText Markup Language (HTML) is used for web pages. Special symbols that are not on a standard keyboard can be put on a web page by using a slight modification of ASCII. A character with ASCII value n (decimal representation) or m (hexadecimal representation) can usually be put on a web page by using

$$\&\#n;$$

or

$$\&\#xm;$$

In addition, many symbols have HTML names. The copyright symbol © can be generated by using any of

$$\&\#169;$$
$$\&\#xA9;$$
$$\©$$

Because of multiple languages and an increased need for more symbols, HTML has outgrown the 128 available ASCII codes. For example, the Euro symbol € can be generated by using any of

$$\&\#8364;$$
$$\&\#x20AC;$$
$$\€$$

Problems 2.2

1. Convert the sequence

   ```
   4E 75 6D 62 65 72 20 74 68 65 6F 72 79 20
   69 73 20 74 68 65 20 71 75 65 65 6E 20 6F
   66 20 6D 61 74 68 65 6D 61 74 69 63 73 2E
   ```

   of ASCII codes into an English sentence.

2. Convert the sequence

   ```
   54 68 65 20 73 65 6D 69 67 72 6F 75 70 20
   74 75 72 6E 65 64 20 6F 75 74 20 74 6F 20
   62 65 20 74 68 65 20 6D 6F 73 74 20 75 73
   65 66 75 6C 20 61 6C 67 65 62 72 61 69 63
   20 6F 62 6A 65 63 74 20 69 6E 20 74 68 65
   6F 72 65 74 69 63 61 6C 20 63 6F 6D 70 75
   74 65 72 20 73 63 69 65 6E 63 65 2E
   ```

---

[1]ASCII values for hex can be found at many sites on the internet. One of these is http://en.wikipedia.org/wiki/ASCII.

of ASCII codes into an English sentence.

3. Convert the sentence

      Mathematics is the queen of the sciences.

into a sequence of ASCII codes.

4. Convert the sentence

      There are aspects of symmetry that are more faithfully represented by a generalization of groups called inverse semigroups.

into a sequence of ASCII codes.

5. Convert the sequence

      41 53 43 49 49 20 69 73 20 70 72 6F 6E 6F
      75 6E 63 65 64 20 61 73 6B 2D 65 65 2E

of ASCII codes into an English sentence.

6. Convert the sentence

      ASCII represents characters as numbers.

into a sequence of ASCII codes

7. Convert the sequence

      55 6E 69 63 6F 64 65 20 63 6F 6E 74 61 69
      6E 73 20 6F 76 65 72 20 74 68 69 72 74 79
      20 74 68 6F 75 73 61 6E 64 20 64 69 73 74
      69 6E 63 74 20 63 6F 64 65 64 20 63 68 61
      72 61 63 74 65 72 73 2E

of ASCII codes into an English sentence.

8. Convert the sentence

      The standard version of ASCII uses 7 bits for each character.

into a sequence of ASCII codes

2.3   Morse Code

Samuel F. B. Morse (1791–1872), an artist by profession, developed the telegraph and Morse code.[2]

---

[2]See http://www.lgny.org/history/morse.html for an informative sketch of Morse's achievements.

Morse code can be thought of as a **ternary code**, based on the three characters dash (—), dot (•), and space. Table 2.5 shows how the letters of the alphabet, and some other symbols, are written in Morse code.

A	•—	N	—•	1	•————
B	—•••	O	———	2	••———
C	—•—•	P	•——•	3	•••——
D	—••	Q	——•—	4	••••—
E	•	R	•—•	5	•••••
F	••—•	S	•••	6	—••••
G	——•	T	—	7	——•••
H	••••	U	••—	8	———••
I	••	V	•••—	9	————•
J	•———	W	•——	.	•—•—•—
K	—•—	X	—••—	,	——••——
L	•—••	Y	—•——	?	••——••
M	——	Z	——••	:	———•••

Table 2.5 Morse code

The dot and space are each one unit long; the dash and the space between letters are each three units; the space between words is equal to seven units. The most commonly used letters in the English language have the shortest codes: E is the most common letter, followed by T.

Perhaps the best known message in Morse code is the universal distress signal ••• ——— •••. This can be thought of as the letters SOS all run together. Morse code can be represented as a binary code by replacing each short space with a 0 and each short sound with a 1. With this scheme, "SOS" would be converted to 1010101110111011011010101.

The Morse code is a variable-length code, and errors can occur because it is difficult to distinguish between such things as •• (I) and • • (EE). Most of the codes that are now in common use are fixed-length binary codes.

Problems 2.3

1. Read the message

2. Read the message

```
—··· —·—— ·—————— ———·· ··· ··—·
—— ——— ·—·· ··· ·
·———· ·—·· ——— —·· ·— —··· ·—·· —·——
—··· ·— —·· ···· ·· ··· ··—· ·· ·—· ··· —
— · ·—·· · ——· ·—· ·— ——· ····
—— ——— —·· · ·—··
·—— ——— ·—· —·— ·· —· ——·
```

3. Write Morse code for the sentence

At Yale College, Morse delighted
in painting miniature portraits.

4. Calculate the number of units required for Morse code for each letter A–Z.

5. The expected relative frequencies are shown in the table. Use the results of Problem 4 to calculate the expected number of time units per letter for Morse code.

	A	B	C	D	E	F	G	H	I	J	K	L	M
%	7.3	0.9	3.0	4.4	13.0	2.8	1.6	3.5	7.4	0.2	0.3	3.5	2.5
	N	O	P	Q	R	S	T	U	V	W	X	Y	Z
%	7.8	7.4	2.7	0.3	7.7	6.3	9.3	2.7	1.3	1.6	0.5	1.9	0.1

6. Read the message

```
··· ·— —— ··— · ·—··
·· — ·— ·—· — · —·· —— ——— ·—· ··· ·
··· — ·— ·—· — · —·· —··· ·—· ·— ——— ··
—· ——·· ·—— —— ·· —··· — —— ·—· · ···
·—— ···· · — ···· —— — ··· ··—·— ·—·—·—
```

7. Read the message

```
··· ··— —··· —··· · —· ·—· —·—— ——·—— —
··· — ·— ·—· — · —·· ···· ·— —·· ·— —·
·· —··· · ·— ·—·—·— —— ·— —·—— —···· ·
· ·—·· · —·—· — ·—· ·· —·—· — — —·——
—·—· ——— —·· · ·—·· —··
— ·—· ·— —· ··· —— ·· —
—— · ···— ··· ·— —— · ··· ·—·—·—
```

8. Read the message

2.4 Braille

Louis Braille was born in France in 1809. When he was 3 years old, he used to play in his father's saddle shop. During one of his playful adventures, Louis accidentally punctured his eye with an awl, a sharp tool used to punch holes in leather. Infection set in and spread to his other eye, leaving him completely blind. He developed the Braille system by the time he was fifteen.[3]

Over 150 years after Louis Braille worked out his basic 6-dot system, its specific benefits remain unmatched by any later technology—though some, computers being a prime example, both complement and contribute to braille.

Joe Sullivan

Braille uses groups of dots in a 3×2 matrix to represent letters, numbers, and punctuation. There are $2^6 = 64$ such groups. Patterns that have dots in only one column normally use the left column. The first few letters use dots only in the top two rows. Generally speaking, the more common letters use fewer dots; less common letters contain four or five dots. Neither the blank nor the six-dot pattern is used (see Table 2.6, where raised dots are indicated by the larger dots and unused matrix positions by small dots).

---

[3]See http://www.cnib.ca/en/living/braille/louis-braille/Default.aspx for more about Louis Braille's contributions.

A	B	C	D	E	F	G
● ○	● ·	● ●	● ●	● ·	● ●	● ●
· ·	● ·	· ·	· ●	· ●	● ·	● ●
· ·	· ·	· ·	· ·	· ·	· ·	· ·

H	I	J	K	L	M	N
● ·	· ●	· ●	● ·	● ·	● ●	● ●
● ●	● ·	● ●	· ·	● ·	· ·	· ●
· ·	· ·	· ·	● ·	● ·	● ·	● ·

O	P	Q	R	S	T	U
● ·	● ●	● ●	● ·	· ●	· ●	● ·
· ●	● ·	● ●	● ●	● ·	● ●	· ·
● ·	● ·	● ·	● ·	● ·	● ·	● ●

V	W	X	Y	Z
● ·	· ●	● ●	● ●	● ·
● ·	● ●	· ·	· ●	· ●
● ●	· ●	● ●	● ●	● ●

Table 2.6   Braille code for A–Z

The digits 1, 2, 3, 4, 5, 6, 7, 8, 9, 0 use a special prefix, followed by the codes for the letters A–J. These are given in Table 2.7.

1				2				3				4			
·	●	●	·	·	●	●	·	·	●	●	●	·	●	●	●
·	●	·	·	·	●	●	·	·	●	·	·	·	●	·	●
●	●	·	·	●	●	·	·	●	●	·	·	●	●	·	·

5				6				7				8			
·	●	●	·	·	●	●	●	·	●	●	●	·	●	●	·
·	●	·	●	·	●	●	·	·	●	●	●	·	●	●	●
●	●	·	·	●	●	·	·	●	●	·	·	●	●	·	·

9				0			
·	●	·	●	·	●	·	●
·	●	●	·	·	●	●	●
●	●	·	·	●	●	·	·

Table 2.7   Braille code for digits 0–9

Punctuation marks use dots in rows 2 and 3 (see Table 2.8).

,		;		:		.		!		(		)	
·	·	·	·	·	·	·	·	·	·	·	·	·	·
●	·	●	·	●	●	●	●	●	●	●	●	●	●
·	·	●	·	·	·	·	●	●	·	●	●	●	●

Table 2.8   Braille code for punctuation

Problems 2.4

1. Create Braille code for the sentence

   At the age of ten he attended a school for blind boys in Paris.

2. Create Braille code for the sentence

   A soldier named Barbier invented night writing for trench warfare.

3. Translate the phrase

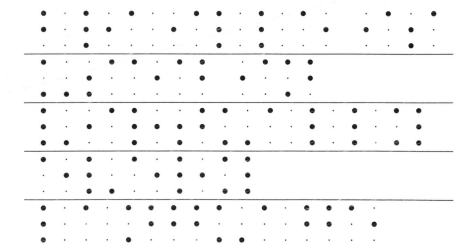

4. Translate the phrase

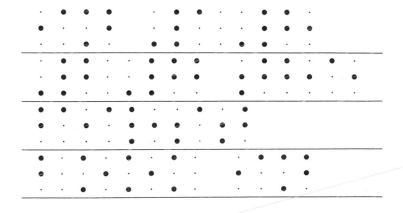

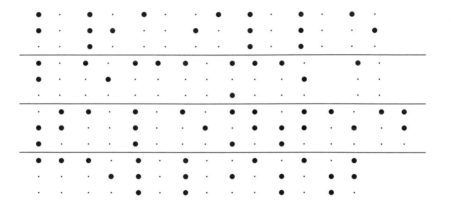

5. Translate the phrase

6. Translate the phrase

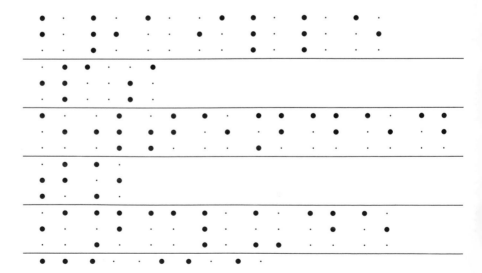

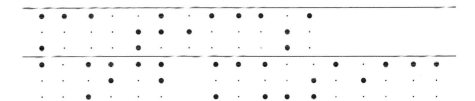

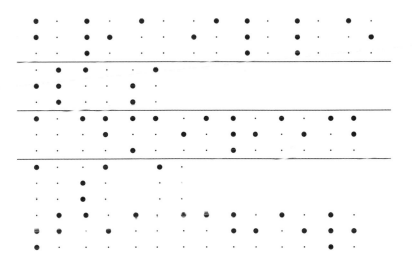

7. Translate the phrase

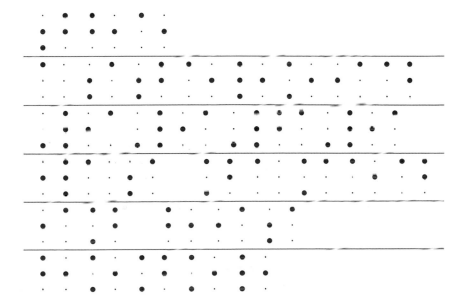

8. Translate the phrase

2.5   Two-out-of-Five Code

When programmers stored data on paper tape, a popular code was the **two-out-of-five code**. This paper tape was about one inch wide, with two holes cut out of five possible locations (see Figure 2.1). Thus there were $\binom{5}{2} = 10$ possible patterns, so the two-out-of-five code could code the ten digits 0, 1,..., 9.

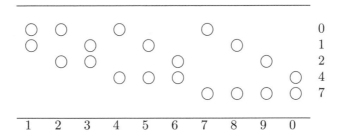

Figure 2.1   Two-out-of-five paper tape

By assigning the values 0, 1, 2, 4, and 7 to the five possible locations for the holes, it is an easy calculation to convert from holes to digits. This is illustrated in Figure 2.2, where the locations of the holes are represented by 1's.

1	1		1		1					0
1		1		1		1				1
	1	1			1		1			2
		1	1	1				1		4
					1	1	1	1		7

| 0+1 | 0+2 | 1+2 | 0+4 | 1+4 | 2+4 | 0+7 | 1+7 | 2+7 | 4+7 |

Figure 2.2   Column sums for two-out-of-five paper tape

Notice how well the column sums match with the decimal representations. If we consider the remainder of the column sums after division by 11, then the correspondence is exact. The two-out-of-five code was very reliable, because errors were usually easy to recognize. The most likely errors were when the reader thought it saw an extra hole (three altogether) or missed a hole (saw only one hole). Since no patterns correspond to one or three holes, the reader would recognize the error. Thus the two-out-of-five code is an **error-detecting code**. We will see several additional examples of such codes in *Chapter 5:Error Control Codes*. Many of these codes are relatives of the two-out-of-five code.

Example 2.10 The two-out-of-five code can be used for phone numbers, zip codes, and Social Security numbers.

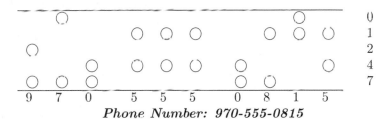

Phone Number: 970-555-0815

Problems 2.5

1. How many symbols can be represented by a two-out-of-four code?

2. Suppose you wanted to design an x-out-of-six code. What integer x would code the largest set of symbols? How many symbols could be represented?

3. List all of the arrangements for a three-out-of-six code.

4. Design an n-out-of-m code that could represent the letters A–Z plus a space character and the usual punctuation symbols (. , ; : ? !).

5. Design an n-out-of-m code that could code nearly the entire 128-character ASCII set. What advantage(s) would this code have over the usual 7-bit binary code? What makes your code better than other possible n-out-of-m codes?

6. Explain why a three-out-of-five code is essentially the same as a two-out-of-five code. In general, why is an n-out-of-m code equivalent to an $(m - n)$-out-of-m code?

7. Explain why a set with m elements has exactly 2^m subsets. Use this to show why

$$\sum_{k=0}^{m} \binom{m}{k} = 2^m$$

8. Draw a picture that shows why

$$\binom{m + 1}{k} = \binom{m}{k} + \binom{m}{k - 1}$$

2.6   Hollerith Codes

Herman Hollerith (1860–1929) developed a system of punched cards for the
1890 United States Census. The punched cards allowed the census data to be
tabulated in three months instead of the expected two years. A company that
he formed later changed its name to International Business Machines (IBM).[4]

Libraries once used punched cards for checking out books, using the holes
for sorting purposes. Punched cards have been used for tabulating votes. To get
an idea of how punched cards can be used for sorting and tabulating, imagine
a set of cards with holes or slots along one edge, as illustrated in Figure 2.3.

Figure 2.3   Binary sort

Suppose a hole represents 0 and a slot represents 1. To sort the cards, place
them in a stack and insert a straightened paper clip through the cards, starting
at the right edge. Carefully pull the clip, removing all of the cards with a hole
at the current location, and place those cards on top. Repeat this at all five
locations. Note that five steps are sufficient for up to 32 cards indexed by the
32 binary numbers $(00000)_2$ to $(11111)_2$. By using a similar scheme (binary
sorting), 1000 cards could be sorted in 10 steps, or $1000\,000$ cards could be
sorted in 20 steps.

Hollerith's punched cards were used until the 1970s, when magnetic tape
took over most of the functions for what had become known as IBM cards. The
IBM cards had 80 columns and 12 rows. Each column coded a single character,
using one, two, or three punches (see Figure 2.4).

The 80-column card shown here was designed for FORTRAN coding. The
first five columns contained the statement numbers, column 6 was used as a
continuation indicator for multiple line statements, and columns 73–80 was used
for identification (in case you stumbled and the cards were scattered on your
way to the card reader).

---

[4]See http://www-groups.dcs.st-and.ac.uk/~history/Mathematicians/Hollerith.html for an
interesting account of the life and contributions of Herman Hollerith.

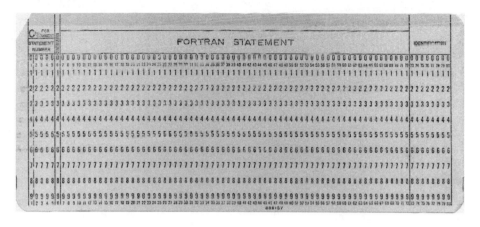

Figure 2.4   80-column card

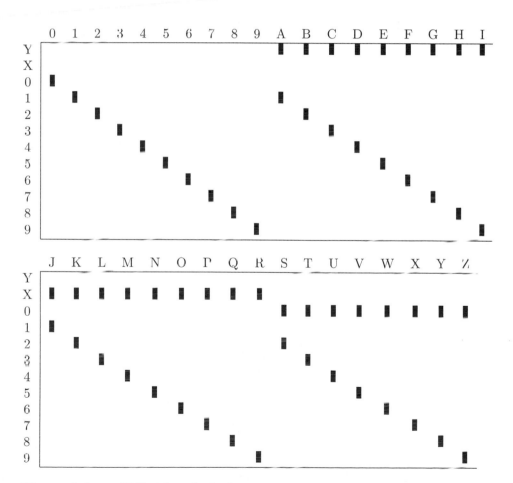

Figure 2.5   Hollerith code for letters

The dictionary in Figure 2.5 describes the Hollerith code used by Control Data Corporation (CDC). Row labels are given along the left edge. Single punches were used for the digits 0–9, with two punches for the letters A–Z.

The special characters used two or three punches, with a special bias for row 8 (see Figure 2.6).[5]

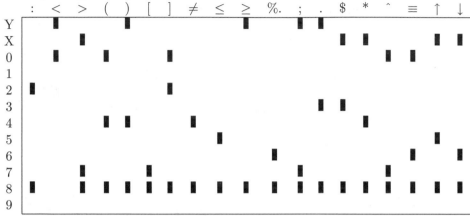

Figure 2.6   Hollerith code for symbols

Example 2.11 Here is some FORTRAN code. Each line is punched on a separate card. The first five columns are for statement numbers, but typing 'C' in the first column indicates a comment. Program statements begin at column 7. Columns 73–80 are ignored by the FORTRAN compiler. These columns are often used for sequence numbers so the cards can be put back in their proper order if the deck of cards is dropped.

```
                     PROGRAM BINOMIAL
        C            COMMENTS LOOK LIKE THIS
                     INTEGER N,K,M,L,I
                     N=5
                     K=2
                     M=1
                     L=N
                     DO 10 I=1,K
                        M=M*L/I
                        L=L-1
        10           CONTINUE
                     PRINT M
                     END
```

This program computes the binomial coefficient $\binom{n}{k} = \binom{5}{2}$. The output of the program is the number 10. Note that in Algorithm 1.1, the expression pt/b is an integer because b divides pt. Thus in this program, M*L/I is always an integer.

---

[5]See http://www.cwi.nl/~dik/english/codes/80col.html for a description of several variations on codes for 80-column cards.

Problems 2.6

1. Why is there a slant cut in the upper left corner of the IBM cards?

2. Give three examples of a pair of punches in a column of an IBM card which does not represent a character in the Hollerith code used by CDC.

3. How many characters could be encoded using at most two punches per column?

4. How many characters could be encoded using at most three punches per column?

5. How many characters could be encoded using exactly four punches per column?

6. How many characters could be encoded using exactly five punches per column?

Chapter 3

Euclidean Algorithm

The Euclidean algorithm was stated by Euclid in his *Elements* over 2000 years ago. It is still the most efficient way to find the greatest common divisor of two integers.

Before investigating the Euclidean algorithm, we take a look at the mod function. This function can be used to define modular arithmetic, which is used extensively in applied algebra.

3.1   The Mod Function

When you look at an analog clock, you can't tell how many times the hour hand has gone around the clock—you only see where the hand is currently pointing. The clock uses mod 12 arithmetic. If it is now 9:00, then 5 hours later it will be 2:00. Thus, on a clock, $9 + 5 = 2$.

We describe this by saying that

$$9 + 5 \bmod 12 = 2$$

We also do this for integers other than 12.

Definition 3.1 (The mod function) *If n and m are integers with $m > 0$, then we define*

$$n \bmod m = n - \lfloor n/m \rfloor m$$

If we rearrange this equation, we see that each integer n can be written as an integer multiple of m plus a remainder which is one of the numbers $0, 1, 2, \ldots, m - 1$:

$$n = \lfloor n/m \rfloor m + n \bmod m$$

We say that when we divide m into n, we get a **quotient** $q = \lfloor n/m \rfloor$ and a **remainder** $r = n \bmod m$. Do you see why the $n \bmod m$ is one of the numbers $0, 1, 2, \ldots, m - 1$?

The computation of the quotient, $\lfloor n/m \rfloor$, and remainder, $n \bmod m$, is the **division algorithm**.

We can write the quotient in terms of the remainder

$$\lfloor n/m \rfloor = \frac{n - n \bmod m}{m}$$

so if a programming language implements the function $n \bmod m$, we can compute the quotient $\lfloor n/m \rfloor$ without having to form the floating point number n/m. On the other hand, the definition of $n \bmod m$ given above is easy to execute on any hand calculator where everything floats.

Example 3.2 Let $n = 23$ and $m = 7$. Then

$$
\begin{aligned}
23 \bmod 7 &= 23 - \lfloor 23/7 \rfloor \, 7 \\
&= 23 - 3 \cdot 7 \\
&= 23 - 21 \\
&= 2
\end{aligned}
$$

On a calculator, you would form $23/7 = 3.285\ldots$, drop the decimal part to get 3, multiply that by 7 and subtract from 23 to get 2.

To compute the base b representation of a positive integer n, we modify Algorithm 2.1 slightly (see Algorithm 3.1).

Algorithm 3.1   Base b representation

Input: Positive integers b and n, where $b \geq 2$

Output: The base b representation of $n = (a_k a_{k-1} \ldots a_2 a_1 a_0)_b$

  Set $i = 0$

  While $n > 0$ **do**

      Set $a_i = n \bmod b$

      Set $n = (n - a_i)/b$

      Set $i = i + 1$

  End While

  Set $k = i - 1$

  Return k, a_0, a_1, $\ldots$, a_k

Example 3.3 For the base 3 expansion of 74, we use the calculations

$$
\begin{array}{ll}
a_0 = 74 \bmod 3 = 2 \qquad & 24 = (74 - 2)/3 \\
a_1 = 24 \bmod 3 = 0 \qquad & 8 = 24/3 \\
a_2 = 8 \bmod 3 = 2 \qquad & 2 = (8 - 2)/3 \\
a_3 = 2 \bmod 3 = 2 \qquad & 0 = (2 - 2)/3
\end{array}
$$

to find that $(2202)_3 = 74$. Indeed,

$$2 \cdot 3^3 + 2 \cdot 3^2 + 0 \cdot 3 + 2 \cdot 3^0 = 74$$

The coefficient a_j can be calculated directly using

$$a_j = \left\lfloor \frac{n}{b^j} \right\rfloor \bmod b = \left\lfloor \frac{n}{b^j} \right\rfloor - \left\lfloor \frac{n}{b^{j+1}} \right\rfloor b$$

Example 3.4 For the base 3 expansion of 74, evaluate the sum

$$\sum_{j=0}^{74} \left(\left\lfloor \frac{74}{3^j} \right\rfloor - \left\lfloor \frac{74}{3^{j+1}} \right\rfloor 3 \right) T^j = 2T^3 + 2T^2 + 2$$

and replace the "T" by "3".

On the other hand, the sum

$$\sum_{j=0}^{74} \left(\left\lfloor \frac{74}{3^j} \right\rfloor - \left\lfloor \frac{74}{3^{j+1}} \right\rfloor 3 \right) 10^j = 2202$$

which is the base 3 representation of 74. Why is that?

The mod function lets us define a new addition and multiplication on the sets $\{0, 1, 2, 3, \ldots, m-1\}$.

Example 3.5 For $m = 5$ we define $\oplus$ and $\otimes$ on $\{0, 1, 2, 3, 4\}$ by

$$a \oplus b = (a + b) \bmod 5$$
$$a \otimes b = ab \bmod 5$$

The addition and multiplication tables are given in Table 3.1.

$\oplus$	0	1	2	3	4		$\otimes$	0	1	2	3	4
0	0	1	2	3	4		0	0	0	0	0	0
1	1	2	3	4	0		1	0	1	2	3	4
2	2	3	4	0	1		2	0	2	4	1	3
3	3	4	0	1	2		3	0	3	1	4	2
4	4	0	1	2	3		4	0	4	3	2	1

Table 3.1   Addition and multiplication modulo 5

Problems 3.1

1. Find the base 5 representation of the decimal number 9374.

2. Give the addition and multiplication tables for the integers modulo 3, where $a \oplus b = (a + b) \bmod 3$ and $a \otimes b = ab \bmod 3$. Use the tables to solve the equations $2 \oplus x = 1$ and $2 \otimes x = 1$.

3. Give the addition and multiplication tables for the integers modulo 4, where $a \oplus b = (a + b) \bmod 4$ and $a \otimes b = ab \bmod 4$. Can you use the tables to solve the equations $2 \oplus x = 1$ and $2 \otimes x = 1$? Why or why not?

4. Give the addition and multiplication tables for the integers modulo 6, where $a \oplus b = (a + b) \bmod 6$ and $a \otimes b = ab \bmod 6$. If a and b are in the set $\{0, 1, 2, 3, 4, 5\}$, can you always solve the equation $a \oplus x = b$? For which choices of a and b can you solve the equation $a \otimes x = b$?

5. Give the addition and multiplication tables for the integers modulo 7, where $a \oplus b = (a + b) \bmod 7$ and $a \otimes b = ab \bmod 7$. If a and b are in the set $\{0, 1, 2, 3, 4, 5, 6\}$, can you always solve the equation $a \oplus x = b$? For which choices of a and b can you solve the equation $a \otimes x = b$?

6. Give the addition and multiplication tables for the integers modulo 13, omitting 0 from the multiplication table. Describe the patterns that appear in the two tables. How are the patterns similar? How are they different?

7. Consider the alphabet as represented by the integers modulo 26, using the conversion table

A	B	C	D	E	F	G	H	I	J	K	L	M
0	1	2	3	4	5	6	7	8	9	10	11	12

N	O	P	Q	R	S	T	U	V	W	X	Y	Z
13	14	15	16	17	18	19	20	21	22	23	24	25

Describe how you would design a word scramble that is based upon addition and/or multiplication modulo 26.

8. Solve the equation $4x + 3 = 7$ in the integers modulo 11.

9. Solve the equation $5x + 8 = 4$ in the integers modulo 11.

10. Solve the system

$$2x + 3y = 5$$
$$3x + 4y = 2$$

of linear equations in the integers modulo 11.

11. Solve the equation $x^2 + 9x + 9 = 0$ in the integers modulo 11.

3.2   Greatest Common Divisors

Every integer a is a divisor of 0 because $0 = 0 \cdot a$. However, a nonzero integer n has only a finite number of divisors because any divisor of n must lie between $-|n|$ and $|n|$.

Definition 3.6 *An integer d is called a **common divisor** of a and b if it divides both a and b; that is, if $d|a$ and $d|b$.*

If either a or b is nonzero, then a and b have only a finite number of common divisors.

Definition 3.7 *If a and b are integers that are not both zero, then the **greatest common divisor** d of a and b is the largest of the common divisors of a and b. We write the greatest common divisor of a and b as*

$$d = \gcd(a, b)$$

Since 1 divides any integer, the greatest common divisor is always positive. It is convenient to set $\gcd(0,0) = 0$. Note that because every number divides 0, there is, strictly speaking, no greatest common divisor of 0 and 0.

Example 3.8 To compute $\gcd(24, 32)$, we can look at the divisors of 24

$$\pm 1, \ \pm 2, \ \pm 3, \ \pm 4, \ \pm 6, \ \pm 8, \ \pm 12, \ \pm 24$$

and the divisors of 32

$$\pm 1, \ \pm 2, \ \pm 4, \ \pm 8, \ \pm 16, \ \pm 32$$

The common divisors of 24 and 32 are the numbers that are in both those sets, namely

$$+1, \ \pm 2, \ \pm 4, \ \pm 8$$

It is easily seen that 8 is the greatest common divisor of 24 and 32. Thus,

$$8 = \gcd(24, 32)$$

Examining all the divisors of a and b is a way to find the greatest common divisor of small integers, but in cryptography we deal with integers that may be hundreds of digits long. We will present an efficient method for finding greatest common divisors of large numbers. First a few observations.

Definition 3.9 *The **absolute value** of a real number x is*

$$|x| = \begin{cases} x & \text{if } x \geq 0 \\ -x & \text{if } x < 0 \end{cases}$$

Theorem 3.10 *If a and b are integers, then $\gcd(a, b) = \gcd(|a|, |b|)$.*

Proof. This is obviously true if $a = b = 0$. Otherwise, note that the divisors of a are the same as the divisors of $|a|$, and the divisors of b are the same as the divisors of $|b|$. So the greatest common divisor of a and b is the same as the greatest common divisor of $|a|$ and $|b|$. ∎

It follows that to compute $\gcd(a, b)$, we may as well assume that $a \geq 0$ and $b \geq 0$.

Theorem 3.11 *If $a > 0$, then $\gcd(a, a) = a$ and $\gcd(a, 0) = a$.*

Proof. The largest divisor of a is a.   ∎

Theorem 3.12 *If a and b are integers, then* $\gcd(a, b) = \gcd(b, a)$.

Proof. The common divisors of a and b are the same as the common divisors of b and a.   ∎

Theorem 3.13 *If a, b, and k are integers, then*

$$\gcd(a, b) = \gcd(a + kb, b)$$

Proof. We will show that the common divisors of a and b are the same as the common divisors of $a + kb$ and b.

If $d|a$ and $d|b$, then $a = xd$ and $b = yd$ for some integers x and y. So

$$a + kb = xd + kyd = (x + ky)d$$

which means that d is a divisor of $a + kb$, so d is a common divisor of $a + kb$ and b.

Conversely, if c is a common divisor of $a + kb$ and b, then $a + kb = xc$ and $b = yc$ for some integers x and y. So

$$a = xc - kb = xc - kyc = (x - ky)c$$

so c is a common divisor of a and b.

This shows that the set of common divisors of a and b is the same as the set of common divisors of $a + kb$ and b, so $\gcd(a, b) = \gcd(a + kb, b)$.   ∎

The following corollary leads to an efficient method for computing greatest common divisors.

Corollary 3.14 *If a and b are integers with $b > 0$, then*

$$\gcd(a, b) = \gcd(a \bmod b, b)$$

Proof. Recall that $a \bmod b = a - \lfloor a/b \rfloor b$, so that $a \bmod b = a + kb$ for $k = -\lfloor a/b \rfloor$. The result now follows from Theorems 3.12 and 3.13.   ∎

As $\gcd(m, n)$ is always equal to $\gcd(n, m)$, we can write the preceding equation as

$$\gcd(a, b) = \gcd(b, a \bmod b)$$

which is the form we will use.

Example 3.15 To compute $\gcd(24, 32)$, we proceed as follows:

$$\begin{aligned}
\gcd(32, 24) &= \gcd(24, 32 \bmod 24) \\
&= \gcd(24, 8) \\
&= \gcd(8, 24 \bmod 8) \\
&= \gcd(8, 0) \\
&= 8
\end{aligned}$$

You get a less cluttered display of the running of this algorithm by simply printing out the sequence $a, b, r_0, r_1, r_2, \ldots$ where each term in the sequence is obtained by applying the mod function to the previous two terms. In this example, the sequence is $24, 32, 24, 8, 0$ so the gcd is equal to 8. The third term is 24 because we are taking 24 mod 32.

Example 3.16 The calculation of gcd $(31899744, 44216928)$ requires more steps. Repeated use of the mod function yields the sequence $31\,899\,744$, $44\,216\,928$, $31\,899\,744$, $12\,317\,184$, $7\,265\,376$, $5\,051\,808$, $2\,213\,568$, $624\,672$, $339\,552$, $285\,120$, $54\,432$, $12\,960$, 2592, 0, so the gcd is 2592. The first two terms in the sequence are the input. The computations for the third, fourth, and fifth terms are

$$31899744 \bmod 44216928 = 31\,899\,744$$
$$44216928 \bmod 31\,899\,744 = 12\,317\,184$$
$$31\,899\,744 \bmod 12\,317\,184 = 7\,265\,376$$

Euclid gave an algorithm to compute the greatest common divisor over 2000 years ago:

Algorithm 3.2   Euclidean algorithm

Input: Integers a and b

Output: $d = \gcd(a, b)$

  Set $b = |b|$

  While $b > 0$ **do**

    Set $c = b$

    Set $b = a \bmod b$

    Set $a = c$

  End While

  Return $|a|$

The Euclidean algorithm produces a sequence of remainders $r_0, r_1, r_2, \ldots$:

$$r_0 = a \bmod b$$
$$r_1 = b \bmod r_0$$
$$r_2 = r_0 \bmod r_1$$
$$r_3 = r_1 \bmod r_2$$
$$\vdots$$

For example, if $a = 34$ and $b = 13$, then the sequence is $8, 5, 3, 2, 1, 0$. In general, if $r_n \neq 0$, then $r_{n+1} = r_{n-1} \bmod r_n$.

Since gcd $(r_n, r_{n+1}) = \gcd(r_{n-1}, r_n)$ (see problem 9), the number gcd (r_n, r_{n+1}) is a **loop invariant** in the Euclidean algorithm. The last nonzero remainder r_m is the gcd because gcd $(r_m, r_{m+1}) = \gcd(r_m, 0) = r_m$.

There is also a recursive version of the Euclidean algorithm (see Algorithm 3.3). A recursive algorithm is one that calls on itself.

Algorithm 3.3	Euclidean algorithm (recursive)

Input: Nonnegative integers a and b
Output: $d = \gcd(a, b)$
    If $b = 0$
        Then Set $d = a$
    Else
        Set $d = \gcd(b, a \bmod b)$
    End If
    Return d

The Euclidean algorithm computes $\gcd(a, b)$ in very few steps. Let $r_0, r_1, r_2, \ldots$ be the sequence of remainders:

$$
\begin{aligned}
r_0 &= a \bmod b \\
r_1 &= b \bmod r_0 \\
r_{n+1} &= r_{n-1} \bmod r_n \text{ for } n \geq 1
\end{aligned}
$$

We might as well assume that $a > b > 0$. The algorithm stops when $r_m = 0$. How big can m be? We will give a fairly crude bound that is enough to show that the algorithm is quite fast.

First we show that $r_0 < a/2$. Indeed, if $b \leq a/2$, then $r_0 < b \leq a/2$, while if $b > a/2$, then $r_0 \leq a - b < a/2$. For the same reason, $r_{2i} < r_{2i-2}/2$ for $i \geq 1$.

Thus, each term in the sequence $a, r_0, r_2, r_4, \ldots$ is less than half of the preceding one. So $r_{2i} < a \cdot (1/2)^{i+1}$. Choose the smallest i so that $a \leq 2^{i+1}$, so $r_{2i} < 1$. Either $r_{2i} = 0$, or $r_m = 0$ for some $m < 2i$. Now $a \leq 2^{i+1}$ exactly when $\log_2 a \leq i + 1$. So i is the smallest integer such that $2 \log_2 a \leq 2i + 2$ whence $2 \log_2 a > 2i$. Also $r_m = 0$ for some $m \leq 2i$. Thus $r_m = 0$ for some $m < 2 \log_2 a$.

For example, if $a = 32$, then the algorithm must stop at some $m < 10 = 2 \log_2 32$. If a is a one-hundred digit number, then $\log_2 a$ is less than 336 so we know that the algorithm takes fewer than 672 steps. That's a pretty small number when you think about how many steps would be required to factor a one-hundred digit number by trying to divide it by smaller numbers. You would have to try to divide it by all numbers up to fifty digits, which would require over 10^{50} steps.

The usual terminology for this situation is that the number of steps is $O(\log a)$. That's pronounced "Big O." It means that you can bound the number of steps by a constant times the logarithm of a. (For this notion it doesn't matter what logarithm you use because each is a constant times $\log_2$.)

Problems 3.2

1. Compute $\gcd(48, 72)$ by writing out all the divisors of 48 and all the divisors of 72.

2. Compute $\gcd(168, 245)$ using Example 3.15 as a guide.

3. Compute $\gcd(55\,440, 48\,000)$ by factoring $55\,440$ and $48\,000$ into prime powers.

4. Compute $\gcd(40\,768, 13\,689)$ using a computer algebra system and the mod function.

5. Compute $\gcd(29\,432\,403, 22\,254\,869)$ by computing a sequence of quotients $q_0, q_1, \ldots$ and a sequence of remainders $r_0, r_1, \ldots$, where $r_{n-1} = r_n q_{n+1} + r_{n+1}$.

6. Compute $\gcd(2456513580, 2324849811)$.

7. Given two integers a and b that differ by 5, show that

$$\gcd(a, b) = 1 \quad \text{or} \quad \gcd(a, b) = 5$$

8. Explain the role of the integer c in the Euclidean algorithm.

9. Verify that $\gcd(a, b) = \gcd(r_k, r_{k-1})$, where $r_0 = a \bmod b$, $r_1 = b \bmod r_0$, and $r_{n+1} = r_{n-1} \bmod r_n$ for $n \geq 1$.

10. Let $a \,\mathrm{MOD}\, m = r$, where

$$r = \begin{cases} a \bmod m & \text{if } a \bmod m \leq m/2 \\ a \bmod m - m & \text{if } a \bmod m > m/2 \end{cases}$$

Note that

$$-\frac{m}{2} < r \leq \frac{m}{2}$$

The number r is called the **absolutely least residue** of a modulo m. Compare the number of steps required to calculate

$$\gcd(1784, 2392)$$

using the two functions $a \bmod m$ and $a \,\mathrm{MOD}\, m$.

11. The worst case for the Euclidean algorithm is when a and b are two consecutive Fibonacci numbers (see page 14). How many steps are required to compute $\gcd(233, 144)$? How many steps are required to compute $\gcd(F_{21}, F_{20})$?

12. How many steps does the Euclidean algorithm take to calculate the greatest common divisor of F_{n+1} and F_n?

3.3 Extended Euclidean Algorithm

In this section we show how to find integers x and y such that $\gcd(a, b) = ax + by$. One way is to keep track of all the steps in the Euclidean algorithm and then use back substitution.

Example 3.17 Let $a = 41$ and $b = 54$. Then successive applications of the division algorithm give the equations

$$54 = 1 \cdot 41 + 13$$
$$41 = 3 \cdot 13 + 2$$
$$13 = 6 \cdot 2 + 1$$
$$2 = 2 \cdot 1 + 0$$

Forget the fourth equation—it just tells you that $1 = \gcd(41, 54)$. What we want to do is write 1 in terms of 41 and 54 using the first three equations.

Solve the first equation for 13, the second for 2, and the third for 1

$$13 = 54 - 1 \cdot 41$$
$$2 = 41 - 3 \cdot 13$$
$$1 = 13 - 6 \cdot 2$$

Start with the third of these equations and substitute the expression for 2 from the second equation

$$1 = 13 - 6 \cdot 2$$
$$= 13 - 6 \cdot (41 - 3 \cdot 13)$$
$$= 19 \cdot 13 - 6 \cdot 41$$

Now substitute the expression for 13 from the first equation into this

$$1 = 19 \cdot 13 - 6 \cdot 41$$
$$= 19 \cdot (54 - 1 \cdot 41) - 6 \cdot 41$$
$$= 19 \cdot 54 - 25 \cdot 41$$
$$= 54 \cdot 19 + 41 \cdot (-25)$$

Theorem 3.18 *The greatest common divisor of two integers a and b can be written as $ax + by$ for some integers x and y.*

Proof. We will compute sequences $x_0, x_1, x_2, \ldots$ and $y_0, y_1, y_2, \ldots$ so that

$$r_k = ax_k + by_k$$

where $r_0, r_1, r_2, \ldots$ is the sequence of remainders in the Euclidean algorithm. Recall that $r_0 = a$ and $r_1 = b$.

We start by setting $(x_0, y_0) = (1, 0)$ and $(x_1, y_1) = (0, 1)$. Trivially

$$r_0 = a = a \cdot 1 + b \cdot 0 = ax_0 + by_0$$
$$r_1 = b = a \cdot 0 + b \cdot 1 = ax_1 + by_1$$

For $i > 0$, set $q_{i+1} = \lfloor r_{i-1}/r_i \rfloor$ and

$$r_{i+1} = r_{i-1} - r_i q_{i+1}$$

so r_{i+1} is indeed the next remainder. Now set

$$x_{i+1} = x_{i-1} - x_i q_{i+1}$$
$$y_{i+1} = y_{i-1} - y_i q_{i+1}$$

Each of these last three equations has the same pattern. It remains to show that $r_k = ax_k + by_k$ for each nonnegative integer k. We have seen that this is trivially true for $k = 0$ and $k = 1$, and we will show that it is true for $k = i+1$ if it is true for $k = i-1$ and $k = i$.

If $r_{i-1} = ax_{i-1} + by_{i-1}$ and $r_i = ax_i + by_i$, then

$$
\begin{aligned}
r_{i+1} &= r_{i-1} - r_i q_{i+1} \\
&= ax_{i-1} + by_{i-1} - (ax_i + by_i) q_{i+1} \\
&= a(x_{i-1} - x_i q_{i+1}) + b(y_{i-1} - y_i q_{i+1}) \\
&= ax_{i+1} + by_{i+1}
\end{aligned}
$$

Thus by the principle of mathematical induction,

$$r_k = ax_k + by_k$$

for each nonnegative integer k. Eventually,

$$r_{n+1} = 0$$

and

$$r_n = \gcd(a, b) = ax_n + by_n$$

Thus the greatest common divisor of a and b can be written in the form $ax + by$. The expression $ax + by$ is called a **linear combination** of a and b.   ∎

The proof of Theorem 3.18 gives an algorithm for finding the integers x and y (see Algorithm 3.4).

Algorithm 3.4   Extended Euclidean algorithm

Input: Integers a and b

Output: Integers x, y, and d, where $d = \gcd(a, b) = ax + by$

  Set $d_0 = a$ Set $x_0 = 1$ Set $y_0 = 0$
  Set $d_1 = b$ Set $x_1 = 0$ Set $y_1 = 1$
  While $d_1 \neq 0$ **Do**
      Set $q = \lfloor d_0/d_1 \rfloor$
      Set $d_2 = d_1$ Set $x_2 = x_1$ Set $y_2 = y_1$
      Set $d_1 = d_0 - qd_1$
      Set $x_1 = x_0 - qx_1$
      Set $y_1 = y_0 - qy_1$
      Set $d_0 = d_2$ Set $x_0 = x_2$ Set $y_0 = y_2$
  End While

  Return $[d, x, y] = [d_0, x_0, y_0]$

Example 3.19 Let $a = 52$ and $b = 96$. The extended Euclidean algorithm produces the numbers in the following table.

d_0	x_0	y_0	d_1	x_1	y_1	q
52	1	0	96	0	1	0
96	0	1	52	1	0	-1
52	1	0	44	-1	1	1
44	-1	1	8	2	-1	-5
8	2	-1	4	-11	6	-2
4	-11	6	0	24	-13	

Table 3.2   Extended Euclidean algorithm

From this table, it follows that $4 = \gcd(52, 96) = (-11) \cdot 52 + 6 \cdot 96$.

The algorithm can be stated succinctly using matrices (see Algorithm 3.5).

Algorithm 3.5   Extended Euclidean algorithm (matrix version)

Input: Integers a and b, not both zero
Output: Integers x, y, and d where $d = \gcd(a, b) = ax + by$

$$\textbf{Set } \begin{bmatrix} d_0 & x_0 & y_0 \\ d_1 & x_1 & y_1 \end{bmatrix} = \begin{bmatrix} a & 1 & 0 \\ b & 0 & 1 \end{bmatrix}$$

While $d_1 \neq 0$ **Do**

$$\textbf{Set } \begin{bmatrix} d_0 & x_0 & y_0 \\ d_1 & x_1 & y_1 \end{bmatrix} = \begin{bmatrix} 0 & 1 \\ 1 & -\lfloor d_0/d_1 \rfloor \end{bmatrix} \begin{bmatrix} d_0 & x_0 & y_0 \\ d_1 & x_1 & y_1 \end{bmatrix}$$

End While

Return $[d, x, y] = [d_0, x_0, y_0]$

Example 3.20 Let $a = 52$ and $b = 96$. The steps of Algorithm 3.5 are

$$\begin{bmatrix} 0 & 1 \\ 1 & -\lfloor 52/96 \rfloor \end{bmatrix} \begin{bmatrix} 52 & 1 & 0 \\ 96 & 0 & 1 \end{bmatrix} = \begin{bmatrix} 96 & 0 & 1 \\ 52 & 1 & 0 \end{bmatrix}$$

$$\begin{bmatrix} 0 & 1 \\ 1 & -\lfloor 96/52 \rfloor \end{bmatrix} \begin{bmatrix} 96 & 0 & 1 \\ 52 & 1 & 0 \end{bmatrix} = \begin{bmatrix} 52 & 1 & 0 \\ 44 & -1 & 1 \end{bmatrix}$$

$$\begin{bmatrix} 0 & 1 \\ 1 & -\lfloor 52/44 \rfloor \end{bmatrix} \begin{bmatrix} 52 & 1 & 0 \\ 44 & -1 & 1 \end{bmatrix} = \begin{bmatrix} 44 & -1 & 1 \\ 8 & 2 & -1 \end{bmatrix}$$

$$\begin{bmatrix} 0 & 1 \\ 1 & -\lfloor 44/8 \rfloor \end{bmatrix} \begin{bmatrix} 44 & -1 & 1 \\ 8 & 2 & -1 \end{bmatrix} = \begin{bmatrix} 8 & 2 & -1 \\ 4 & -11 & 6 \end{bmatrix}$$

$$\begin{bmatrix} 0 & 1 \\ 1 & -\lfloor 8/4 \rfloor \end{bmatrix} \begin{bmatrix} 8 & 2 & -1 \\ 4 & -11 & 6 \end{bmatrix} = \begin{bmatrix} 4 & -11 & 6 \\ 0 & 24 & -13 \end{bmatrix}$$

It follows that

$$\gcd(52, 96) = 4 = (-11) \cdot 52 + 6 \cdot 96$$

The case where $\gcd(a, b) = 1$ is of special interest.

Definition 3.21 *If* $\gcd(a, b) = 1$, *then we say that* a *and* b *are* **relatively prime** *and write* $a \perp b$.

So 12 and 35 are relatively prime, but 12 and 34 are not. If a and b are relatively prime, then $\gcd(a, b) = 1$ so we can find integers x and y such that $ax + by = 1$. The converse statement is also true.

Theorem 3.22 *If* $ax + by = 1$ *for some integers* x *and* y, *then* a *and* b *are relatively prime.*

Proof. Any positive common divisor of a and b must be a divisor of 1, so $\gcd(a, b) = 1$.

Theorem 3.23 *The greatest common divisor of two integers* a *and* b *is the smallest positive integer* d *that can be written in the form* $d = ax + by$ *where* x *and* y *are integers.*

Proof. (See problem 8 at the end of this section.)

Theorem 3.24 *If* $a|bc$ *and* $\gcd(a, b) = 1$, *then* $a|c$.

Proof. (See problem 9 at the end of this section.) ∎

As a corollary to this theorem, we have a result often called Euclid's lemma. It is Proposition 30 of Book VII of Euclid.

Corollary 3.25 (Euclid's lemma) *If* p *is a prime and* $p|ab$, *then* $p|a$ *or* $p|b$.

Proof. Assume p is a (positive) prime and $p|ab$. If $p|a$ there is nothing to prove, so assume $p \nmid a$. Since the only positive divisors of p are 1 and p, it follows that $\gcd(p, a) = 1$. Thus $p|b$ by Theorem 3.24. ∎

Euclid's lemma extends to products of more than two factors.

Corollary 3.26 *If* p *is a prime and* $p|a_1 a_2 \cdots a_k$, *then* $p|a_i$ *for some* i.

Proof. If p is a prime such that

$$p|a_1 a_2 \cdots a_k = (a_1 a_2 \cdots a_{k-1})a_k$$

Then by the previous corollary, $p|a_1 a_1 \cdots a_{k-1}$ or $p|a_k$. If $p|p_k$ we are done. Otherwise, repeat this argument on $p|(a_1 a_2 \cdots a_{k-2})a_{k-1}$. Eventually, we get $p|a_i$ for some i. ∎

Problems 3.3

1. Use back substitution to find $d = \gcd(43, 56)$ and integers x and y such that $d = 43x + 56y$.

2. Use back substitution to find $d = \gcd(27, 68)$ and integers x and y such that $d = 27x + 68y$.

3. Use the extended Euclidean algorithm to find $d = \gcd(43, 56)$ and integers x and y such that $d = 43x + 56y$.

4. Use the extended Euclidean algorithm to find $d = \gcd(27, 68)$ and integers x and y such that $d = 27x + 68y$.

5. Use a computer algebra system to find integers d, x, and y such that

$$d = \gcd(742789479, 9587374758) = 742789479x + 9587374758y$$

6. Show that the greatest common divisor of two consecutive integers is 1.

7. Find $d = \gcd(4, 6)$ and integers x and y such that $d = 4x + 6y$ by creating a table of values of $4x + 6y$ and picking the smallest positive value.

8. Prove Theorem 3.23.

9. Prove Theorem 3.24.

10. The **least common multiple** of two positive integers a and b is the smallest positive integer that is divisible by both a and b. Find an equation relating ab, the least common multiple of a and b, and $\gcd(a, b)$.

3.4   The Fundamental Theorem of Arithmetic

The set
$$\mathbb{M} = \{1, 4, 7, 10, 13, 16, 19, 22, 25, \ldots\}$$
consists of those positive integers of the form $3k + 1$. It is the set of positive integers a such that $a \bmod 3 = 1$. Call an element p of $\mathbb{M}$ a **prime** in $\mathbb{M}$ if $p > 1$ and the only factors of p in $\mathbb{M}$ are 1 and p. The set of primes in $\mathbb{M}$ is $\{4, 7, 10, 13, 19, 22, 25, \ldots\}$. Note that 1 is a unit and $16 = 4 \cdot 4$ is composite in $\mathbb{M}$.

The surprise here is that $100 = 10 \cdot 10 = 4 \cdot 25$, so 100 can be written as a product of primes in $\mathbb{M}$ in two essentially different ways. Can this happen in the set of all the positive integers?

The fundamental theorem of arithmetic says that factorization of positive integers into primes is unique. Of course $2 \cdot 2 \cdot 3$ and $2 \cdot 3 \cdot 2$ are both ways of writing 12 as a product of primes, but they are not essentially different. We will show that the product is unique if we arrange the primes in order of size.

Theorem 3.27 *(Fundamental Theorem of Arithmetic) Every integer greater than 1 is either a prime or can be written uniquely as a product of primes.*

Proof. We leave it to the reader to show that every integer greater than 1 is either a prime or a product of primes. We will show that an integer a can be written in only one way as a product of primes

$$a = p_1 p_2 \cdots p_k$$

with $p_1 \leq p_2 \leq \cdots \leq p_k$. We do this by describing what those primes $p_1, p_2, \ldots, p_k$ must be.

The prime p_1 is the smallest prime that divides a. Indeed, let q be the smallest prime that divides a. The extension of Euclid's lemma (the second corollary to Theorem 3.24) says that q divides one of the primes p_j. Since p_j is prime its only divisors are 1 and p_j, so $q = p_j$. But since q is the *smallest* prime that divides a, we must have $q = p_1$.

Now we repeat the argument on $a/p_1 = p_2 p_3 \cdots p_k$. The prime p_2 is the smallest prime that divides a/p_1. Similarly, the prime p_3 is the smallest prime that divides $a/(p_1 p_2)$ and so on. ∎

Edsger Dijkstra, a noted computer scientist, said this was a common but bad way to state the fundamental theorem of arithmetic. First of all, he said, you should consider a prime to be a product of (one) primes. So you can eliminate the phrase "is either a prime or." Second, you should consider the number 1 to be a product of zero primes (note that $3^0 = 1$), so you can replace the phrase "Every integer greater than 1" by "Every positive integer."

It is often convenient to write the prime factorization as a product of prime powers, with the primes written in ascending order. For example,

$$10! = 2^8 \cdot 3^4 \cdot 5^2 \cdot 7^1$$

and

$$20! = 2^{18} \cdot 3^8 \cdot 5^4 \cdot 7^2 \cdot 11 \cdot 13 \cdot 17 \cdot 19$$

The fundamental theorem of arithmetic gives a way to picture the greatest common divisor. We can write any two integers as products of prime powers for the same primes: $a = p_1^{e_1} p_2^{e_2} \cdots p_k^{e_k}$ and $b = p_1^{f_1} p_2^{f_2} \cdots p_k^{f_k}$, where some of the exponents e_i or f_j may be zero. Then

$$\gcd(a, b) = p_1^{g_1} p_2^{g_2} \cdots p_k^{g_k}$$

where $g_i = \min(e_i, f_i)$; that is, the exponent of each prime in $\gcd(a, b)$ is the smaller of the exponents in a and b.

Example 3.28 Consider $35640 = 2^3 \cdot 3^4 \cdot 5 \cdot 11$ and $7409556 = 2^9 \cdot 3^7 \cdot 7 \cdot 11^2$. The greatest common divisor is $3564 = 2^2 \cdot 3^4 \cdot 5^0 \cdot 7^0 \cdot 11^1$.

This picture does not usually make the greatest common divisor easier to compute because the Euclidean algorithm is the world's greatest algorithm while factoring large numbers into primes is generally quite difficult.

Definition 3.29 *The **least common multiple** $\text{lcm}(a, b)$ of two positive integers a and b is the smallest positive integer m such that $a|m$ and $b|m$.*

Example 3.30 Given the integers 8625 and $14\,835$, we have

$$\gcd\,(8625, 14\,835) = 345$$
$$\text{lcm}(8625, 14\,835) = 370\,875$$

Note that

$$\gcd\,(8625, 14\,835) \cdot \text{lcm}(8625, 14\,835) = 345 \cdot 370\,875$$
$$= 127\,951\,875$$
$$= 8625 \cdot 14\,835$$

The next theorem shows that this is no accident.

Theorem 3.31 *The least common multiple of two positive integers a and b is given by*
$$\text{lcm}\,(a, b) = \frac{ab}{\gcd\,(a, b)}$$

Proof. Let $d = \gcd\,(a, b)$ and $z = ab/d$. We will show that z is the least common multiple of a and b. As d divides both a and b, we have $z = a\,(b/d) = (a/d)\,b$ is a multiple of both a and b. Now write d as $ax + by$ and suppose that m is a common multiple of a and b, say $m = sa = tb$. Then

$$m = s\frac{zd}{b} = t\frac{zd}{a}$$

so

$$ma = ztd \quad \text{and} \quad mb = zsd$$

and so

$$md = m(ax + by) = max + mby = ztdx + zsdy = z\,(tx + sy)\,d$$

Cancelling the d on both sides gives $m = z\,(tx + sy)$ so m is a multiple of z and therefore is at least as big as z. So z is the least common multiple of a and b. ∎

Problems 3.4

In problems 1-7, $\mathbb{M} = \{1, 4, 7, 10, 13, 16, 19, 22, 25, \ldots\}$.

1. List the next six elements after 25 in $\mathbb{M}$.

2. List the next six primes after 25 in $\mathbb{M}$.

3. List the first six composites in $\mathbb{M}$.

4. Show that if $a = 3k + 2$ and $b = 3m + 2$, then ab is an element of $\mathbb{M}$.

5. Show that if $p = 3k + 2$ and $q = 3m + 2$ are ordinary primes, then pq is a prime in $\mathbb{M}$.

6. Find three distinct factorizations of 1870 as a product of 2 primes in $\mathbb{M}$.

7. Show that if $a = 3s + 2$, $b = 3t + 2$, $c = 3u + 2$, and $d = 3v + 2$ are distinct ordinary primes, then $(ab)(cd) = (ac)(bd) = (ad)(bc)$ is an example of an element of $\mathbb{M}$ that has three essentially different prime factorizations.

8. Use a computer algebra system to write 1000! as a product of powers of primes. Use a direct argument to verify that the exponent on the prime 2 is correct.

9. Use a direct argument to explain why 1000! ends in 249 zeros.

10. Use a computer algebra system to write $2^{20} + 20!$ as a product of powers of primes.

11. Prove that if $a \mid n$ and $b \mid n$, then $\mathrm{lcm}\,(a, b) \mid n$.

3.5 Modular Arithmetic

Earlier in this chapter, we looked at ideas related to constructing remainders modulo m. In this section we will build on those ideas. Given any positive integer m, we get a **congruence relation** on the integers as follows

Definition 3.32 *Given two integers a and b and a positive integer m, we say that a is **congruent** to b modulo m and write*

$$a \equiv b \pmod{m}$$

if $m|(a - b)$.

Theorem 3.33 *The condition $a \equiv b \pmod{m}$ is equivalent to $a \bmod m = b \bmod m$.*

Proof. Suppose that $m|(a - b)$, say $a - b = mk$ for some integer k. Then $a = b + mk$, so

$$a \bmod m = (b + mk) \bmod m$$

$$= b + mk - \left\lfloor \frac{b + mk}{m} \right\rfloor m$$

$$= b + mk - \left\lfloor \frac{b}{m} + k \right\rfloor m$$

$$= b + mk - \left(\left\lfloor \frac{b}{m} \right\rfloor + k \right) m$$

$$= b + mk - \left\lfloor \frac{b}{m} \right\rfloor m - km$$

$$= b - \left\lfloor \frac{b}{m} \right\rfloor m = b \bmod m$$

Conversely, suppose that $a \bmod m = b \bmod m$. Then $a - \lfloor a/m \rfloor m = b - \lfloor b/m \rfloor m$, which can be rewritten as $a - b = \lfloor a/m \rfloor m - \lfloor b/m \rfloor m = (\lfloor a/m \rfloor - \lfloor b/m \rfloor)m$. Since $\lfloor a/m \rfloor - \lfloor b/m \rfloor$ is an integer, it follows that $m | (a - b)$. $\blacksquare$

Why the new notation $a \equiv b \pmod m$? The point here is to make things easier to remember. With this notation, which goes back to Gauss, many properties of divisibility look like familiar properties of equality.

Definition 3.34 *A set C of integers is a m if for each integer a there is a unique c in C such that $a \equiv c \pmod m$.*

The set
$$\{0, 1, 2, 3, \ldots, m - 1\}$$

is the most common example of a complete residue system. It is called the **least nonnegative residue system** modulo m. Note that every complete residue system has exactly m elements.

Theorem 3.35 *Let m and d be positive integers and let a, b, c be any integers. The following hold.*

i. $a \equiv a \pmod m$.

ii. If $a \equiv b \pmod m$, then $b \equiv a \pmod m$.

iii. If $a \equiv b \pmod m$ and $b \equiv c \pmod m$, then $a \equiv c \pmod m$.

iv. If $a \equiv b \pmod m$, then $a + c \equiv b + c \pmod m$ and $ac \equiv bc \pmod m$.

v. If $a \equiv b \pmod m$, then $a^d \equiv b^d \pmod m$.

vi. If $a \perp m$ and $ab \equiv ac \pmod m$, then $b \equiv c \pmod m$.

vii. If $\gcd(a, b) = d$, then $(a/d) \perp (b/d)$.

viii. If $\gcd(a, m) = d$ and $ab \equiv ac \pmod m$, then $b \equiv c \pmod{m/d}$.

Proof. i. As $a - a = 0$, it follows that $m | (a - a)$.

ii. As $b - a = -(a - b)$, if $m | (a - b)$, then $m | (b - a)$.

iii. If $m | (a - b)$ and $m | (b - c)$, then $m | ((a - b) + (b - c))$. But $(a - b) + (b - c) = a - c$.

iv. As $(a + c) - (b + c) = a - b$, and $ac - bc = (a - b)c$, if $m | (a - b)$, then $m | ((a + c) - (b + c))$ and $m | (ac - bc)$.

v. We use induction on d. If $a \equiv b \pmod m$, then $a^1 \equiv b^1 \pmod m$. Assuming $a^k \equiv b^k \pmod m$, we have

$$a^{k+1} = a^k a \equiv a^k b \equiv b^k b \equiv b^{k+1} \pmod m$$

vi. If $a \perp m$ and $ab \equiv ac \pmod{m}$, then $m \mid (a(b-c))$ so $m \mid (b-c)$ which means that $b \equiv c \pmod{m}$.

vii. Assume $c \mid (a/d)$ and $c \mid (b/d)$, say $a/d = cx$ and $b/d = cy$. Then $a = cdx$ and $b = cdy$, so cd is a common divisor of a and b. But d is the greatest common divisor of a and b, so $cd \leq d$. But this means that $c = \pm 1$. It follows that $(a/d) \perp (b/d)$.

viii. Let $d = ax + my$. Then $m \mid (ab - ac)$ implies $mk = a(b-c)$. Dividing by d gives

$$(m/d)k = (a/d)(b-c)$$

where m/d and a/d are integers, so $(m/d) \mid (a/d)(b-c)$. But $(m/d) \perp (a/d)$, so $(m/d) \mid (b-c)$; that is, $b \equiv c \pmod{m/d}$. ∎

We will often be faced with the problem of solving the congruence

$$ax \equiv b \pmod{m}$$

for some integer x. The key to solving such a problem is the idea of an inverse modulo m.

Definition 3.36 (Inverse modulo m) *If $ab \equiv 1 \pmod{m}$, then b is called an **inverse** of a modulo m and $a^{-1} \bmod m$ is the smallest positive integer b such that $ab \equiv 1 \pmod{m}$.*

Theorem 3.37 *The integer a has an inverse modulo m if and only if $a \perp m$.*

Proof. Suppose a has such an inverse, say $ab \equiv 1 \pmod{m}$. Then $m \mid (ab - 1)$ implies $mx = ab - 1$. We can write this as $1 = ab - mx$. It follows that any common factor of a and m must divide 1, so $a \perp m$.

On the other hand, if $a \perp m$, then $1 = \gcd(a, m) = as + mt$ means that $1 - as = mt$ and therefore $m \mid (1 - as)$. That is,

$$1 \equiv as \pmod{m}$$

so s is an inverse of a modulo m. ∎

If $\gcd(a, m) = 1$, then the extended Euclidean algorithm will produce integers s and t such that $1 = as + mt$, and s will be an inverse of a modulo m. If $\gcd(a, m) > 1$, then the extended Euclidean algorithm will reveal this, and a has no inverse modulo m.

The mod functions can be extended to fractions with denominators relatively prime to the modulus. So $(2/9) \bmod 35 = 8$ because the inverse of 9 modulo 35 is 4 and $(4 \cdot 2) \bmod 35 = 8$. In particular, if b is an inverse of a modulo m, and $1 \leq b < m$, then we write $b = a^{-1} \bmod m$.

Algorithm 3.6 is a variation of Algorithm 3.5 that computes the inverse of a modulo m.

Algorithm 3.6   Inverse modulo m

Input: Integers a and m with $m > 0$
Output: Integer $a^{-1} \bmod m$ or the message "Inverse does not exist"

Set $\begin{bmatrix} d_0 & x_0 \\ d_1 & x_1 \end{bmatrix} = \begin{bmatrix} a & 1 \\ m & 0 \end{bmatrix}$

While $d_1 \neq 0$ **Do**

$\quad$ Set $\begin{bmatrix} d_0 & x_0 \\ d_1 & x_1 \end{bmatrix} = \begin{bmatrix} 0 & 1 \\ 1 & -\lfloor d_0/d_1 \rfloor \end{bmatrix} \begin{bmatrix} d_0 & x_0 \\ d_1 & x_1 \end{bmatrix}$

End While
If $d_0 = 1$ **Then**
$\quad$ **Return** $a^{-1} = x_0 \bmod m$
Else
$\quad$ **Return** 'Inverse does not exist'
End If

Example 3.38 To compute the inverse of 13 modulo 29, we use the following matrix calculations.

$$\begin{bmatrix} 0 & 1 \\ 1 & -\lfloor 13/29 \rfloor \end{bmatrix} \begin{bmatrix} 13 & 1 \\ 29 & 0 \end{bmatrix} = \begin{bmatrix} 29 & 0 \\ 13 & 1 \end{bmatrix}$$

$$\begin{bmatrix} 0 & 1 \\ 1 & -\lfloor 29/13 \rfloor \end{bmatrix} \begin{bmatrix} 29 & 0 \\ 13 & 1 \end{bmatrix} = \begin{bmatrix} 13 & 1 \\ 3 & -2 \end{bmatrix}$$

$$\begin{bmatrix} 0 & 1 \\ 1 & -\lfloor 13/3 \rfloor \end{bmatrix} \begin{bmatrix} 13 & 1 \\ 3 & -2 \end{bmatrix} = \begin{bmatrix} 3 & -2 \\ 1 & 9 \end{bmatrix}$$

$$\begin{bmatrix} 0 & 1 \\ 1 & -\lfloor 3/1 \rfloor \end{bmatrix} \begin{bmatrix} 3 & -2 \\ 1 & 9 \end{bmatrix} = \begin{bmatrix} 1 & 9 \\ 0 & -29 \end{bmatrix}$$

It follows that $13^{-1} \bmod 29 = 9$. Checking, we see that

$$13 \cdot 9 = 117$$
$$= 29 \cdot 4 + 1$$
$$\equiv 1 \pmod{29}$$

Since the first row of the new matrix is always the second row of the previous matrix, these calculations can be simplified slightly for hand calculations (see Algorithm 3.7).

Algorithm 3.7   Inverse modulo m

Input: Integers a and m with $m > 0$
Output: The inverse of a modulo m or the message that $\gcd(a, m) > 1$
$\quad$ Set $d_0 = a$
$\quad$ Set $d_1 = m$
$\quad$ Set $x_0 = 1$
$\quad$ Set $x_1 = 0$
$\quad$ While $d_1 \neq 0$ **Do**
$\quad\quad$ Set $q = \lfloor d_0/d_1 \rfloor$
$\quad\quad$ Set $d_2 = d_0 - qd_1$
$\quad\quad$ Set $x_2 = x_0 - qx_1$
$\quad\quad$ Set $x_0 = x_1$
$\quad\quad$ Set $x_1 = x_2$
$\quad\quad$ Set $d_0 = d_1$
$\quad\quad$ Set $d_1 = d_2$
$\quad$ **End While**
$\quad$ **If** $d_0 = 1$ **Then**
$\quad\quad$ Return x_0
$\quad$ **Else**
$\quad\quad$ Return '$\gcd(a, m) > 1$'
$\quad$ **End If**

Example 3.39 Table 3.3 can be used to construct the inverse of 13 modulo 29.

$q = \lfloor d_0/d_1 \rfloor$	$d_2 = d_0 - qd_1$	$x_2 = x_0 - qx_1$
	$d_0 = 13$	$x_0 = 1$
	$d_1 = 29$	$x_1 = 0$
$q = 0 = \lfloor 13/29 \rfloor$	$d_2 = 13 - 29 \cdot 0 = 13$	$x_2 = 1 - 0 \cdot 0 = 1$
$q = 2 = \lfloor 29/13 \rfloor$	$d_2 = 29 - 2 \cdot 13 = 3$	$x_2 = 0 - 2 \cdot 1 = -2$
$q = 4 = \lfloor 13/3 \rfloor$	$d_2 = 13 - 4 \cdot 3 = 1$	$x_2 = 1 - 4 \cdot (-2) = 9$
$q = 3 = \lfloor 3/1 \rfloor$	$d_2 = 3 - 3 \cdot 1 = 0$	

Table 3.3   Calculating the inverse of 13 *modulo* 29

As soon as a zero appears in the d-column, the inverse is the previous entry in the x-column. Thus $9 = 13^{-1} \bmod 29$.

Example 3.40 Solve the congruence $13x \equiv 19 \pmod{29}$.
$\quad$ Solution. Multiply both sides of the congruence by 9 to get $x \equiv 9 \cdot 19 \equiv 171 \equiv 26 \pmod{29}$. As a check, you can multiply $26 \cdot 13$ to get $338 \equiv 19 \pmod{29}$.

Problems 3.5

1. Show that the congruence $10x \equiv 14 \pmod 8$ has the same set of solutions as the congruence $5x \equiv 7 \pmod 4$.

2. Find the inverse of 9 modulo 23 using the methods of this section. Check your answer by hand.

3. Compute $3/8 \bmod 13$ and $-14/3 \bmod 17$.

4. Compute $3/2 \bmod 35$ and $-3/14 \bmod 55$.

5. Find the inverse of 34 modulo 113 using a sequence of matrix products. Check your answer by multiplication.

6. Use a computer algebra system to calculate the inverse of 28394325 modulo 849289528 and to check your answer.

7. Solve the congruence $7382784739x \equiv 1727372727 \pmod{2783479827}$. Check your answer by plugging your answer back into the congruence and using a computer algebra system to do the arithmetic.

8. Let n be an odd integer greater than 1. Show that

$$\left\{ -\frac{n-1}{2}, \ldots, -1, 0, 1, \ldots, \frac{n-1}{2} \right\}$$

is a complete residue system modulo n.

9. Construct a multiplication table for the integers modulo 5 that uses the residue system

$$\left\{ -\frac{n-1}{2}, \ldots, -1, 0, 1, \ldots, \frac{n-1}{2} \right\}$$

(see Problem 8). Find two solutions to the equation $x^2 = -1$.

10. Solve the equation $93x + 47 = 61$ in the integers modulo 101.

11. Solve the system

$$23x + 37y = 14$$
$$53x + 17y = 25$$

of linear equations in the integers modulo 101.

12. Solve the equation $x^2 + x + 10 = 0$ in the integers modulo 11.

13. For which primes p does -1 have a square root modulo p?

14. Does the matrix

$$\begin{pmatrix} 1 & 2 \\ 3 & 3 \end{pmatrix}$$

have an inverse in the integers modulo 5? If so, find it. If not, why not?

Chapter 4

Ciphers

The history of America and of secret communications includes many examples of enterprising men and women who, with little in the way of resources, developed innovative devices and systems that have become a part of this cryptologic legacy of freedom. One of the most inspiring stories is the creation of slave quilts in the early and mid-1800s. The secret messages embedded in the quilts, some say, assisted slaves from the South in their efforts to escape to freedom in the North. Each quilt contained a specific code or message that conveyed important information to those who were attempting the dangerous journey from the southern regions of the nation to the free states and Canada.

National Security Agency web site

People have disguised messages for as long as written languages have been used. A **cipher** is a method for disguising messages by replacing each letter by another letter, by a number, or by some other symbol.

4.1 Cryptography

Definition 4.1 *Cryptography is the art of disguising messages so that only friendly eyes can read them. The original message is called the **plaintext**; the disguised text is called the **ciphertext**.*

Encryption is going from plaintext to ciphertext. **Decryption** is going from ciphertext to plaintext.

The goals of a cryptographer are:

1. To provide an easy and inexpensive way for an authorized user to encrypt and decrypt messages

2. To make it difficult and expensive for an unauthorized user to decrypt the ciphertext

We will normally write plaintext in lowercase letters and ciphertext in uppercase letters.

Caesar Cipher

In the method of encryption attributed to the Roman emperor Julius Caesar, the lower case plaintext letters are replaced by upper case ciphertext letters according to the following scheme:

a	b	c	d	e	f	g	h	i	j	k	l	m
↓	↓	↓	↓	↓	↓	↓	↓	↓	↓	↓	↓	↓
D	E	F	G	H	I	J	K	L	M	N	O	P

n	o	p	q	r	s	t	u	v	w	x	y	z
↓	↓	↓	↓	↓	↓	↓	↓	↓	↓	↓	↓	↓
Q	R	S	T	U	V	W	X	Y	Z	A	B	C

Table 4.1   Caesar cipher

Note that the ciphertext alphabet is shifted to the left by three, and the letters A, B, and C are put on the end. We can give a formula for this cipher by thinking of the letters A through Z (or a through z) as corresponding to the numbers 0 through 25 as shown in Table 4.2.

A	B	C	D	E	F	G	H	I	J	K	L	M
↕	↕	↕	↕	↕	↕	↕	↕	↕	↕	↕	↕	↕
0	1	2	3	4	5	6	7	8	9	10	11	12

N	O	P	Q	R	S	T	U	V	W	X	Y	Z
↕	↕	↕	↕	↕	↕	↕	↕	↕	↕	↕	↕	↕
13	14	15	16	17	18	19	20	21	22	23	24	25

Table 4.2   Letters are numbers

In the **Caesar cipher**, the formula

$$y = (x + 3) \bmod 26$$

is used to encrypt and the formula

$$x = (y - 3) \bmod 26$$

to decrypt. To encrypt the letter f, which corresponds to 5, we form the sum $5 + 3 = 8$, which corresponds to the letter I. To encrypt the letter y, which corresponds to 24, we form the sum $24 + 3 = 27$ and reduce modulo 26 to get 1, which corresponds to the letter B. To decrypt the letter A, we form $0 - 3 = -3$ which is 23 modulo 26, so A decrypts to x.

The arithmetic of encrypting with the Caesar cipher is shown by the arrows in Table 4.3.

a	b	c	d	e	f	g	h	i	j	k	l	m
↓	↓	↓	↓	↓	↓	↓	↓	↓	↓	↓	↓	↓
0	1	2	3	4	5	6	7	8	9	10	11	12
↓	↓	↓	↓	↓	↓	↓	↓	↓	↓	↓	↓	↓
3	4	5	6	7	8	9	10	11	12	13	14	15
↓	↓	↓	↓	↓	↓	↓	↓	↓	↓	↓	↓	↓
D	E	F	G	H	I	J	K	L	M	N	O	P

n	o	p	q	r	s	t	u	v	w	x	y	z
↓	↓	↓	↓	↓	↓	↓	↓	↓	↓	↓	↓	↓
13	14	15	16	17	18	19	20	21	22	23	24	25
↓	↓	↓	↓	↓	↓	↓	↓	↓	↓	↓	↓	↓
16	17	18	19	20	21	22	23	24	25	0	1	2
↓	↓	↓	↓	↓	↓	↓	↓	↓	↓		↓	↓
Q	R	S	T	U	V	W	X	Y	Z	A	B	C

Table 4.3 Caesar cipher arithmetic

t	h	e	c	a	e	s	a	r	c	i
↓	↓	↓	↓	↓	↓	↓	↓	↓	↓	↓
19	7	4	2	0	4	18	0	17	2	8
↓	↓	↓	↓	↓	↓	↓	↓	↓	↓	↓
22	10	7	5	3	7	21	3	20	5	11
↓	↓	↓	↓	↓	↓	↓	↓		↓	↓
W	K	H	F	D	H	V	D	U	F	L

p	h	e	r	i	s	e	a	s	y
↓	↓		↓	↓	↓	↓	↓	↓	↓
15	7	4	17	8	18	4	0	18	24
↓	↓	↓	↓	↓	↓	↓	↓	↓	↓
18	10	7	20	11	21	7	3	21	1
↓	↓	↓	↓	↓	↓	↓	↓		↓
S	K	H	U	L	V	H	D	V	B

Table 4.4 Caesar encrypting

Example 4.2 The plaintext *"the caesar cipher is easy"* becomes the ciphertext *"WKH FDHVDU FLSKHU LV HDVB"* as shown in Table 4.4 from top to bottom. Each letter in the plaintext is replaced by its corresponding number, the numbers are shifted by the formula $y = (x + 3) \bmod 26$, and finally each number is replaced by its corresponding letter. Decryption is done in the reverse order, from bottom to top.

Vigenère Cipher

The Caesar cipher is completely described by the string of letters

$$\text{D E F G H I J K L M N O P Q R S T U V W X Y Z A B C}$$

which is the ciphertext for the plaintext abcdefghijklmnopqrstuvwxyz:

$$
\begin{array}{cccccccccccc}
a & b & c & d & e & f & g & h & i & j & k & l & m \\
D & E & F & G & H & I & J & K & L & M & N & O & P
\end{array}
$$

$$
\begin{array}{cccccccccccc}
n & o & p & q & r & s & t & u & v & w & x & y & z \\
Q & R & S & T & U & V & W & X & Y & Z & A & B & C
\end{array}
$$

A rearrangement of the twenty-six letters is called an **alphabet**. The Caesar cipher is one of the twenty-six **shift ciphers**. Each shift cipher has an alphabet starting with a different letter. Here are the alphabets for two other shift ciphers, shifting by 10 and shifting by 24:

$$\text{K L M N O P Q R S T U V W X Y Z A B C D E F G H I J}$$

$$\text{Y Z A B C D E F G H I J K L M N O P Q R S T U V W X}$$

A **monoalphabetic cipher** uses just one alphabet. A **polyalphabetic cipher** uses different alphabets for different positions in the text. The simplest polyalphabetic cipher is the **Vigenère cipher** which uses the twenty-six shift ciphers. Often the alphabets used in a Vigenère cipher are described by a **key word**. For example, if the key word is CLINT, then we use the alphabets starting with those letters:

$$\text{C D E F G H I J K L M N O P Q R S T U V W X Y Z A B}$$

$$\text{L M N O P Q R S T U V W X Y Z A B C D E F G H I J K}$$

$$\text{I J K L M N O P Q R S T U V W X Y Z A B C D E F G H}$$

$$\text{N O P Q R S T U V W X Y Z A B C D E F G H I J K L M}$$

$$\text{T U V W X Y Z A B C D E F G H I J K L M N O P Q R S}$$

The first letter of plaintext is enciphered using the first alphabet, the second letter using the second alphabet, until we reach the sixth letter of plaintext which is again enciphered using the first alphabet and so on. So the word **attack** would be enciphered as CFCNVM. Notice that the first and third letters of the ciphertext are the same, but the first and third letters of the plaintext are different. Also the second and third letters of the plaintext are the same, but the second and third letters of the ciphertext are different.[1]

Affine Ciphers

The Caesar cipher is a special kind of affine cipher.

Definition 4.3 *An **affine cipher** is given by $y = (kx + s) \bmod 26$, where x is the plaintext integer, k is the multiplier, and s is the shift. If $k = 1$, the cipher is called a **shift cipher**.*

For the Caesar cipher, $k = 1$ and $s = 3$.
If $\gcd(k, 26) = 1$, then the equation

$$y = (kx + s) \bmod 26$$

can be solved uniquely for x

$$x = j(y - s) \bmod 26$$

where $jk \equiv 1 \pmod{m}$. If $\gcd(k, 26) > 1$, then different plaintexts will give the same ciphertext, so the ciphertext cannot be uniquely decrypted. For example, if $k = 8$ and $s = 0$, then we get the encryption shown in Table 4.5. Notice the ambiguity holds for friend and foe alike.

A	B	C	D	E	F	G	H	I	J	K	L	M
↕	↕	↕	↕	↕	↕	↕	↕	↕	↕	↕	↕	↕
0	8	16	24	6	14	22	4	12	20	2	10	18

N	O	P	Q	R	S	T	U	V	W	X	Y	Z
↕	↕	↕	↕	↕	↕	↕	↕	↕	↕	↕	↕	↕
0	8	16	24	6	14	22	4	12	20	2	10	18

Table 4.5   Ambiguous encryption

Example 4.4 To encrypt the message

Short ciphers are hard to break

using the affine cipher $y = (9x + 7) \bmod 26$, we break the message up into five-letter groups and replace letters as in the following table:

SHORT	CIPHE	RSARE	HARDT	OBREA	K
NSDEW	ZBMSR	ENHER	SHEIW	DQERH	T

---

[1]To play with Vigenère ciphers, go to http://www.math.fau.edu/Richman/viginere.htm. You will be able to encipher there using a key word. Can you figure out how to decipher there also?

In particular, $18 \cdot 9 + 7 = 169 = 26 \cdot 6 + 13 \equiv 13 \pmod{m}$, so $S \to 18 \to 13 \to N$.

To decrypt, it is necessary to solve $y \equiv 9x + 7 \pmod{26}$ for x in terms of y. First, we subtract 7 from both sides to get $y - 7 \equiv 9x \pmod{26}$. Since $9 \cdot 3 = 27 = 26 + 1 \equiv 1 \pmod{26}$, it follows that $3 = 9^{-1} \bmod 26$ and hence $x = 3(y - 7) \bmod 26$. In particular, $3(13 - 7) = 3 \cdot 6 = 18$ and hence $N \to 13 \to 18 \to S$.

There are 26 shift ciphers. The number of affine ciphers $y = (kx + s) \bmod 26$ is $12 \cdot 26 = 312$ because there are 12 choices of k with $\gcd(k, 26) = 1$, and 26 choices of s.

Polyalphabetic Ciphers

Two or more affine ciphers can be used to construct a **polyalphabetic cipher**. For example, two affine ciphers can be alternated, as in the following example.

Example 4.5 Encrypt the plaintext

The British are coming!

using the affine cipher $y = 5x + 7 \bmod 26$ for the odd-numbered letters and $y = 3x + 4 \bmod 26$ for the even-numbered letters. So

$$T \longrightarrow 19 \longrightarrow 5 \cdot 19 + 7 \bmod 26 = 24 \longrightarrow Y$$

and

$$H \longrightarrow 7 \longrightarrow 3 \cdot 7 + 4 \bmod 26 = 25 \longrightarrow Z$$

Table 4.6 yields the ciphertext YZB HOCYCTZ HDB KZOVRL.

t	h	e	b	r	i	t	i	s	h	a	r	e	c	o	m	i	n	g
↓	↓	↓	↓	↓	↓	↓	↓	↓	↓	↓	↓	↓	↓	↓	↓	↓	↓	↓
19	*7*	*4*	*1*	*17*	*8*	*19*	*8*	*18*	*7*	*0*	*17*	*4*	*2*	*14*	*12*	*8*	*13*	*6*
↓	↓	↓	↓	↓	↓	↓	↓	↓	↓	↓	↓	↓	↓	↓	↓	↓	↓	↓
24	*25*	*1*	*7*	*14*	*2*	*24*	*2*	*19*	*25*	*7*	*3*	*1*	*10*	*25*	*14*	*21*	*17*	*11*
↓	↓	↓	↓	↓	↓	↓	↓	↓	↓	↓	↓	↓	↓	↓	↓	↓	↓	↓
Y	*Z*	*B*	*H*	*O*	*C*	*Y*	*C*	*T*	*Z*	*H*	*D*	*B*	*K*	*Z*	*O*	*V*	*R*	*L*

Table 4.6   Polyalphabetic encryption

Problems 4.1

1. Use the Caesar cipher to encrypt the plaintext

Hello.

2. Use the Caesar cipher to decrypt the ciphertext

   ZOVMQ LDOXM EVFPQ EBPZF BKZBL CPBZO
   BQTOF QFKD

3. Use the shift cipher $y = x + 6$ to encrypt the plaintext

   *Encryption products with less than sixty four bits
   are freely exportable.*

4. Use the affine cipher $y = 5x + 7 \bmod 26$ to encrypt the plaintext

   *The width of a complete filled rectangle
   must be a divisor of the length of the message.*

5. Use the Caesar cipher to decrypt the ciphertext

   JRRGE BH

6. Use the Caesar cipher to unscramble the ciphertext

   LDPJR LQJWR VSDLQ WRILJ KWDQD UPBZL
   WKRXW DJHQH UDODQ GWKHQ FHWRW KLIHDV
   WWRIL JKWDJ HQHUD OZLWK RXWDQ DUPB

   This statement is ascribed to Julius Caesar himself.

7. Unscramble the following ciphertext, which was encrypted using the affine
   cipher $y = x + 5 \bmod 26$.

   HFJXF WNXHT SXNIJ WJIYT GJTSJ TKYMJ
   KNWXY UJWXT SXYTM FAJJA JWJRU QTDJI
   JSHWD UYNTS KTWYM JXFPJ TKXJH ZWNSL
   RJXXF LJX

8. Use the Vigenère cipher with keyword SING to encrypt the plaintext

   There are two kinds of music: country and western.

9. Use the Vigenère cipher with keyword GOLF to decrypt the ciphertext

   JFTAKTZWYVZBVIEYLCCIUIRM

10. Decrypt the ciphertext

    HEJGI JTTPU WHBDH UHPDII AMTLH SB1UF
    1ZUFT 1ZUJS IHVHU B

    which was encrypted using an affine cipher $y = mx + b \bmod 26$, knowing
    that the plaintext begins with *el*.

11. Encrypt the message

> You should be aware that encrypted
> communications are illegal in some
> parts of the world.

using a polyalphabetic cipher that alternates the use of the three affine ciphers

$$f(x) = 11x + 2 \bmod 26$$
$$g(x) = 15x + 5 \bmod 26$$
$$h(x) = 19x + 7 \bmod 26$$

12. Decrypt the ciphertext

    DGFEH LDJNE DNPOF DEFHV LU

    encrypted using a polyalphabetic cipher that alternated the use of the three affine ciphers

$$f(x) = 11x + 2 \bmod 26$$
$$g(x) = 15x + 5 \bmod 26$$
$$h(x) = 19x + 7 \bmod 26$$

13. Plaintext is encrypted using the affine cipher $y = 3x + 5 \bmod 26$; then the ciphertext in encrypted again using the affine cipher $y = 15x + 4 \bmod 26$. Give a simple equivalent to the compound cipher.

14. The affine cipher $y = mx + b \bmod 26$ has an inverse cipher for only 12 different choices of m. What is the effect of increasing the alphabet size from 26 to 27? How about 29? 30?

4.2   Cryptanalysis

Cryptanalysis is the art of breaking codes. For every coded message, there might be several unauthorized persons trying to learn what the message says. This could involve industrial espionage, electronic eavesdropping, or simple curiosity.

The letter count for the first paragraph of this section is given in Table 4.7. The second column is the number of occurrences of each letter, while the third column gives the relative frequency.

Are these relative frequencies typical? The fourth column gives the relative frequencies of letters from a large sample of written English. As you can see from the table, letters such as 'Y' are over-represented in Paragraph 1, while letters such as 'X' are under-represented. Although 'E' has the highest relative frequency in both lists, 'S' is second in one list while 'T' is second in the other.

However, the relative frequencies of letters in the small sample are in general agreement with that of the large sample.

As the amount of text increases, we normally get better agreement between the relative frequencies of letters in the text and their expected relative frequencies. This phenomenon is used by cryptanalysts to break codes. Very short messages are usually much harder to break than long messages.

Letter	Frequency	Relative Frequency	Expected Relative Frequency
A	17	7.2%	7.3%
B	1	0.4%	0.9%
C	8	3.6%	3.0%
D	6	2.7%	4.4%
E	24	10.3%	13.0%
F	4	1.8%	2.8%
G	9	4.0%	1.6%
H	8	3.6%	3.5%
I	18	8.1%	7.4%
J	0	0.0%	0.2%
K	1	0.4%	0.3%
L	12	4.9%	3.5%
M	2	0.9%	2.5%
N	16	6.7%	7.8%
O	18	7.6%	7.4%
P	8	3.6%	2.7%
Q	0	0.0%	0.3%
R	21	9.0%	7.7%
S	17	7.6%	6.3%
T	20	8.5%	9.3%
U	4	1.8%	2.7%
V	3	1.3%	1.3%
W	2	0.9%	1.6%
X	0	0.0%	0.5%
Y	11	4.9%	1.9%
Z	0	0.0%	0.1%

Table 4.7   Letter count from selected text

Example 4.6 If we suspect that a simple shift cipher, $y = (x + b) \mod 26$, was used, we count the frequencies of each letter and shift the left side of the table up until we get a good match with the expected frequencies.[2]

---

[2]You can get frequency counts for any text you enter at http://www.math.fau.edu/Richman/Liberal/freqs.htm.

Sample					Expected
‖‖‖‖‖‖‖	14	I	A	12.191	‖‖‖‖‖‖,
‖	2	J	B	1.503	‖
‖‖‖	5	K	C	5.01	‖‖‖
‖‖‖	5	L	D	7.348	‖‖‖‖,
‖‖‖‖‖‖‖‖‖	19	M	E	21.71	‖‖‖‖‖‖‖‖‖‖‖
‖‖	4	N	F	4.676	‖‖‖
‖‖	4	O	G	2.672	‖
‖‖‖‖	8	P	H	5.845	‖‖‖
‖‖‖‖‖	10	Q	I	12.358	‖‖‖‖‖,
	0	R	J	0.334	,
‖	1	S	K	0.501	‖
‖‖‖‖	9	T	L	5.845	‖‖‖
‖‖‖	6	U	M	4.175	‖‖,
‖‖‖‖‖‖	13	V	N	13.026	‖‖‖‖‖‖
‖‖‖‖‖‖	12	W	O	12.358	‖‖‖‖‖,
‖	1	X	P	4.509	‖‖,
	0	Y	Q	0.501	‖
‖‖‖‖‖	11	Z	R	12.859	‖‖‖‖‖‖
‖‖‖‖‖	11	A	S	10.521	‖‖‖‖,
‖‖‖‖‖‖‖	17	B	T	15.531	‖‖‖‖‖‖‖,
‖‖	4	C	U	4.509	‖‖,
‖‖	4	D	V	2.171	‖,
‖‖‖	5	E	W	2.672	‖
	0	F	X	0.835	‖
‖‖	4	G	Y	3.173	‖‖,
	0	H	Z	0.167	,

Table 4.8   Frequencies for shifted ciphertext

Table 4.8 was obtained from the ciphertext

BEMVB   GGMIZ   AIOWB   ZQITL   QDQAQ   WVEIA   MAAMV

BQITT   GBPMN   IABMA   BNIKB   WZQVO   UMBPW   LSVWE

VAQVK   MBPMV   QUXZW   DMLIT   OWZQB   PUAPI   DMJMM

VQVDM   VBMLB   PIBIT   TWECA   BWNIK   BWZUC   KPTIZ

OMZVC   UJMZA   BPIVE   MKWCT   LNWZU   MZTG

This matching scheme corresponds to the shift cipher $y = (x + 8) \bmod 26$. We use the inverse shift $y = (x - 8) \bmod 26$ to get the plaintext

```
TWENT   YYEAR   SAGOT   RIALD   IVISI   ONWAS   COSEN
TIALL   YTHEF   ASTES   TFACT   ORING   METHO   DKNOW
NSINC   ETHEN   IMPRO   VEDAL   GORIT   HMSHA   VEBEE
NINVE   NTEDT   HATAL   LOWUS   TOFAC   TORMU   CHLAR
GERNU   MBERS   THANW   ECOUL   DFORM   ERLY
```

It's not hard to insert punctuation and spaces to get the message:

> *Twenty years ago trial division was essentially the fastest fac-*
> *toring method known. Since then improved algorithms have been*
> *invented that allow us to factor much larger numbers than we could*
> *formerly.*

Example 4.7 Table 4.9 gives the letter frequencies for the ciphertext

```
RZDTZ   ECATR   TBSPZ   GLCAD   RLOYZ   SYTVN
LCTRZ   KALRC   LXBCT   IDBCA   TDBCL   XRBRL
OCATX   BRTZE   CATTS   SLKCL   XXNGY   TDTCA
ZIZEA   TOIGL   HSTOR   CGB
```

This time, sliding the left-hand side up or down never gives a good match, so we try the assumption that this is an affine cipher $y = (kx + s) \bmod 26$ rather than a shift cipher. The two letters that appear most frequently in the ciphertext are likely to correspond to plaintext letters such as e and t that we expect to see often. In this case, T appears 16 times and C appears 12 times, so let us assume that in this cipher, e $\to$ T and t $\to$ C; that is, $19 = (4k + s) \bmod 26$ and $2 = (19k + s) \bmod 26$ (since C $\leftrightarrow$ 2, e $\leftrightarrow$ 4, and T $\leftrightarrow$ 19 $\leftrightarrow$ t).

To solve the system

$$19 = 4k + s \bmod 26$$
$$2 = 19k + s \bmod 26$$

first eliminate s by subtracting the second equation from the first to get

$$17 = 11k \bmod 26$$

Multiply this equation by 19, which is the inverse of 11 modulo 26, to get

$$k = 17 \cdot 19 \bmod 26 = 11$$

Substituting $k = 11$ into the equation $19 = (4k + s) \bmod 26$ gives

$$s = 19 - 4 \cdot 11 \bmod 26 = 1$$

If the ciphertext is given by $y = 11x + 1 \bmod 26$, then to decrypt the ciphertext we must solve this equation for x in terms of y. We have $11x = y - 1 \bmod 26$ and 19 is the inverse of 11 modulo 26, so

$$x = 19(y - 1) \bmod 26 = 19y + 7 \bmod 26$$

Sample				Expected
‖‖‖‖	8	A	7.884	‖‖‖
‖‖‖	7	B	0.972	‖
‖‖‖‖‖	12	C	3.24	‖‖.
‖‖	5	D	4.752	‖‖
‖	3	E	14.04	‖‖‖‖‖‖
	0	F	3.024	‖‖
‖‖	4	G	1.728	‖
‖	1	H	3.78	‖‖
‖	3	I	7.992	‖‖‖‖
	0	J	0.216	.
‖	2	K	0.324	,
‖‖‖‖	10	L	3.78	‖‖
	0	M	2.7	‖
‖	2	N	8.424	‖‖‖.
‖‖	4	O	7.992	‖‖‖
‖	1	P	2.916	‖
	0	Q	0.324	,
‖‖‖	9	R	8.316	‖‖‖.
‖‖	5	S	6.804	‖‖
‖‖‖‖‖	15	T	10.044	‖‖‖‖
	0	U	2.916	‖
‖	1	V	1.404	‖
	0	W	1.728	‖
‖‖	5	X	0.54	,
‖	3	Y	2.052	‖
‖‖‖	8	Z	0.108	.

Table 4.9   Simple shift fails

Applying this to the ciphertext, and then inserting the appropriate spaces and punctuation, we get

Some of these algorithms involve quite sophisticated mathematics, as in the case of the elliptic curve method of Hendrik Lenstra.

Example 4.8 Matrix algebra can also be used to solve the system

$$19 = 4k + s \bmod 26$$
$$2 = 19k + s \bmod 26$$

of congruences. The system is equivalent to the matrix equation

$$\begin{pmatrix} 4 & 1 \\ 19 & 1 \end{pmatrix} \begin{pmatrix} k \\ s \end{pmatrix} \bmod 26 = \begin{pmatrix} 19 \\ 2 \end{pmatrix}$$

or

$$\begin{pmatrix} k \\ s \end{pmatrix} = \begin{pmatrix} 4 & 1 \\ 19 & 1 \end{pmatrix}^{-1} \begin{pmatrix} 19 \\ 2 \end{pmatrix} \bmod 26 = \begin{pmatrix} 11 \\ 1 \end{pmatrix}$$

Problems 4.2

1. The ciphertext

    ZNKUR JKYZQ TUCTK TIXEV ZOUTJ KBOIK
    OYZNK YIEZG RK

   was encrypted using a shift

   $$y = x + a \bmod 26$$

   Determine a and decipher the message.

2. The ciphertext

    DROBO KBODG YWKSX DIZOC YPMSZ ROBCK
    CELCD SDEDS YXMSZ ROBKX NKDBK XCZYC
    SDSYX MSZRO B

   was encrypted using a shift

   $$y = x + a \bmod 26$$

   Determine a and decipher the message.

3. The ciphertext

    RJUMK QRADU KSNMO MRUPS ZRGSH SWNPX
    OUKUM SZGSS PGJOK JJAPU LAKRD QRJUM
    AIUPO IUNMO SNM

   was encrypted using an affine cipher

   $$y = kx \bmod 26$$

   Determine k and decipher the message.

4. The ciphertext

    AOEBX CPEWG UGUAZ BXAHC DEOEJ ANMZC
    DDCPU JDXCA ZBXAH CDEWA ZYAMW CNOEB
    XCPCV HMDXC WAGCO EBXCP DCFDZ CDDCP

   was encrypted using an affine cipher

   $$y = kx \bmod 26$$

   Determine k and decipher the message.

5. The ciphertext

$$
\begin{array}{llllll}
\text{GWUUE} & \text{SWUMW} & \text{JWWRA} & \text{CLWLP} & \text{IMORL} & \text{ORSEN} \\
\text{OWAWC} & \text{HNEJA} & \text{DGWRF} & \text{EJCQR} & \text{LFDWA} & \text{IVORL} \\
\text{WJORE} & \text{UNOJE} & \text{VERLM} & \text{JOFOR} & \text{SFDWG} & \text{WUUES} \\
\text{WEVCR} & \text{SFDWV} & \text{WRSFD} & \text{CHFDW} & \text{AIVOR} & \text{LWJ} \\
\end{array}
$$

was encrypted using an affine cipher

$$y = kx + s \bmod 26$$

Determine k and s, then decipher the message.

6. The two shift ciphers

$$
\begin{aligned}
f\left(x\right) &= x + a \bmod 26 \\
g\left(x\right) &= x + b \bmod 26
\end{aligned}
$$

were combined in a polyalphabetic cipher to construct the ciphertext

$$
\begin{array}{llllllll}
\text{VMGFT} & \text{YKXVH} & \text{CWNLQ} & \text{WOFPB} & \text{CXQSG} & \text{THYJJ} & \text{VBGSV} & \text{DPNPJ} \\
\text{QWKLK} & \text{SCQPF} & \text{XFLTE} & \text{TFJVF} & \text{NPGWU} & \text{FUJNJ} & \text{EYIWQ} & \text{ZRTHN} \\
\text{PIKFP} & \text{XTJEW} & \text{WNVJF} & \text{GAYJJ} & \text{OFTNP} & \text{JETTU} & \text{UNPBQ} & \text{WNIYF} \\
\text{TYYTV} & \text{TUJPI} & \text{OJUXC} & \text{LGXKS} & \text{VMGSC} & \text{ACOQQ} & \text{CSIZC} & \text{LG} \\
\end{array}
$$

Decrypt the message.

7. The two shift ciphers

$$
\begin{aligned}
f\left(x\right) &= x + a \bmod 26 \\
g\left(x\right) &= x + b \bmod 26
\end{aligned}
$$

were combined in a polyalphabetic cipher to construct the ciphertext

$$
\begin{array}{llllllll}
\text{ASPCA} & \text{JMTCP} & \text{JZKPA} & \text{LSVLC} & \text{ZLAEL} & \text{YKPKE} & \text{OPKPK} & \text{TJLAT} \\
\text{VYVQA} & \text{SLYHG} & \text{HUVNV} & \text{OLEHW} & \text{RPYPE} & \text{SPMPE} & \text{ASLPE} & \text{SPMPE} \\
\text{PYJWB} & \text{OLDHO} & \text{PDWWH} & \text{JVQWS} & \text{VEVRY} & \text{LWSZP} & \text{XFPAT} & \text{PUEHY} \\
\text{KEOPV} & \text{CPRPY} & \text{HWJZK} & \text{PHWVY} & \text{NHPEO} & \text{LUPEA} & \text{SLULA} & \text{TVYVQ} \\
\text{OZDEO} & \text{PJZKP} & \text{DZYVL} & \text{OKPKT} & \text{JLATV} & \text{YJPYP} & \text{TZUTL} & \text{DPYJW} \\
\text{BOLOZ} & \text{ALPJS} & \text{LDIJA} & \text{SLEOP} & \text{UOLAB} & \text{EFDLN} & \text{YPALY} & \text{JVQKP} \\
\text{MPUDL} & \text{OVYHW} & \text{KLAHV} & \text{ZKFZD} & \text{LYHEV} & \text{CQZOY} & \text{TNJLP} & \text{YVQHC} \\
\text{PKVYH} & \text{LUOUL} & \text{CLQZW} & \text{CLDPO} & \text{LYAAL} & \text{ELCZZ} & \text{UKHSA} & \text{SLYHG} \\
\text{HUVGL} & \text{ELCHY} & \text{ZLUOA} & \text{SLTYQ} & \text{HXPWP} & \text{PZEYL} & \text{CPSPK} & \text{EVEOP} \\
\text{JPYPT} & \text{ZUJMC} & \text{VXASL} & \text{TYSVX} & \text{LDVYA} & \text{SLYHG} & \text{HUVCL} & \text{DLCCL} \\
\text{ATVYD} & \text{SPNOT} & \text{UNSFK} & \text{PZAHC} & \text{ADVQH} & \text{CPKVY} & \text{HYLHT} & \text{PETJZ} \\
\text{HYKFA} & \text{LO} \\
\end{array}
$$

Decrypt the message.

8. The three shift ciphers

$$f(x) = x + a \bmod 26$$
$$g(x) = x + b \bmod 26$$
$$h(x) = x + c \bmod 26$$

were combined in a polyalphabetic cipher to construct the ciphertext

```
DHNMO MOTJV KNBSF RILRE EONCE AUVYW EMKOR NNKBX
ETCRR NOHEX DAODO SFCIM NXWNB ELROB ONOYB TQOIA
ZRXPI NXCHS NWKVJ TOJXD NXGUS SQ
```

Decrypt the message.

4.3 Substitution and Permutation Ciphers

In this section we look at two kinds of ciphers that do not have simple formulas. The first is the general substitution cipher. Shift ciphers and affine ciphers are substitution ciphers.

Substitution Ciphers

Definition 4.9 *A **substitution cipher** is a one-to-one function from the set* $\{A, B, C, \ldots, Z\}$ *onto itself.*

The function tells us what ciphertext letter corresponds to a plaintext letter. How many substitution ciphers are there? There are 26 possibilities for the ciphertext letter corresponding to A. After choosing the ciphertext letter for A, there are 25 possibilities for the ciphertext letter for B, twenty four for C, and so on. Altogether there are

$$26! = 26 \cdot 25 \cdot 24 \cdots 2 \cdot 1 = 403\,291\,461\,126\,605\,635\,584\,000\,000$$

substitution ciphers.

A typical substitution cipher σ is given in Table 4.10, using numbers instead of letters. We denote by x^σ the ciphertext letter that σ associates with the plaintext letter x under the cipher σ.

x =	0	1	2	3	4	5	6	7	8	9	10	11	12
	↓	↓	↓	↓	↓	↓	↓	↓	↓	↓	↓	↓	↓
x^σ =	17	10	22	23	11	1	5	13	7	0	14	3	20

x =	13	14	15	16	17	18	19	20	21	22	23	24	25
	↓	↓	↓	↓	↓	↓	↓	↓	↓	↓	↓	↓	↓
x^σ =	2	16	8	12	15	25	21	9	18	6	24	19	4

Table 4.10 Substitution cipher σ

This table is a clumsy way to present σ. We will investigate more efficient ways.

Substitution ciphers are special cases of permutations.

Definition 4.10 *A **permutation** is a one-to-one function from a finite set onto itself.*

We will develop a more concise notation for a permutation. Here is our current notation for a particular permutation on $\{0, 1, 2, 3, 4, 5, 6, 7, 8, 9\}$:

$$
\begin{array}{cccccccccc}
0 & 1 & 2 & 3 & 4 & 5 & 6 & 7 & 8 & 9 \\
\downarrow & \downarrow & \downarrow & \downarrow & \downarrow & \downarrow & \downarrow & \downarrow & \downarrow & \downarrow \\
5 & 8 & 4 & 1 & 6 & 9 & 2 & 0 & 3 & 7
\end{array}
$$

We write this more concisely by starting with 0 and repeatedly applying the permutation until we return to where we started $0 \rightarrow 5 \rightarrow 9 \rightarrow 7 \rightarrow 0$. This is called a **cycle**, and we write it simply as (0 5 9 7). Now repeat the process starting with the first element that has not yet appeared, so we start with 1 and get $1 \rightarrow 8 \rightarrow 3 \rightarrow 1$ which we represent as (1 8 3). Finally, we get $2 \rightarrow 4 \rightarrow 6 \rightarrow 2$ which we represent as (2 4 6). Note that each integer from 0 through 9 appears in exactly one of these cycles. The first, (0 5 9 7), is called a 4-**cycle** and the other two are called 3-**cycles**. Finally, we write the permutation as

$$(0\ 5\ 9\ 7)(1\ 8\ 3)(2\ 4\ 6)$$

It doesn't matter where you start in a cycle. In fact,

$$(0\ 5\ 9\ 7) = (7\ 0\ 5\ 9) = (9\ 7\ 0\ 5) = (5\ 9\ 7\ 0)$$

since each corresponds to the cycle

$$
\begin{array}{ccc}
0 & \longrightarrow & 5 \\
\uparrow & & \downarrow \\
7 & \longleftarrow & 9
\end{array}
$$

The procedure can be used to write any permutation as a product of disjoint cycles. The cycles are called **disjoint** because no symbol appears in more than one cycle.

A cycle τ can be thought of as a permutation by defining $x^\tau = x$ for each x not in the cycle. The same is true for a product of disjoint cycles. With this interpretation, we state the following theorem.

Theorem 4.11 *Every permutation can be written as a product of disjoint cycles.*

Proof. (See problem 9.)   ∎

Example 4.12 Let $n = 10$. The permutation

$$\sigma = (0\ 3\ 2)(1\ 9\ 6\ 8\ 7\ 5)(4)$$

consists of a 3-cycle

a 6-cycle

$$
\begin{array}{ccccc}
1 & \longrightarrow & 9 & \longrightarrow & 6 \\
\uparrow & & & & \downarrow \\
5 & \longleftarrow & 7 & \longleftarrow & 8
\end{array}
$$

and a 1-cycle

$$4 \circlearrowleft$$

To invert a cycle, just reverse the arrows. The 4-cycle $(2\ 7\ 4\ 5)$ has $(2\ 5\ 4\ 7)$ as its inverse because reversing the arrows below in the left square gives the right square:

$$
\begin{array}{ccc}
2 & \dashrightarrow & 7 \\
\uparrow & & \downarrow \\
5 & \longleftarrow & 4
\end{array}
\qquad
\begin{array}{ccc}
2 & \longleftarrow & 7 \\
\downarrow & & \uparrow \\
5 & \longrightarrow & 4
\end{array}
$$

In particular, a 2-cycle is its own inverse, as is a 1-cycle.

If a permutation σ is written as a product of disjoint cycles then σ^{-1} is the product of the inverses of these cycles. For example, if $\sigma = (0\ 8\ 2\ 3\ 5)(1\ 4)(6\ 9\ 7)$ then $\sigma^{-1} = (0\ 5\ 3\ 2\ 8)(1\ 4)(6\ 7\ 9)$.

Definition 4.13 *To obtain the **product** $\sigma\tau$ of two permutations, first apply σ then apply τ. That is, $x^{\sigma\tau} = (x^{\sigma})^{\tau}$.*

Example 4.14 Let $\sigma = (0\ 5\ 9)(1\ 8\ 4\ 2\ 3\ 7\ 6)$ and $\tau = (0\ 1)(2\ 5\ 8)(3\ 4\ 6\ 9\ 7)$. Then $2^{\sigma\tau} = (2^{\sigma})^{\tau} = 3^{\tau} = 4$ and $5^{\sigma\tau} = (5^{\sigma})^{\tau} = 9^{\tau} = 7$. The product $\sigma\tau$ can be written as a product of disjoint cycles:

$$
\begin{aligned}
\sigma\tau &= (0\ 5\ 9)(1\ 8\ 4\ 2\ 3\ 7\ 6)(0\ 1)(2\ 5\ 8)(3\ 4\ 6\ 9\ 7) \\
&= (0\ 8\ 6)\,(1\ 2\ 4\ 5\ 7\ 9)\,(3) \\
&= (0\ 8\ 6)\,(1\ 2\ 4\ 5\ 7\ 9)
\end{aligned}
$$

Theorem 4.15 *The inverse of a product $\sigma\tau$ can be written as*

$$(\sigma\tau)^{-1} = \tau^{-1}\sigma^{-1}$$

Proof. (See problem 10.) ■

Example 4.16 Let $\sigma = (0\ 5\ 9)(1\ 8\ 4\ 2\ 3\ 7\ 6)$ and $\tau = (0\ 1)(2\ 5\ 8)(3\ 4\ 6\ 9\ 7)$. Then

$$
\begin{aligned}
\tau^{-1}\sigma^{-1} &= (0\ 1)(2\ 8\ 5)(3\ 7\ 9\ 6\ 4)(0\ 9\ 5)(1\ 6\ 7\ 3\ 2\ 4\ 8) \\
&= (0\ 6\ 8)\,(1\ 9\ 7\ 5\ 4\ 2)\,(3) \\
&= (\sigma\tau)^{-1}
\end{aligned}
$$

When working with permutation ciphers, it is convenient to use directly the letters A, B, ..., Z rather than the integers 0, 1, ..., 25.

Example 4.17 Use the substitution cipher

$$
\sigma = (A\ P\ H\ I\ T\ X)(B\ E\ R\ C)(D\ N\ Z\ F\ V\ M)(G\ J\ K\ W\ L\ O\ Y\ Q\ S\ U)
$$

to encipher the plaintext

> If it is correct that the inscription rectangle is completely
> filled, then a solution will be obtained by rearranging
> the column so as to form plaintext.

Note that I$\longrightarrow$ T, F$\longrightarrow$ V, and so forth. Repeating this process for each plaintext character, we get

```
TVTXT   UBYCC   RBXXI   PXXIR   TZUBC   THXTY   ZCRBX
PZJOR   TUBYD   HORXR   OQVTO   ORNXI   RZPUY   OGXTY
ZLTOO   ERYEX   PTZRN   EQCRP   CCPZJ   TZJXI   RBYOG
DZUYP   UXYVY   CDHOP   TZXRA   X
```

The plaintext can be recovered by reading the permutation σ from right to left.

For small n, permutations on $\{1, 2, \ldots, n\}$ can be computed using permutation matrices. The permutation $\sigma = (1\ 4\ 2\ 5\ 3)$ corresponds to the 5×5 Boolean matrix

$$
S = \begin{pmatrix}
0 & 0 & 0 & 1 & 0 \\
0 & 0 & 0 & 0 & 1 \\
1 & 0 & 0 & 0 & 0 \\
0 & 1 & 0 & 0 & 0 \\
0 & 0 & 1 & 0 & 0
\end{pmatrix}
$$

where $S_{ij} = 1$ if and only if $i^\sigma = j$. To evaluate 3^σ you can compute the matrix product

$$(0 \ \ 0 \ \ 1 \ \ 0 \ \ 0) \begin{pmatrix} 0 & 0 & 0 & 1 & 0 \\ 0 & 0 & 0 & 0 & 1 \\ 1 & 0 & 0 & 0 & 0 \\ 0 & 1 & 0 & 0 & 0 \\ 0 & 0 & 1 & 0 & 0 \end{pmatrix} = (1 \ \ 0 \ \ 0 \ \ 0 \ \ 0)$$

This indicates that $3^{\sigma} = 1$.

Given $\tau = (1 \ 3 \ 2 \ 4 \ 5)$ and the corresponding matrix

$$T = \begin{pmatrix} 0 & 0 & 1 & 0 & 0 \\ 0 & 0 & 0 & 1 & 0 \\ 0 & 1 & 0 & 0 & 0 \\ 0 & 0 & 0 & 0 & 1 \\ 1 & 0 & 0 & 0 & 0 \end{pmatrix}$$

note that the permutation product

$$\sigma\tau = (1 \ 4 \ 2 \ 5 \ 3)(1 \ 3 \ 2 \ 4 \ 5) = (1 \ 5 \ 2)(3)(4)$$

corresponds to the matrix product

$$ST = \begin{pmatrix} 0 & 0 & 0 & 1 & 0 \\ 0 & 0 & 0 & 0 & 1 \\ 1 & 0 & 0 & 0 & 0 \\ 0 & 1 & 0 & 0 & 0 \\ 0 & 0 & 1 & 0 & 0 \end{pmatrix} \begin{pmatrix} 0 & 0 & 1 & 0 & 0 \\ 0 & 0 & 0 & 1 & 0 \\ 0 & 1 & 0 & 0 & 0 \\ 0 & 0 & 0 & 0 & 1 \\ 1 & 0 & 0 & 0 & 0 \end{pmatrix} = \begin{pmatrix} 0 & 0 & 0 & 0 & 1 \\ 1 & 0 & 0 & 0 & 0 \\ 0 & 0 & 1 & 0 & 0 \\ 0 & 0 & 0 & 1 & 0 \\ 0 & 1 & 0 & 0 & 0 \end{pmatrix}$$

Permutation Ciphers

In contrast to a substitution cipher, a permutation cipher rearranges the string of plaintext letters according to a fixed permutation. The ciphertext is an anagram of the plaintext.

Definition 4.18 *A **permutation cipher** of length m permutes letters in blocks of length m using a fixed permutation.*

Permutation ciphers cannot be attacked by using character frequencies. However, if the block length is small, say $m = 9$, there are only $9! = 362\,880$ permutations to try in order to break the cipher.

Example 4.19 Let
$$\sigma = (147)(238956)$$
be a fixed permutation of $\{1, 2, 3, 4, 5, 6, 7, 8, 9\}$. First break up the plaintext

Stream ciphers are often used in applications
where high speed and low delay are required

into blocks of length 9 as follows

STREAMCIP HERSAREOF TENUSEDIN APPLICATI
ONSWHEREH IGHSPEEDA NDLOWDELA YAREREQUI
REDEXTRAS

where extra letters are added at the end so that the last block will have nine
letters. Now apply the permutation σ to each block to get

ERICMTSPA SROEREHFA UNIDEETNS LPTACPAII
WSERENOHH SHDEEGIAP OLLEDDNAW ERUQEAYIR
EDARTERSX

The encryption of the first few letters is shown in Table 4.11. Note that the
letters in each block are rearranged, but the letters are unchanged.

S	T	R	E	A	M	C	I	P
↓	↓	↓	↓	↓	↓	↓	↓	↓
1	2	3	4	5	6	7	8	9
↓	↓	↓	↓	↓	↓	↓	↓	↓
4	3	8	7	6	2	1	9	5
↓	↓	↓	↓	↓	↓	↓	↓	↓
E	R	I	C	M	T	S	P	A

Table 4.11   Permutation cipher

Substitution ciphers are often used in combination with other ciphers. For
example, a substitution cipher may be used to give a correspondence between
letters and numbers; then an affine cipher or a block cipher (such as those
introduced in the next section) may be added on top of the substitution.

Although there are

$$26! = 403\,291\,461\,126\,605\,635\,584\,000\,000$$

possible substitution ciphers, these can be broken by methods that involve
counting character frequencies.

Problems 4.3

1. Use the substitution cipher

$$\sigma = (NEW)\,(MXICO)\,(STA)\,(UVRY)\,(BDFGHJKLPQZ)$$

to encrypt the plaintext, *I live in Las Cruces.*

2. The ciphertext

$$VDPJV\ HJLIO\ LRLAD\ CL$$

was encrypted using a substitution cipher with

$$\sigma = (COLRAD)\,(STE)\,(UNIVY)\,(BFGHJKMPQWXZ)$$

What is the plaintext?

3. Use a permutation cipher with

$$\sigma = (1563)\,(24)$$

to encrypt the plaintext

I have a secret.

4. Decrypt the ciphertext

$$ESCROUSEHWI\ WRESOBEIYUT$$

that was encrypted using a permutation cipher

$$\sigma = (1\ 5\ 2\ 4\ 8\ 9)\,(3\ 6\ 7\ 11\ 10)$$

5. Use the substitution cipher

$$\sigma = (A\ R\ P\ I\ H\ N\ C\ W\ G\ F\ B\ K\ O\ Q\ M\ U\ J)(D\ X\ Y\ T\ V\ S\ Z\ E\ L)$$

to encrypt the plaintext

*There is, of course, no difficulty in recognizing
that a cipher is transposition and not substitution.*

6. Verify that a product of disjoint cycles commutes; that is, if σ and τ are disjoint cycles, then $x^{\sigma\tau} = x^{\tau\sigma}$.

7. If $\sigma = (0\ 4\ 7)(1\ 6\ 5\ 3)(2\ 9\ 8)$ and $\tau = (0)(1\ 3\ 5\ 9\ 7)(2\ 8\ 6\ 4)$, express $\sigma\tau$ and $\tau\sigma$ as products of disjoint cycles.

8. Compute σ^{-1}, τ^{-1}, $(\sigma\tau)^{-1}$ and $(\tau\sigma)^{-1}$ for the permutations σ and τ given in Problem 7.

9. Prove Theorem 4.11.

10. Prove that the inverse of a product $\sigma\tau$ of two permutations is given by $(\sigma\tau)^{-1} = \tau^{-1}\sigma^{-1}$. (See Theorem 4.15.)

4.4   Block Ciphers

In this section we look at block ciphers that are determined by matrix multiplication modulo 26. In this way, strings of characters are enciphered as blocks.
Consider an $n \times n$ matrix M whose entries are in $\{0, 1, 2, \ldots, 25\}$, and let

$$\mathbf{x} = \begin{pmatrix} x_1 & x_2 & \ldots & x_n \end{pmatrix}$$

be the row vector that corresponds to a string of n plaintext characters, and let

$$\mathbf{y} = \begin{pmatrix} y_1 & y_2 & \ldots & y_n \end{pmatrix}$$

be the row vector defined by

$$\mathbf{y} = \mathbf{x}M \bmod 26$$

We can use the row vector $\mathbf{y}$ as ciphertext.

If M has a matrix inverse modulo 26, then the process

$$\mathbf{y} = \mathbf{x}M \bmod 26$$

is reversible and we have

$$\mathbf{x} = \mathbf{y}M^{-1} \bmod 26$$

In practice, block ciphers are secure only when the block size is relatively large. In the following examples, we use a small block size to illustrate the ideas.

Example 4.20 Here is a small block cipher determined by matrix multiplication modulo 26. Let $n = 2$ and take

$$M = \begin{pmatrix} 2 & 9 \\ 3 & 5 \end{pmatrix}$$

To encode the plaintext HELLO, first break it up into groups of two and add dummy letters (R in this case) to make it come out even: HE LL OR. Then convert the letter pairs into the number pairs $(7 \quad 4)$, $(11 \quad 11)$, $(14 \quad 17)$ of integers. Next, compute the matrix products

$$\begin{pmatrix} 7 & 4 \end{pmatrix} \begin{pmatrix} 2 & 9 \\ 3 & 5 \end{pmatrix} \bmod 26 = \begin{pmatrix} 0 & 5 \end{pmatrix}$$

$$\begin{pmatrix} 11 & 11 \end{pmatrix} \begin{pmatrix} 2 & 9 \\ 3 & 5 \end{pmatrix} \bmod 26 = \begin{pmatrix} 3 & 24 \end{pmatrix}$$

$$\begin{pmatrix} 14 & 17 \end{pmatrix} \begin{pmatrix} 2 & 9 \\ 3 & 5 \end{pmatrix} \bmod 26 = \begin{pmatrix} 1 & 3 \end{pmatrix}$$

Now convert back to letters. The ciphertext is AFDYBD.

To recover the plaintext, we can compute the inverse of the matrix M:

$$\begin{pmatrix} 2 & 9 \\ 3 & 5 \end{pmatrix}^{-1} \bmod 26 = \begin{pmatrix} 15 & 25 \\ 17 & 6 \end{pmatrix}$$

The plaintext is recovered by multiplying the ciphertext by M^{-1}:

$$\begin{pmatrix} 0 & 5 \end{pmatrix} \begin{pmatrix} 15 & 25 \\ 17 & 6 \end{pmatrix} \bmod 26 = \begin{pmatrix} 7 & 4 \end{pmatrix}$$

$$\begin{pmatrix} 3 & 24 \end{pmatrix} \begin{pmatrix} 15 & 25 \\ 17 & 6 \end{pmatrix} \bmod 26 = \begin{pmatrix} 11 & 11 \end{pmatrix}$$

$$\begin{pmatrix} 1 & 3 \end{pmatrix} \begin{pmatrix} 15 & 25 \\ 17 & 6 \end{pmatrix} \bmod 26 = \begin{pmatrix} 14 & 17 \end{pmatrix}$$

Example 4.21 The plaintext

If a portion of the ciphertext is suspected to yield some specific plaintext then the values obtained for the cipher digraphs involved can be entered wherever those digraphs appear. This may in special situations result in additional text being guessed.

can be transformed into the ciphertext

YTQXE	DNDYJ	MVFYI	UARFF	BOUVN	DESAK	PZSJQ	ZJHQW
CNVNQ	EGWPZ	IGPHM	PRVKH	HJPQF	YSXFY	ELKDS	SYYHA
MOTHV	AEDFY	IUARF	FDMGY	FXNOC	UZOIZ	DBXLA	NRDSX
HJLRB	HFFLR	DBBOH	DWANV	UJQXN	XQXPZ	GHFYY	GCEQW
BZPZI	GKDCU	FLWRS	MBZLR	UCNMK	HYPTP	NDYJK	DHJPQ
RDKHK	GSSES	QZ					

by using the encoding

$$\begin{pmatrix} x_1 & x_2 \end{pmatrix} \begin{pmatrix} 11 & 9 \\ 8 & 5 \end{pmatrix} \bmod 26 = \begin{pmatrix} y_1 & y_2 \end{pmatrix}$$

on pairs of letters. For example, the plaintext letters IF correspond to the number pair $\begin{pmatrix} 8 & 5 \end{pmatrix}$. The product

$$\begin{pmatrix} 8 & 5 \end{pmatrix} \begin{pmatrix} 11 & 9 \\ 8 & 5 \end{pmatrix} \bmod 26 = \begin{pmatrix} 24 & 19 \end{pmatrix}$$

gives the number pair $\begin{pmatrix} 24 & 19 \end{pmatrix}$, which corresponds to the ciphertext YT.

	ciphertext		*plaintext*
	ABCDEFGHIJKLMNOPQRSTUVWXYZ		ABCDEFGHIJKLMNOPQRSTUVWXYZ
A	00000000001001000200000000	A	00010000100301030101000000
B	00000001000000200000000002	B	00002000000000000001000000
C	00001000000001000000200000	C	00000000200000010001000000
D	02000000000010000000000000	D	00101100100000000010001000
E	00020000000100000020000000	E	00220000000102000010010100
F	00000300000100000000000150	F	00000000100000000000000000
G	00000001000000000000001010	G	00000010000000000100000000
H	10010000030000000000000000	H	00003000000000100010000000
I	00000020000000000000200001	I	00000110000003120010000000
J	00000001000000000000000000	J	00000000000000000000000000
K	00030013000000000000000000	K	00000000000000000000000000
L	00000000000000000300000000	L	10000000000000000001000000
M	00000000000000110000010000	M	10000000000000000000000000
N	00030000000010100000010100	N	00001000000000000020010000
O	00000000000000000000000000	O	00000100000112000200000000
P	00000001000000002000000004	P	00004001000000000000000000
Q	00001000000000000000002302	Q	00000000000000000000000000
R	00020000000000000000010000	R	10013000000000000002000000
S	00000000010010000020000200	S	00001000200000100020100000
T	00000001000000010000000000	T	00003005300000100000100000
U	00100000010000000000010000	U	00002000000000000010000000
V	10000000000001000000000000	V	00002000000000000000000000
W	10000000000000000100000000	W	00000000000000000000000000
X	00000000000100000000000000	X	00000000000000000002000000
Y	00000010020000010001000010	Y	00000002000000000000000000
Z	00000000000000100000000000	Z	00000000000000000000000000

Table 4.12   Letter pair (digraph) frequencies

A count of the ciphertext letters gives a distribution that is quite different from the character ciphers. Instead of ranging between typical values of 0.001 and 0.13, the relative frequencies are in a very narrow range between 0.014 and 0.066. To apply cryptanalysis to such problems, it is necessary to look at frequencies of pairs of letters. According to Table 4.12, the pair *FY* appears five times in the ciphertext and *PZ* appears four times in the ciphertext. In the plaintext, *th* appears five times and *pe* appears four times. The letter pair *th* is the most common letter pair in the English language. The pair *pe* is much farther down on the list of popularity. After *th* comes the pairs *he, in, er, re, on, an, en, at,* and *es*.

Making the guess that *th*⟶ *FY* leads to the equation

$$(5 \quad 24) = (19 \quad 7) \begin{pmatrix} a_{11} & a_{12} \\ a_{21} & a_{22} \end{pmatrix}$$

The second most common letter pair is *he*. There are several potential cipher-text pairs that could correspond to *hc*. The lucky guess *he* ⟶ *FF* gives the equation

$$(5 \quad 5) = (7 \quad 4) \begin{pmatrix} a_{11} & a_{12} \\ a_{21} & a_{22} \end{pmatrix}$$

Putting these two equations together

$$\begin{pmatrix} 5 & 24 \\ 5 & 5 \end{pmatrix} = \begin{pmatrix} 19 & 7 \\ 7 & 4 \end{pmatrix} \begin{pmatrix} a_{11} & a_{12} \\ a_{21} & a_{22} \end{pmatrix}$$

so

$$\begin{pmatrix} a_{11} & a_{12} \\ a_{21} & a_{22} \end{pmatrix} = \begin{pmatrix} 19 & 7 \\ 7 & 4 \end{pmatrix}^{-1} \begin{pmatrix} 5 & 24 \\ 5 & 5 \end{pmatrix} \bmod 26$$
$$= \begin{pmatrix} 11 & 9 \\ 8 & 5 \end{pmatrix}$$

The plaintext can be recovered by computing

$$(x_1 \quad x_2) = (y_1 \quad y_2) \begin{pmatrix} 11 & 9 \\ 8 & 5 \end{pmatrix}^{-1} \bmod 26$$

In particular, the first two letters *YT* of the ciphertext correspond to

$$(24 \quad 19) \begin{pmatrix} 11 & 9 \\ 8 & 5 \end{pmatrix}^{-1} \bmod 26 = (8 \quad 5)$$

which yields the two letters *if*. After inserting spaces and punctuation in appropriate places, we recover the original plaintext

If a portion ...

Problems 4.4

1. Use the block cipher $Y = X \begin{pmatrix} 2 & 1 \\ 5 & 3 \end{pmatrix} \bmod 26$ to encrypt the plaintext

   I spy.

2. Assume that the block cipher $Y = X \begin{pmatrix} 7 & 11 \\ 13 & 16 \end{pmatrix} \bmod 26$ was used to produce the ciphertext

   VX XC ZD HG WC RJ AR

   Decrypt the message.

3. Use the block cipher

$$Y = X \begin{pmatrix} 2 & 1 \\ 5 & 3 \end{pmatrix} \bmod 26$$

to encrypt the plaintext

> *A substitution alphabet derived by a linear transformation on the normal sequence introduces at most two unknown quantities.*

4. Use the block cipher

$$Y = X \begin{pmatrix} 0 & 7 & 3 & 6 & 8 \\ 5 & 8 & 1 & 9 & 5 \\ 3 & 7 & 3 & 4 & 5 \\ 6 & 1 & 7 & 9 & 6 \\ 8 & 6 & 9 & 3 & 1 \end{pmatrix} \bmod 26$$

to encrypt the plaintext

> *Praise for their skill, speed, and accuracy accrued throughout the war. At Iwo Jima, Major Howard Connor declared, "Were it not for the Navajos, the Marines would never have taken Iwo Jima." Connor had six Navajo code talkers working around the clock during the first two days of the battle. Those six sent and received over eight hundred messages, all without error.*

5. What is the inverse cipher for the cipher given in Problem 4?

6. Assume a 2×2 block cipher of the form

$$Y = XM$$

was used to produce the ciphertext

```
BIMZU PTTOG VKIIC DBGGJ QCVFQ WXMKL TMANE UNXQR
WEQRV MSNAW XCEIT TCJDO BTZQR GDOEK NEMFD OKWVV
QKWHT TEMGT YEUDK WWVIX MQZDS ZZYGH PPYDY TEDDO
GHNKZ FHOCF CMVCZ XANOU VRDOY DKAAE WULQB HGPSZ
SEYTU OJHCE PESIZ FIZGX QWVPE CQOAE ANEUM IVCLI
GMMDP APUUN HOIYD LIKTT WSJGP OHOTA QBHCZ FHCKG
IZUD
```

Decipher it.

7. Use the affine transformation

$$
X \longrightarrow
\begin{pmatrix}
1 & 9 & 7 & 6 & 0 & 2 & 5 & 5 & 7 \\
0 & 5 & 9 & 3 & 4 & 3 & 3 & 7 & 7 \\
7 & 3 & 3 & 2 & 5 & 7 & 9 & 1 & 4 \\
3 & 7 & 8 & 1 & 0 & 9 & 7 & 2 & 5 \\
6 & 0 & 6 & 3 & 5 & 5 & 7 & 7 & 7 \\
8 & 4 & 0 & 0 & 4 & 1 & 9 & 8 & 2 \\
5 & 5 & 3 & 1 & 9 & 7 & 4 & 2 & 9 \\
8 & 6 & 1 & 8 & 0 & 3 & 3 & 7 & 1 \\
1 & 2 & 4 & 3 & 2 & 3 & 5 & 7 & 4
\end{pmatrix}
X +
\begin{pmatrix}
5 \\ 5 \\ 8 \\ 8 \\ 6 \\ 5 \\ 4 \\ 3 \\ 0
\end{pmatrix}
\mod 10
$$

to encrypt your Social Security number. Compute the inverse transformation and test it on your encrypted Social Security number.

8. Use the 10×10 block cipher

$$
X \longrightarrow X
\begin{pmatrix}
105 & 233 & 85 & 11 & 79 & 235 & 221 & 233 & 118 & 119 \\
246 & 248 & 253 & 192 & 164 & 214 & 160 & 82 & 190 & 195 \\
207 & 79 & 198 & 134 & 82 & 200 & 249 & 132 & 204 & 197 \\
251 & 98 & 212 & 42 & 235 & 243 & 165 & 172 & 113 & 188 \\
104 & 164 & 188 & 213 & 18 & 131 & 109 & 192 & 105 & 41 \\
253 & 89 & 127 & 116 & 205 & 69 & 228 & 57 & 243 & 76 \\
48 & 222 & 249 & 147 & 189 & 218 & 160 & 234 & 124 & 172 \\
147 & 110 & 206 & 223 & 197 & 128 & 154 & 180 & 147 & 178 \\
109 & 203 & 239 & 4 & 207 & 180 & 127 & 36 & 8 & 252 \\
219 & 13 & 54 & 112 & 114 & 112 & 137 & 108 & 38 & 252
\end{pmatrix}
\mod 256
$$

to encrypt the message

Technology and security experts oppose restrictions on encryption, arguing that such restrictions would damage consumer trust.

by breaking the message into blocks of 10 characters and using ASCII values for each character. Pad the plaintext with extra "#" so that the total number of characters is a multiple of 10.

9. The 10×10 block cipher

$$
X \longrightarrow X
\begin{pmatrix}
105 & 233 & 85 & 11 & 79 & 235 & 221 & 233 & 118 & 119 \\
246 & 248 & 253 & 192 & 164 & 214 & 160 & 82 & 190 & 195 \\
207 & 79 & 198 & 134 & 82 & 200 & 249 & 132 & 204 & 197 \\
251 & 98 & 212 & 42 & 235 & 243 & 165 & 172 & 113 & 188 \\
104 & 164 & 188 & 213 & 18 & 131 & 109 & 192 & 105 & 41 \\
253 & 89 & 127 & 116 & 205 & 69 & 228 & 57 & 243 & 76 \\
48 & 222 & 249 & 147 & 189 & 218 & 166 & 234 & 124 & 172 \\
147 & 110 & 206 & 223 & 197 & 128 & 154 & 180 & 147 & 178 \\
109 & 203 & 239 & 4 & 207 & 180 & 127 & 36 & 8 & 252 \\
219 & 13 & 54 & 112 & 114 & 112 & 137 & 108 & 38 & 252
\end{pmatrix}
\mod 256
$$

was used to generate the ciphertext

$$\begin{pmatrix}
102 & 66 & 144 & 98 & 34 & 179 & 118 & 173 & 169 & 92 \\
101 & 139 & 32 & 218 & 201 & 85 & 216 & 114 & 213 & 34 \\
40 & 185 & 54 & 216 & 201 & 141 & 214 & 114 & 84 & 190 \\
163 & 219 & 30 & 135 & 217 & 72 & 243 & 5 & 233 & 231 \\
55 & 88 & 234 & 36 & 119 & 166 & 213 & 71 & 99 & 253 \\
159 & 149 & 32 & 204 & 64 & 159 & 169 & 250 & 240 & 174 \\
112 & 112 & 22 & 248 & 142 & 130 & 220 & 152 & 147 & 91 \\
170 & 99 & 184 & 200 & 81 & 126 & 38 & 130 & 202 & 254 \\
180 & 241 & 182 & 56 & 224 & 229 & 124 & 31 & 208 & 121
\end{pmatrix}$$

as described in problem 8. What is the plaintext?

4.5   The Playfair Cipher

Sir Charles Wheatstone[3] (1802–1875) is remembered by physicists for his contributions to the rheostat and the Wheatstone bridge, by musicians as the developer of the enchanted lyre, and by cryptographers for his development and promotion of the Playfair cipher. This cipher was used by the British in the Boer War and World War I, and by Lt. John F. Kennedy to arrange the rescue of his crew from a Japanese-controlled island after his PT-109 was sunk in the Solomon Islands.

The Playfair cipher is determined by writing the twenty six letters of the alphabet into a 5×5 matrix. To make it fit, the letters I and J are treated as one letter. Normally, one chooses a keyword to aid in this process, such as COLORADOSTATEUNIVERSITY. The keyword is written into the matrix, omitting repeated letters, and is followed by the remaining letters of the alphabet in alphabetical order. Several patterns can be used to enter the letters, including row by row or a spiral pattern. For example, the matrix

C	O	L	R	A
F	G	H	K	D
B	X	Z	M	S
Y	W	Q	P	T
V	IJ	N	U	E

was constructed by starting at the top left and spiraling clockwise in toward the center.

The plaintext is then broken up into two-letter pairs. If the letters in a pair are the same, enter an X between them. If there is only a single letter in the last group, add an X. The plaintext

The Playfair cipher was an immediate success

---

[3]Search the web for summaries of the many contributions of Charles Wheatstone.

is grouped as

$$\text{TH EP LA YF AI RC IP HE RW AS}$$
$$\text{AN IM ME DI AT ES UC CF SX SX}$$

The ciphertext is generated two letters at a time by locating a pair of plaintext letters in the 5×5 matrix. For example, the letters TH appear in the opposite corners of rectangle

H	•	D
•	•	•
Q	•	T

so choose first the letter in the row with T, then the letter in the row with H.
Replace TH with QD. In a similar manner, replace EP with UT. The letters in the pair LA both appear in row

C	•	L	R	A

so choose the letter immediately to the right (with wraparound). The pair LA corresponds to RC. The letters in the pair YF both appear in column

so choose the letter immediately below each plaintext letter. Thus YF corresponds to VB. Continue in this manner to get ciphertext as illustrated in Table 4.13.

TH	EP	LA	YF	AI	RC	IP	HE	RW	AS
↓	↓	↓	↓	↓	↓	↓	↓	↓	↓
QD	UT	RC	VB	OE	AO	UW	DN	OP	DT

AN	IM	ME	DI	AT	ES	UC	CE	SX	SX
↓	↓	↓	↓	↓	↓	↓	↓	↓	↓
LE	UX	SU	GE	DE	AT	VR	AV	BZ	BZ

Table 4.13   Playfair encryption

Reverse the process to decrypt a message. If the letters in a pair are in different rows and columns, pick the opposite corners of the rectangle. If two letters appear in the same row, pick the letters immediately to the left. If two letters appear in the same column, pick the letters immediately above them.

Problems 4.5

1. Create a Playfair cipher using the keyword CIPHERSAREUS (Ciphers are us) by starting in the upper left corner and following the arrows in diagram:

→	↙	→	↙	↓
↓	↗	↙	↗	↙
↗	↙	↗	↙	↓
↓	↗	↙	↗	↙
↗	→	↗	→	○

Encipher the message

Wheatstone named the Playfair cipher after his friend Lyon Playfair.

2. Create a Playfair cipher using the keyword CHARLESWHEATSTONE (Charles Wheatstone) by starting at the center and following the arrows in the diagram:

○	→	→	→	↓
↑	↑	→	↓	↓
↑	↑	↑	↓	↓
↑	↑	←	←	↓
↑	←	←	←	←

(Whenever you enter a new square, go in the direction indicated.) Encipher the message

Wheatstone's work in acoustics won him
a professorship of experimental physics.

3. The keyword RHEOSTAT was used with the pattern

→	→	→	→	↓
↓	←	←	←	←
→	→	→	→	↓
↓	←	←	←	←
→	→	→	→	○

(starting in the upper left corner) to create the ciphertext

```
CSBXE FNTOV MROSK DHCOB
LTASP ODEFB HILEC SSPOB
RMHOP SMARK HOELT LKYBH
```

using a Playfair cipher. Decrypt the message.

4. The key AMMETERS ARE CONNECTED IN SERIES was used with the Playfair pattern

→	→	→	→	↓
→	→	→	↓	↓
↑	→	○	↓	↓
↑	↑	←	←	↓
↑	←	←	←	←

(starting in the upper left corner) to create the Playfair ciphertext

```
OGQME TRTSC MARZN RIRUR
TIDFK EAMYB PMYB
```

Decrypt the message.

5. A variant of the Playfair cipher uses two squares, each generated using its own key. A pair of letters are located in the first square; then opposite corners of the second square are used as ciphertext. Thus IT ⟶ DQ. If a pair of letters appear in the same row on the left, then the characters immediately to the right of the corresponding locations in the right box are used.

•	•	•	•	•		•	•	•	•	•
•	•	I	•	•		•	•	•	•	D
•	•	•	•	•		•	•	•	•	•
•	•	•	•	T		•	•	Q	•	•
•	•	•	•	•		•	•	•	•	•

The keywords SCIENTIFIC NOTEBOOK and MAPLE MUPAD were used to generate the pair

S	C	I	E	N		V	O	G	U	M
T	F	O	B	K		W	Q	H	D	A
A	D	G	H	L		X	R	I	B	P
M	P	Q	R	U		Y	S	K	C	L
V	W	X	Y	Z		Z	T	N	F	E

of squares. Decrypt the ciphertext

DXOCX RTDIV SUOKC DLSBZ BDVKK LWNOB
BUOMK UYXUV KHXWU VKYHZ DVZH

6. Another variant of the Playfair uses rectangles of other sizes and larger alphabets. For example, a 27-character alphabet fills a 3×9 rectangle. The key GADZOOKS was used to fill the rectangle

.	Y	X	W	V	U	T	R	Q
E	F	H	I	J	L	M	N	P
C	B	S	K	O	Z	D	A	G

and generate the ciphertext

RVDVG SMFKX MWNSN RMADJ FUNJ.

HAGME NSCFR VN.L. CHCT

Decrypt the ciphertext.

7. The key WHODONEIT was use to create the Playfair rectangle

W	B	C	R	S	:
H	A	F	Q	U	!
O	T	G	P	V	?
D	I	J	M	X	.
N	E	K	L	Y	Z

and produce the ciphertext

HOTOW DPIZA

What is the plaintext?

4.6   Unbreakable Ciphers

We can extend the idea of a shift cipher by using different shifts at different letter positions. Instead of choosing a shift s, we choose a vector of shifts $\mathbf{s} = (s_1, s_2, \ldots, s_k)$. Given plaintext $\mathbf{x} = (x_1, x_2, \ldots, x_k)$, the associated ciphertext $\mathbf{y} = (y_1, y_2, \ldots, y_k)$ is

$$y_i = (x_i + s_i) \bmod 26$$

which can be expressed concisely as

$$(\mathbf{y} = \mathbf{x} + \mathbf{s}) \bmod 26$$

The vector **s** is called the *key*. If the key is as long as the plaintext, and if it is used only once (and is sufficiently random), then the ciphertext is essentially unbreakable. This is called a **one-time pad**. A friend with the same key can recover the plaintext by computing

$$\mathbf{x} = \mathbf{y} - \mathbf{s} \bmod 26$$

This scheme is occasionally used for highly sensitive documents. However, it requires the generation and distribution of potentially large numbers of keys. These keys are traditionally hand carried in diplomatic pouches to embassies around the world.

Example 4.22 If the plaintext *THEBU YISGO INGDO WNATN OON* is added to the key *OCSJZ VOCKP UNBLN PDVNU MRJ*, the resulting ciphertext is *HJWKT TWUQD CAHOB LQVGH AFW*. For example, since $T \to 19$ and $O \to 14$, the first ciphertext character is given by $19 + 14 \bmod 26 = 7 \to H$.

If the key is used more than once, then the first character of the key is subject to attack by a statistical analysis of the first letters being sent in the various messages. This is similar to the way that simple shift ciphers are broken. The other letters in the key are subject to the same attack.

Most messages do not require absolute security. It is enough to make the cost of reading the message higher than its potential value to the eavesdropper. An easy modification of the one-time pad is to create keys that are short enough to be remembered. This is idea behind the Vigenère cipher.

Example 4.23 Using a key *OCSJZ* of length 5, it is possible to encipher the plaintext *THEBU YISGO INGDO WNATN OON* by adding the extended key *OCSJZ OCSJZ OCSJZ OCSJZ OCS* to generate the ciphertext *HJWKT MKKPN WPYMN KPSCM CQF*.

How secure is such a system based on short keys? The problem is the same as when you use a key more than once. Each message now becomes a sequence of messages each enciphered by the same key.

Problems 4.6

1. Use the one-time key

        SHORTCIPHERTEXTMESSAGES
        AREEXTREMELYHARDTOBREAK

to encrypt the message

The longer the message the easier the decryption

using $\mathbf{y} = \mathbf{x} + \mathbf{s} \bmod 26$ where **x** is the plaintext vector and **s** is the key vector.

2. The ciphertext

$$\text{LLGMV GEVGX VQMAZ KWASK WWUHT}$$
$$\text{TSEMS NSNOX AFPSR I}$$

   was encrypted using the one-time key

$$\text{SECUREKEYEXCHANGEIS}$$
$$\text{ESSENTIALFORONETIMEKEYS}$$

   Decrypt the message.

3. The short key MILLIONAIRE was used to produce the ciphertext

$$\text{IPZTA HUEEV EWMDE TWAK}$$

   Decrypt the message.

4. The short key PLAYFAIR was used to produce the ciphertext

$$\text{HSOPY KMPHN ALJAA ZAJBC}$$
$$\text{WEKFV YIXJD}$$

   Decrypt the message.

5. The short key CODETALKER was use to generate the ciphertext

$$\text{YVBTE AJPEZ TKKIG TSOSK JSUKN YDKVV}$$
$$\text{WGLRZ RDK}$$

   Decrypt the message.

6. A pseudo key of longer length can be generated by using two short keys. Thus the two keys CODE and ENCRYPT can be used as a two-stage cipher

	p	l	a	i	n	t	e	x	t	⋯	⋯
+	C	O	D	E	C	O	D	E	C	O	⋯
+	E	N	C	R	Y	P	T	E	N	C	⋯
	c	i	p	h	e	r	t	e	x	t	⋯

   where the ith ciphertext character is the mod 26 sum of the three characters above it. What is the length of the pseudo key?

7. Encryption can also be done at the bit level. Given a byte $(1, 0, 0, 0, 1, 0, 1, 1)$ and a key $(1, 1, 0, 1, 1, 0, 0, 1)$, the vector sum modulo 2 is given by

$$(1, 0, 0, 0, 1, 0, 1, 1) + (1, 1, 0, 1, 1, 0, 0, 1) \bmod 2 = (0, 1, 0, 1, 0, 0, 1, 0)$$

   Show that decryption is the same as encryption.

4.7 Enigma Machine

The Enigma machine (see Figure 4.1) is an electromechanical cipher device that was used by Germany during World War II. It was placed in submarines, ships, and other locations to transmit and receive sensitive information.

Mathematicians and cryptanalysts from Poland, Great Britain, and the United States were partially successful in breaking the Enigma ciphers, and the information gained was used to help the Allied war effort. The British and American efforts took place at Bletchley Park under the leadership of the British mathematician Alan Turing[4] (1912–1954) who is best known for his formulation of an abstract computational device now called a Turing machine.

The following statement by the Bletchley Park Board of Trustees speaks to the significance of this operation.

Bletchley Park during the Second World War seethed with life, intellectual stimulus, individuality and eccentricity. It was a hotbed of revolutionary thinking; ideas whose practical application in time of crisis preserved freedoms, saved lives and changed the way the world communicated. The work of Bletchley Park's pioneers secretly affected the fate of nations during the course of the war and helped shorten it by at least two years. Since then, millions of people have been influenced by what happened on and beyond this site.

Bletchley Park Board of Trustees

The Enigma machine[5] used a keyboard to enter plaintext. This keyboard was connected with wires and gears to light bulbs that indicated the corresponding ciphertext. The gears, or rotors, were configured much like automobile odometers used to be, so that one rotor moved with each keystroke, and as it moved from position 25 to position 0, it forced the second rotor to move one notch. Each machine contained at least three rotors plus a fixed reflector (see Figure 4.1).

The rotors could be removed and replaced in any order and in any initial position, so a great many variations were possible. The daily key consisted of the initial positions of the rotors. Each rotor was connected with a series of wires that created a permutation cipher. An electric charge from the keyboard would enter each rotor at one position, then the permutation would connect the charge to a different position on the far side of the rotor.

The reflector rotor at the end was a fixed product of disjoint 2-cycles, say

$$\rho = (0\ 12)(1\ 9)(2\ 3)(4\ 24)(5\ 18)(6\ 23)(7\ 8)(10\ 25)$$
$$\cdot (11\ 13)(14\ 21)(15\ 17)(16\ 19)(20\ 22)$$

---

[4]See http://www.turing.org.uk/turing/ for interesting detail about the life and contributions of Alan Turing.

[5]See http://www.bletchleypark.org.uk for a history of Bletchley Park and the Enigma machine.

Such a permutation has the property that $x^{\rho\rho} = x$ for each x, so that $\rho^2 = \iota$, the identity function.

The rotor allowed the charge to travel back through the set of rotors to the keyboard, where one of the keys would be illuminated. The ciphertext character would be noted by the keyboard operator, and the resulting ciphertext would be transmitted by a radio operator.

It is easier to describe the Enigma machine mathematically than it is to describe it physically. There are permutations σ_1, σ_2, and σ_3 that correspond to the rotors. For each plain-text character x, a cipher-text character y is generated by

$$y = x^{\sigma_1 \sigma_2 \sigma_3 \rho \sigma_3^{-1} \sigma_2^{-1} \sigma_1^{-1}}$$

The kicker is that σ_1 changes each time a new character is typed in, σ_2 changes occasionally, and once in a while σ_3 also changes.

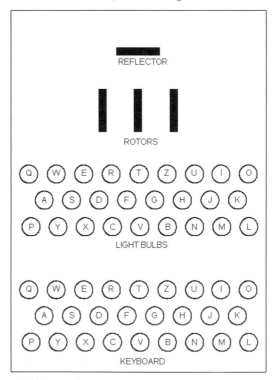

Figure 4.1   Enigma machine

To keep track of what σ_1 does in a certain state, we introduce a shift variable for each rotor, so that s_1, s_2, and s_3 keep track of what the top of each rotor is marked. Instead of computing x^{σ_1}, we actually compute

$$y = (x + s_1 \bmod 26)^{\sigma_1}$$

then compute

$$z = (y + s_2 \bmod 26)^{\sigma_2}$$

and so forth.

To compute the inverse of the function

$$y = (x + s_1 \bmod 26)^{\sigma_1}$$

we must solve for x in terms of y. Applying σ_1^{-1} to both sides of the equation, we get $x + s_1 \bmod 26 = y^{\sigma_1^{-1}}$ or $x = y^{\sigma_1^{-1}} - s_1 \bmod 26$.

The actual machines had a few additional complications. In particular, a plugboard was added when the machine was modified for military use. The plugboard allows the user to pair up ten pairs of letters with ten plug-in wires. The number of possible plugboard settings is $150,738,274,937,250$ which greatly increases the difficulty of deciphering Enigma when you do not know the plugboard setting.

Algorithm 4.1   Enigma Code

Do for i from 1 to 3
    Set $s(i)$ = initial shift of rotor i
    Do for x from 0 to 25
        Set $e(i, x) = x^{\sigma_i}$
        Set $d(i, c(i, x)) = x$
    Loop
Loop
Do for x from 0 to 25
    Set $r(x) = x^{\rho}$
Loop
Do for k from 1 to message length
    Set $x = m_k$
    Do for i from 1 to 3
        Set $x = e(i, x + s(i) \bmod 26)$
    Loop
    Set $x = r(x)$
    Do for i from 3 down to 1
        Set $x = d(i, x) - s(i) \bmod 26$
    Loop
    Print ASCII(x)
    Set $j = 0$
    Do while $j < 3$
        Set $s(j) = s(j) + 1 \bmod 26$
    Loop until $s(j) \neq 0$
Loop

To use Algorithm 4.1, assume permutations σ_1, σ_2, σ_3, and ρ have been defined, and m_k represents the number equivalent of the kth alphabet character in either plaintext or ciphertext. The number $e(i, x)$ is the image of x under the permutation σ_i and $d(i, y) = x$ if and only if $e(i, x) = y$ and hence $d(i, y) = y^{\sigma_i^{-1}}$.

Example 4.24 For concreteness, let

$$\sigma_1 = (0\ 15\ 7\ 8\ 19\ 23)\,(1\ 4\ 17\ 2)\,(3\ 13\ 25\ 5\ 21\ 12)\,(6\ 9\ 10\ 22\ 11\ 14\ 24\ 16\ 18\ 20)$$
$$\sigma_2 = (0\ 5\ 9)\,(1\ 2\ 3\ 7\ 6\ 8\ 4)\,(10\ 11\ 12\ 15\ 18\ 13\ 14\ 16\ 19\ 17)\,(20\ 25\ 22\ 21\ 24\ 23)$$
$$\sigma_3 = (0\ 10\ 20\ 3\ 13\ 23\ 7\ 17\ 5\ 15\ 25\ 1\ 11\ 21)\,(2\ 12\ 22\ 4\ 14\ 24\ 8\ 18\ 9\ 19\ 6\ 16)$$
$$\rho = (0\ 12)\,(1\ 9)\,(2\ 3)\,(4\ 24)\,(5\ 18)\,(6\ 23)\,(7\ 8)\,(10\ 25)\,(11\ 13)\,(14\ 21)\,(15\ 17)\cdot$$
$$(16\ 19)\,(20\ 22)$$
$$s_1 = 24$$
$$s_2 = 10$$
$$s_3 = 5$$

We will use these initial settings to encipher the message

$$\text{Get your forces into position to attack.}$$

We take $x = 6 \leftarrow \text{G}$ and compute

$$y = (x + s_1 \bmod 26)^{\sigma_1} = (6 + 24 \bmod 26)^{\sigma_1} = 4^{\sigma_1} = 17$$

Next, we compute z by

$$z = (y + s_2 \bmod 26)^{\sigma_2} = (17 + 10 \bmod 26)^{\sigma_2} = 1^{\sigma_2} = 2$$

and compute w by

$$w = (z + s_3 \bmod 26)^{\sigma_3} = (2 + 5 \bmod 26)^{\sigma_3} = 7^{\sigma_3} = 17$$

The reflector ρ then yields

$$u = (w)^\rho = 17^\rho = 15$$

Now redefine z by

$$z = \left(u^{\sigma_3^{-1}} - s_3 \bmod 26\right) = 15^{\sigma_3^{-1}} - 5 \bmod 26 = 5 - 5 = 0$$

redefine y by

$$y = (z^{\sigma_2^{-1}} - s_2 \bmod 26) = 0^{\sigma_2^{-1}} - 10 \bmod 26 = 9 - 10 \bmod 26 = 25$$

and redefine x by

$$x = (y^{\sigma_1^{-1}} - s_1 \bmod 26) = 25^{\sigma_1^{-1}} - 24 \bmod 26 = 13 - 24 \bmod 26 = 15$$

and hence the first ciphertext character is $15 \to$ P. Finally, we update the shift constants:

$$s_1 := s_1 + 1 \bmod 26 = 25$$

(If $s_1 = 0$, then we would update s_2; if also $s_2 = 0$, then we would update s_3.) Continuing in this way, we see that the ciphertext is

$$\text{PIIAY BLRQH YTDXP XZIHU YKTQP DFCAV QET}$$

One feature of Enigma is that the enciphering procedure is exactly the same as the deciphering procedure. To see this, let $\mu = \sigma_1\sigma_2\sigma_3\rho\sigma_3^{-1}\sigma_2^{-1}\sigma_1^{-1}$. Then

$$
\begin{aligned}
\mu^2 &= \left(\sigma_1\sigma_2\sigma_3\rho\sigma_3^{-1}\sigma_2^{-1}\sigma_1^{-1}\right)\left(\sigma_1\sigma_2\sigma_3\rho\sigma_3^{-1}\sigma_2^{-1}\sigma_1^{-1}\right) \\
&= \sigma_1\left(\sigma_2\left(\sigma_3\left(\rho\left(\sigma_3^{-1}\left(\sigma_2^{-1}\left(\sigma_1^{-1}\sigma_1\right)\sigma_2\right)\sigma_3\right)\rho\right)\sigma_3^{-1}\right)\sigma_2^{-1}\right)\sigma_1^{-1} \\
&= \sigma_1\left(\sigma_2\left(\sigma_3\left(\rho\left(\sigma_3^{-1}\left(\sigma_2^{-1}\sigma_2\right)\sigma_3\right)\rho\right)\sigma_3^{-1}\right)\sigma_2^{-1}\right)\sigma_1^{-1} \\
&= \sigma_1\left(\sigma_2\left(\sigma_3\left(\rho\left(\sigma_3^{-1}\sigma_3\right)\rho\right)\sigma_3^{-1}\right)\sigma_2^{-1}\right)\sigma_1^{-1} \\
&= \sigma_1\left(\sigma_2\left(\sigma_3\left(\rho\rho\right)\sigma_3^{-1}\right)\sigma_2^{-1}\right)\sigma_1^{-1} \\
&= \sigma_1\left(\sigma_2\left(\sigma_3\sigma_3^{-1}\right)\sigma_2^{-1}\right)\sigma_1^{-1} \\
&= \sigma_1\left(\sigma_2\sigma_2^{-1}\right)\sigma_1^{-1} \\
&= \sigma_1\sigma_1^{-1} \\
&= \iota
\end{aligned}
$$

This means that if $y = x^\mu$, then $x = x^{\mu\mu} = y^\mu$. Thus with all the rotors in the initial daily settings, encrypting $y \leftarrow x^\mu$ is exactly the same as decrypting $x \leftarrow y^\mu$.

Problems 4.7

1. Use rotor settings $s_1 = 25$, $s_2 = 13$, and $s_3 = 4$. Encrypt the message

$$Hi$$

2. The rotor settings $s_1 = 24$, $s_2 = 25$, and $s_3 = 19$ were used to produce the ciphertext

$$\text{QAV}$$

Decrypt the message.

3. Rotors 1 and 3 were interchanged with the rotor settings $s_1 = 23$, $s_2 = 3$, and $s_3 = 7$ to produce the ciphertext

$$\text{SKNSL BOWU}$$

Decrypt the message.

4. Interchange rotors 1 and 2 and set the initial rotor placements as $s_1 = 20$, $s_2 = 25$, and $s_3 = 4$ to encrypt the plaintext

   Combinatorial mathematics is the study of the arrangements of objects.

5. There are 26 letters on the Enigma plugboard. A plugboard setting consists of ten (unordered) pairs of letters. So, for example, one plugboard setting is $\{A, Z\}, \{B, Y\}, \{C, X\}, \{D, W\}, \{E, V\}, \{F, U\}, \{G, T\}, \{H, S\}, \{I, R\}, \{J, Q\}$. Verify that there are $150, 738, 274, 937, 250$ plugboard settings.

6. Given the permutations

$$\sigma = \begin{pmatrix} 1 & 2 & 3 & 4 & 5 & 6 & 7 \\ 6 & 3 & 5 & 7 & 4 & 1 & 2 \end{pmatrix} \text{ and } \tau = \begin{pmatrix} 1 & 2 & 3 & 4 & 5 & 6 & 7 \\ 4 & 1 & 2 & 6 & 5 & 7 & 3 \end{pmatrix}$$

   what are $\sigma\tau$ and $\tau\sigma$?

7. Given the permutations

$$\sigma = \begin{pmatrix} 1 & 2 & 3 & 4 & 5 & 6 & 7 \\ 6 & 3 & 5 & 7 & 4 & 1 & 2 \end{pmatrix} \text{ and } \tau = \begin{pmatrix} 1 & 2 & 3 & 4 & 5 & 6 & 7 \\ 4 & 1 & 2 & 6 & 5 & 7 & 3 \end{pmatrix}$$

   what are σ^{-1} and τ^{-1}?

8. Write the permutations

$$\sigma = \begin{pmatrix} 1 & 2 & 3 & 4 & 5 & 6 & 7 \\ 6 & 3 & 5 & 7 & 4 & 1 & 2 \end{pmatrix} \text{ and } \tau = \begin{pmatrix} 1 & 2 & 3 & 4 & 5 & 6 & 7 \\ 4 & 1 & 2 & 6 & 5 & 7 & 3 \end{pmatrix}$$

   as a products of disjoint cycles.

9. If ρ is a k-cycle and σ is a permutation, show that $\sigma\rho\sigma^{-1}$ is also a k-cycle. Give a formula for this k-cycle in terms of σ.

10. Show that if $\rho = (a_1, a_2)(a_3, a_4) \cdots (a_{25}, a_{26})$ is a product of disjoint 2-cycles, σ is a permutation, and $\tau = \sigma^{-1}$, then

$$\sigma\rho\tau = (a_1^\tau, a_2^\tau)(a_3^\tau, a_4^\tau) \cdots (a_{25}^\tau, a_{26}^\tau)$$

   is also a product of disjoint 2-cycles.

Chapter 5

Error-Control Codes

In this chapter, we see how transmission errors can be detected and even corrected. There are numerous applications, including the use of error detection and correction, for bar code scanning.

5.1 Weights and Hamming Distance

Errors occur when transmitting data. They can result from noisy telephone lines, sun spots, thunder storms, or faulty solder joints. Even under ideal circumstances, random gamma radiation can alter the data. Since errors are unavoidable, we acknowledge this fact and take steps to minimize their harmful effects.

By analogy, consider the problem of minimizing the harmful effects of automobile accidents. There are two approaches. We can make accidents less likely to occur (improve brakes and handling, decrease the number of drunk drivers), or we can make cars safer in the event of an accident (energy-absorbing frames, seat belts, air bags).

Much has been done to make errors in data transmission less likely to occur. Shielded cable, better product testing, and clean rooms all help. In this chapter, we will look at mathematical tools that mitigate the harmful effects of those errors that do occur.

The subject of *information theory*, created by Claude Shannon (1916–2001), is one of the great intellectual achievements of the twentieth century. In an historic paper published in 1948, Shannon formulated a model of a communications channel as illustrated in Figure 5.1.

Figure 5.1   Communications channel

The channel is noisy, so errors in the message might be introduced between the time that it leaves the encoder and the time it is received by the decoder. The idea of an error-control code is to add enough redundant information to the message so that if errors occur they will be detected (error-detecting codes) or even corrected (error-correcting codes).

Here are two examples of how the sentence

```
When transmitting data, there is
always the possibility of error.
```

might look if it is transmitted over a channel with an error rate of 1/10.

```
When hransmitling zvta, thdre is
alwayt the possibility ofrvrrmr.
Whet transuiorirg data, thire is
always the pojsibiqity of error.
```

Redundancy, in this case transmitting the entire sentence a second time, makes it relatively easy for a human to interpret the garbled message.

A simple example of an error-correcting code is the **triple-repetition code**, where each bit is repeated three times. Thus the word 01101 is encoded as 000111111000111. The received message is decoded by majority rule. Three errors are made in the example shown in Figure 5.2, yet the decoder gives the correct message.

Original	0	1	1	0	1
	↓	↓	↓	↓	↓
Encoded	000	111	111	000	111
	↓	↓	↓	↓	↓
Received	010	111	110	000	101
	↓	↓	↓	↓	↓
Decoded	0	1	1	0	1

Figure 5.2   Triple-repetition code

In the triple-repetition code the two codewords, 000 and 111, encode the words 0 and 1. It is called a (3,1) code because a codeword of length 3 is used to encode a word of length 1. The codewords are illustrated in the following figure. If one of the other three-bit words is received, then at least one error has occurred. If we think of these words as being the vertices of a cube, then vertices that are a distance 1 apart are connected by an edge. Majority rule decoding takes the top four points to 1 and the bottom four points to 0.

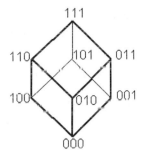

Figure 5.3   Majority rule decoding

Although this simple scheme will allow some garbled messages to be correctly decoded, it is not very efficient. The coded message is three times as long as the original message, and takes three times as long to transmit. We would like more efficient codes—ones that do a good job of controlling errors but require only a few additional bits.

The messages that we have in mind will be messages between computers or between a computer and a storage device, so they will be in binary form. We will assume that the noisy channel is a **binary symmetric channel**. This means that the only errors that can occur are replacing a zero by a one or a one by a zero, and that the probabilities of the two types of error are the same. If the probability of a change is p, then the probability of no change is $1 - p$. Table 5.1 lists the four possible events in a binary symmetric channel together with their probabilities.

$$\text{no error} \quad \begin{cases} 0 \xrightarrow{1-p} 0 \\ 1 \xrightarrow{1-p} 1 \end{cases}$$

$$\text{error} \quad \begin{cases} 0 \xrightarrow{p} 1 \\ 1 \xrightarrow{p} 0 \end{cases}$$

Table 5.1   Binary symmetric channel

In general, p will be a small positive number, so that $1 - p$ is close to 1.

Example 5.1 Suppose that the probability of an error in a binary symmetric channel is $p = 0.01$. What is the probability that a single error occurs if a string of 50 bits are transmitted through the channel? The probability of an error in the first bit is 0.01 and the probability that the remaining 49 bits are transmitted correctly is 0.99^{49}. There are exactly 50 ways that such a single error could occur, so the probability of a single error is $(50) \cdot (0.01) \cdot (0.99)^{49} \approx 0.30250$.

What is the probability of at least one error? To compute this probability, compute the probability of zero errors, and subtract this probability from 1. The probability of zero errors is $0.99^{50} \approx 0.605006$, so the probability of at least one error is approximately $1 - 0.605006 = 0.394994$.

In general, if p is the probability of a single error in a binary symmetric channel, then the probability of exactly k errors when n bits are transmitted is given by

$$P = \binom{n}{k} p^k (1-p)^{n-k}$$

This example illustrates the central challenge of error-control codes. Small probabilities of single errors can translate into large probabilities when large amounts of data are transmitted.

The encoders will work on binary words $\mathbf{u} = (u_1, u_2, \ldots, u_n)$, where each u_i is either 0 or 1.

Definition 5.2 *The **weight** of a binary word is its number of nonzero coordinates.*

For example, the weight of the word $\mathbf{u} = 01101000$ is 3. We write $w(\mathbf{u}) = 3$.

Definition 5.3 *The **Hamming distance** between two words of equal length is the number of coordinates where the two words differ.*

The Hamming distance between the English words "lemon" and "lemma" is two. The Hamming distance between the binary words 01101000 and 01010101 is five. This distance is named after Richard Hamming (1915–1998) whose seminal paper on error-detecting and error-correcting codes appeared in 1950. Hamming codes (see page 120) are of fundamental importance in coding theory.

Definition 5.4 *The **sum** of two binary words of equal length is their vector sum modulo 2.*

The Hamming distance $d(\mathbf{u}, \mathbf{v})$ between two binary words $\mathbf{u}$ and $\mathbf{v}$ is the weight of their sum, so that

$$d(\mathbf{u}, \mathbf{v}) = w(\mathbf{u} + \mathbf{v})$$

Example 5.5 If

$$\mathbf{u} = (0,0,1,0,1,1,0,1,0,0,1) \quad and \quad \mathbf{v} = (1,1,1,0,1,1,0,1,0,0,0)$$

then

$$\mathbf{u} + \mathbf{v} = (1,1,0,0,0,0,0,0,0,0,1)$$

and hence $w(\mathbf{u} + \mathbf{v}) = 3$. The distance $d(\mathbf{u}, \mathbf{v}) = 3$ because $\mathbf{u}$ and $\mathbf{v}$ differ in exactly three coordinates (the first, second, and eleventh).

The weight of a binary vector is its dot product with itself (not reduced modulo 2). The calculations

$$(0,0,1,0,1) + (1,1,1,0,1) \bmod 2 = (1,1,0,0,0)$$
$$(1,1,0,0,0) \cdot (1,1,0,0,0) = 2$$

show that the distance between the two vectors $(0, 0, 1, 0, 1)$ and $(1, 1, 1, 0, 1)$ is 2, as is the distance between the two words 00101 and 11101.

A **parity check bit** is used on personal computers in order to control memory errors. An 8-bit word $\mathbf{u} = (u_1, u_2, \dots, u_8)$ is stored using the 9-bit codeword

$$\mathbf{v} = (u_1, u_2, \dots, u_8, \sum_{i=1}^{8} u_i \bmod 2)$$

so this is a $(9, 8)$ code. The weight of the vector $\mathbf{v}$ is always even. If a memory error occurs in a single bit, then the weight of the new vector is odd, and an error message is displayed on the screen.

Encoding can be thought of as a matrix product, with operations modulo 2. The matrix is

$$P = \begin{pmatrix} 1 & 0 & 0 & 0 & 0 & 0 & 0 & 0 & 1 \\ 0 & 1 & 0 & 0 & 0 & 0 & 0 & 0 & 1 \\ 0 & 0 & 1 & 0 & 0 & 0 & 0 & 0 & 1 \\ 0 & 0 & 0 & 1 & 0 & 0 & 0 & 0 & 1 \\ 0 & 0 & 0 & 0 & 1 & 0 & 0 & 0 & 1 \\ 0 & 0 & 0 & 0 & 0 & 1 & 0 & 0 & 1 \\ 0 & 0 & 0 & 0 & 0 & 0 & 1 & 0 & 1 \\ 0 & 0 & 0 & 0 & 0 & 0 & 0 & 1 & 1 \end{pmatrix}$$

and the codeword $\mathbf{v}$ is given by $\mathbf{v} = \mathbf{u}P \bmod 2$, for $\mathbf{u}$ an eight-bit word.

We say this is a **linear code** because the sum of any two codewords is another codeword: if $\mathbf{v}_1 = \mathbf{u}_1 P$ and $\mathbf{v}_2 = \mathbf{u}_2 P$, then $\mathbf{v}_1 + \mathbf{v}_2 = (\mathbf{u}_1 + \mathbf{u}_2) P$. Since the codewords all have even weight, the sum of any two codewords has even weight. It follows that the minimum distance between two distinct codewords is 2.

Problems 5.1

1. If you use pure guessing on a 10-question true/false exam, what is the probability that you get them all right?

2. If you use pure guessing on a 10-question true/false exam, what is the probability that you get at least 7 out of 10?

3. What is the minimum distance between codewords for the triple repetition code?

4. What is the probability of no errors if 3 bits are transmitted through a binary symmetric channel with $p = 0.05$?

5. What is the probability of exactly one error if 200 bits are transmitted through a binary symmetric channel with $p = 0.0001$?

6. What is the probability of at least two errors if 1000 bits are transmitted through a binary symmetric channel with $p = 0.001$?

7. What is the expected number of errors if 1000 bits are transmitted through a binary symmetric channel with $p = 0.001$?

8. What is the minimum distance between the following collection of garbled sentences?

```
When trcnsmitthngddata, thery is always
           tte pfssibhlity of error.
When transmittxng datae therj is zlwaps
           the possibwlity of xrror.
When trznspitting data, there is always
           the possibility oe evyorg
Whfn tranymyttgnc data, nheri is always
           the hossibility of error.
When transmitting data, dherexis alwahs
           the jossicility of error.
```

9. The garbled sentences in problem 8 were generated by assuming that the error rate was 1/10. Explain why a minimum distance of 11 is reasonable.

5.2   Bar Codes Based on Two-out-of-Five Code

Bar codes come in many flavors. The bar codes presented in this chapter are called **one dimensional** since a linear beam of light can be used for scanning. Two-dimensional bar codes appear in Chapter 10. So-called three-dimensional codes are actually one-dimensional codes that are engraved or forged onto a solid object such as an engine block.[1]

Two-out-of-Five Bar Code

Two-out-of-five bar code is based on the two-out-of-five code used on paper tape. The bar code adaptation uses five bars per digit, with two wide bars and three narrow bars. The columns can be thought of as having values 0, 1, 2, 4, and 7, with the sum of the wide bars taken modulo 11. Notice that the minimum distance between codewords in Table 5.2 is 2, and hence two-out-of-five bar code detects single errors.

---

[1]See C. K. Harmon and R. Adams, *Reading Between the Lines*, Helmers Publishing, Peterborough, NH, 1989, for issues concerning printing and scanning.

See http://www.cwi.nl/~dik/english/codes/barcodes.html#start for additional examples of one-dimensional bar codes.

1		6	
2		7	
3		8	
4		9	
5		0	

Table 5.2   Two-out-of-five bar code

Three-out-of-nine (Code 39)

Code 39 or three-out-of-nine bar code is a popular bar code that is based on two-out-of-five code. Code 39 also uses a sequence of five bars. By replacing one of the four interior spaces by a wide space, $4 \cdot 10 = 40$ characters can be represented. The asterisk * is used on both ends as a start/stop character, so 39 characters can be represented, thus the name **code 39**.

The name **three-out-of-nine bar code** comes from that fact that there are always three wide elements out of a total of nine elements (five bars plus the four interior spaces). An additional four characters have been added by using five narrow bars, with three wide spaces and one narrow space in the interior.

As with Morse code, the ratio between wide and narrow elements is typically three to one but can be as small as two to one. This makes it natural to represent the bar code as a binary code. For example,

$$A = \text{\rule{1pt}{8pt}} = 111010100010111$$

where wide and narrow bars are represented by 111 and 1, respectively, and wide and narrow spaces are represented by 000 and 0, respectively.

Adding a narrow space between characters, code 39 can be thought of as a 16-bit binary code. Parity check is implemented at the bit level by using the fact that there are an odd number of 0s or 1s in a row. At the character level, each block of length 16 should contain the substring 000 an odd number of times (1 or 3), and the substring 111 an even number of times (0 or 2).

Because of the multiple levels of parity checking, code 39 is one of the most reliable of all the bar codes. The code 39 dictionary is given in Table 5.3. The column order has been altered slightly from two-out-of-five bar code; the values are 1, 2, 4, 7, and 0. To decode, break up the message into blocks of length five, and then find the pattern in the code 39 dictionary. Notice that the wide space locates the column, and the two wide bars locate the row in the dictionary.

1		A		K		U		$	
2		B		L		V		/	
3		C		M		W		+	
4		D		N		X		%	
5		E		O		Y			
6		F		P		Z			
7		G		Q		-			
8		H		R		.			
9		I		S					
0		J		T		*			

Table 5.3   Code 39

Example 5.6 Here is an example. To decode

first break the message up into groups of five, then decode.

This Code 39 is taken from a map distributed by the American Automobile Association.

Note that the initial and final * do not appear below the bar code.

Interleaved Two out of Five

Another interesting variation of two-out-of-five bar code is the **interleaved two out of five** (see Figure 5.4). This code takes advantage of the spaces (including the spaces between characters) as well as the bars to give a code that is nearly twice as dense as two-out-of-five bar code. Interleaved two out of five is used to encode the ten digits 0, 1, ... , 9, just like two-out-of-five bar code. The first digit is encoded using the first five bars, and the second digit is encoded by using the first five spaces. Thus each group of 10 elements (bars and spaces) has four wide elements (two wide bars and two wide spaces). Interleaved two out of five codes an even number of digits.

1 2 3 4 5 6 7 8 9 0

Figure 5.4   Interleaved two out of five

Interleaved two-out-of-five bar codes begin with the start pattern of a narrow bar, narrow space, narrow bar, narrow space and ends with a wide bar, narrow space, narrow bar. The columns have values 1, 2, 4, 7, 0. With n representing "narrow" and w representing "wide," the patterns for both bars and spaces are given in Table 5.4. In particular, the integer 58 is represented by ▐▌▌▌ = wWnNwNnWnN, where lowercase represents bars and uppercase represents spaces.

1	wnnnw	6	nwwnn
2	nwnnw	7	nnnww
3	wwnnn	8	wnnwn
4	nnwnw	9	nwnwn
5	wnwnn	0	nnwwn

Table 5.4   Interleaved two out of five

U.S. Postal Code

Another variation of two-out-of-five code is the U.S. Postal Code, used at the bottom edge of envelopes to encode ZIP codes. This code uses tall and short bars instead of wide and narrow bars, and seems to tolerate a wide range of print quality. The weights of the columns are 7, 4, 2, 1, and 0 from left to right. A check digit is added so that the sum is always a multiple of 10, and one tall bar is added at the beginning and at the end.

The dictionary for the postal code is given in Table 5.5. The ZIP code 80523 is written

‖ıııld‖ııııldıııddıdıı‖ıııldıl

and the 9-digit ZIP code 80521-9900 is written

‖ııldl‖ıııdldıııldlııı‖ldıııldldııllıııdlıııllııl

To read the last Postal Code, for example, separate the code into groups of 5 and decode each group using the Postal Code Dictionary. Thus,

| 8 | 0 | 5 | 2 | 1 | 9 | 9 | 0 | 0 | 6 |

so that the ZIP code is 80521-9900 and the check sum is 6. Indeed,

$$8 + 0 + 5 + 2 + 1 + 9 + 9 + 0 + 0 + 6 = 40$$

is congruent to zero modulo 10.

1	ııı‖	6	ı‖ıı
2	ııdı	7	‖ııı‖
3	ıı‖ı	8	‖ııdı
4	ıdııl	9	lıdıı
5	ıdıdı	0	‖ıııı

Table 5.5   U.S. Postal Code

Problems 5.2

1. Read the Code 2 of 5 message

2. Read the Code 39 message

3. Read the Interleaved 2 of 5 message

4. Read the Postal Code

What is the check sum?

5. Identify and read the bar code

6. Identify and read the bar code

7. Identify and read the bar code

8. Read the Postal Code

What is the check sum?

9. Identify and read the bar code

10. Identify and read the bar code

11. Identify and read the bar code

5.3   Other Commercial Codes

Bar codes have been designed for many special uses. Those described in this section are among the most widely used.

UPC-A

The Universal Product Code (UPC) is widely used by supermarkets and mass market retailers for cash register checkout (see Figure 5.5). Each digit 0–9 is represented by a space-bar-space-bar pattern for characters that appear to the left of center, and bar-space-bar-space for characters that appear on the right of center.

0 1 2 3 4 5   6 7 8 9 0 5

Figure 5.5   UPC-A

The width of each element is from 1 to 4 units wide, and the total width of the pattern is 7 units. Each pattern is conveniently represented by a binary word of length 7, with spaces represented by 0s and bars represented by 1s, as in Table 5.6.

Digit	Left code	Binary	Digit	Left code	Binary
0		0001101	5		0110001
1		0011001	6		0101111
2		0010011	7		0111011
3		0111101	8		0110111
4		0100011	9		0001011

Table 5.6   UPC-A

An extra narrow bar has been added at the left of each bar pattern to show where the left space begins. The right patterns are white/black reversals, with 0s and 1s interchanged. The parity for the left patterns is odd and for the right is even.

To count the number of possible left patterns, consider strings of length 7 that start with 0 and end with 1, say $0xxxxx1$. The pattern of two spaces and two bars means that we can mark the locations where the pattern changes between 0s and 1s with some symbol, say.

Thus $0x - x - xx - x1$ would indicate the pattern 0010011. There are 6 interior places, of which 3 are chosen, so there are $\binom{6}{3} = 20$ possible choices. Half of these have even parity and the other half odd, so there are exactly 10 with odd parity. Thus the patterns given in Table 5.6 give a complete list of the possible patterns with the stated criteria. The minimum distance between these patterns is 2, so single errors can be detected.

The UPC code is designed to be read from either the left or from the right. A total of ten digits are encoded, with five on the left of center and five to the right. The sequence 01010 = ⦀ is placed in the center to separate the two halves, and the string 101 = ‖ is placed at the far left and again at the far right. The digit zero 0001101 = ∣‖ is placed at the left just inside the 101 = ‖ and the sum of the digits modulo ten is placed at the right just inside the 101 = ‖ (see Figure 5.6).

Bar codes are read by reflected light, and a tall bar code makes it more likely that a beam of light will catch all the bars. If the reader does not see a complete pattern that it recognizes, including start/stop characters and the mod 10 check sum, then the entire pattern must be scanned again. On the seventh pass, the checker usually gives up and types the code in by hand on a keypad. It is the job of the error control code to make reasonably sure that if and when a pattern is accepted, then it is actually correct. It is much better to repeat the scan than to interpret something incorrectly. It causes a lot of ill will to charge someone $53.75 for a $1.39 item.

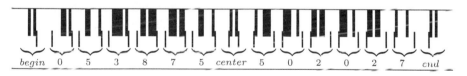

begin 0 5 3 8 7 5 *center* 5 0 2 0 2 7 *end*

Figure 5.6   UPC-A

We have

$$0 + 5 + 3 + 8 + 7 + 5 + 5 + 0 + 2 + 0 + 2 \bmod 10 \equiv 7$$

and the 7 serves as a parity check byte. Thus there is built in error control at two levels. The minimum distance is 2, so single errors are detected at the bit level. At the character level, the sum is checked modulo 10.

ISBN

Error control of a different type is used for the **International Standard Book Number (ISBN)**. Until 2007, an ISBN number was a string of ten digits printed on the back of a book:

$$\text{ISBN } 0\text{-}471\text{-}62546\text{-}9$$

In this case the leading 0 represents the country (United States), 471 the publisher, 62546 a publisher sequence number, and 9 was a computed check-sum digit which is equal to

$$0 \cdot 1 + 4 \cdot 2 + 7 \cdot 3 + 1 \cdot 4 + 6 \cdot 5 + 2 \cdot 6 + 5 \cdot 7 + 4 \cdot 8 + 6 \cdot 9 \bmod 11$$

If this sum is equal to 10 modulo 11, then the character X is used as the last symbol in the ISBN code.

Because of the check-sum digit, the distance between any two such ISBN numbers is at least 2. To see that, consider the ISBN number

$$a_1 - a_2 a_3 a_4 - a_5 a_6 a_7 a_8 a_9 - a_{10}$$

where

$$a_{10} = \sum_{i=1}^{9} i a_i \bmod 11$$

Since $10 = -1 \bmod 11$, it follows that $0 = \sum_{i=1}^{10} i a_i \bmod 11$. If the digit a_i is changed to b_i, then the sum $\sum_{i=1}^{10} i a_i \bmod 11$ changes from 0 to $i\,(b_i - a_i) \bmod 11$ which is only 0 if $b_i = a_i$ because 11 does not divide $i = 1, \dots, 10$. Thus a single error would result in an invalid ISBN number, so the minimum distance between ISBN numbers is at least 2.

In 2007 the ISBN code was changed to a thirteen-digit number. The immediate result is that the new ISBN numbers now start with the three digits 978 or 979, and the computation of the check digit is different. The check digit a_{13} is computed so that

$$a_1 + 3a_2 + a_3 + 3a_4 + a_5 + 3a_6 + a_7 + 3a_8 + a_9 + 3a_{10} + a_{11} + 3a_{12} + a_{13} \equiv 0 \ (\bmod 10)$$

Codabar

Codabar has been in use by libraries for a number of years. It is also known as Code 2 of 7 since most of the characters use two wide elements out of seven elements. Each character is represented by an odd number (1 or 3) of wide bars, and the characters 0–9, hyphen (-), and dollar sign ($) are represented with one wide bar and one wide space. (There is a choice of 1 out of 4 bars and 1 out of 3 spaces for a total of 12 possible sequences.) The remaining eight characters are represented with three wide elements and four narrow elements (see Figure 5.7).

a 1 2 3 4 5 6 7 8 9 0 b

Figure 5.7   Codabar

Unlike most other bar codes, Codabar is not a fixed-width code. The codes
for 0–9 all have two wide elements and five narrow elements, whereas some of
the codes have three wide elements and four narrow elements. The start/stop
characters are A, B, C, and D. Each character has a value assigned to it. Bars
have width 1 or 3 whereas spaces have widths 2 or 4 (see Table 5.7).

Data	Code	Value	Data	Code	Value
0		0	–		10
1		1	$		11
2		2	.		12
3		3	/		13
4		4	.		14
5		5	+		15
6		6	A		16
7		7	B		17
8		8	C		18
9		9	D		19

Table 5.7   Codabar

Usually a parity check character is added so that the sum of the values is
zero modulo 16. Codabar has been replaced at many locations by Code 39.

Code 128

Code 128 is flexible because it allows a full set of 7-bit ASCII characters to be
scanned.

Each of the Code 128 characters consists of three bars and three spaces. The bars and spaces may be one, two, three, or four units wide. The total length of each code 128 character is eleven units, with the total length of the bar units odd, and the total length of the space units even.

To count the number of patterns, note that we are counting strings of length 11 that start with 1 and end with 0 such that there are 3 substrings of all 1s and 3 substrings of all 0s. Placing a dash to indicate where a change takes place, we see that we need 5 dashes out of 10 locations. Thus there are

$$\binom{10}{5} = 252$$

possible patterns. However, the bars and spaces can each be of width at most 4. There are $\binom{6}{1} = 6$ patterns that have a bar or space of width 6 (see Table 5.8).

111111-0-1-0-1-0
1-000000-1-0-1-0
1-0-111111-0-1-0
1-0-1-000000-1-0
1-0-1-0-111111-0
1-0-1-0-1-000000

Table 5.8   Width 6 patterns

There are $6 \cdot 5 = 30$ patterns that have a bar or space of width 5 (and hence a bar or space of width 2). This leaves $252 - 6 - 30 = 216$ that have a bar or space of width at most 4. By symmetry, half of these will have even total bar width, half with odd total bar width. (To see that there is a one-to-one correspondence between patterns with odd bar width and even bar width, consider the mapping that reverses the sequence of 0s and 1s and then interchanges 0s and 1s.) Thus there are 108 patterns with even total bar width. There are three special start characters and one stop character, which leaves a total of 104 other characters that can be used. The special stop character has a total width of 13, including four bars and three spaces.

The even total bar width assures that the minimum distance between codewords is 2.

With the "A" start character, all uppercase alphanumeric characters plus all of the ASCII control characters can be used. With the "B" start character, all upper- and lowercase alphanumeric characters can be used (see Figure 5.8). The "C" start character allows double density numeric characters from 00 to 99 to be used.[2]

---

[2]See http://www.csse.monash.edu.au/~cavram/Cot2030/Barcodes/code_128.html for a character table of Code 128.

123456789

Figure 5.8   Code 128

U. K. Royal Mail

Postal codes in the United Kingdom have letters as well as digits.

A		J		S		1	
B		K		T		2	
C		L		U		3	
D		M		V		4	
E		N		W		5	
F		O		X		6	
G		P		Y		7	
H		Q		Z		8	
I		R		0		9	

Table 5.9   U.K. Royal Mail

The postal code uses a sequence of four elements chosen from High $|$, Low $|$,

Short $|$, and Tall $|$ bars with the property that each sequence contains two elements that extend up and two that extend down. The character set is given in Table 5.9.

How many such sequences are there? There are $4! = 24$ rearrangements of $\left\{ |,|,|,| \right\}$, with $\binom{4}{2} = 6$ rearrangements of $\left\{ |,|,|,| \right\}$, and $\binom{4}{2} = 6$ rearrangements of $\left\{ |,|,|,| \right\}$, for a total of $24 + 6 + 6 = 36$ sequences of length 4 that have two elements that extend up and two that extend down.

Problems 5.3

1. What are the numbers associated with the following UPC-A bar code? Verify that the check sum is correct.

2. An ISBN number is 0-966-x6563-9, where the fifth digit is unreadable. What is the missing digit, assuming the other digits are correct?

3. Show that if two (different) digits in a ten-digit ISBN number are interchanged, then the result is not a valid ISBN number.

4. Referring to the preceding problem, is this still true for the thirteen-digit ISBN numbers?

5. Decode the following Codabar.

6. Compare the strengths and weaknesses of Code 39 and the UPC code. Which is more compact? Which is easier to decode by hand?

7. In the ISBN number 0-471-89614-4, what does the first digit (0) represent? What does the second part (471) represent?

8. What bar code does the Fort Collins Public Library use?

9. Give another example of a commercial product that uses some type of error-control code. Explain how it uses minimum distance, check sums, and so on.

10. The address for Bletchley Park, where mathematicians and logicians worked to decrypt secret messages during World War II, is

> The Bletchley Park Trust
> The Mansion
> Bletchley Park
> Milton Keynes MK3 6EB
> United Kingdom

Give the U.K. Royal Mail code for this address.

11. If the U.K. Royal Mail code were expanded from four to five elements with the requirement of two upward extensions and two downward extensions in each group of five, how many characters could be represented?

5.4   Hamming (7, 4) Code

---

*Since the representation, transmission, and transformation of infor-
mation are fundamental to many other fields as well as to computer
science, it is time to make the theories easily available.*

Richard Hamming

In the Hamming (7, 4) code, the codewords have length 7 for message words
of length 4. This code can not only detect errors but correct them. It uses three
parity-check bits rather than one.

Message words are encoded by multiplying them, modulo 2, by the matrix

$$H = \begin{pmatrix} 1 & 0 & 0 & 0 & 0 & 1 & 1 \\ 0 & 1 & 0 & 0 & 1 & 0 & 1 \\ 0 & 0 & 1 & 0 & 1 & 1 & 0 \\ 0 & 0 & 0 & 1 & 1 & 1 & 1 \end{pmatrix}$$

For $\mathbf{u}$ a 4-bit message word, the 7-bit codeword is $\mathbf{v} = \mathbf{u}H$. The first 4 bits of
the codeword $\mathbf{v}$ are the 4 bits in the word $\mathbf{u}$. These are the message bits. The
last three bits are the parity check bits.

Example 5.7 If the message word is $\begin{pmatrix} 0 & 0 & 1 & 1 \end{pmatrix}$, then the codeword is
given by

$$\begin{pmatrix} 0 & 0 & 1 & 1 \end{pmatrix} \begin{pmatrix} 1 & 0 & 0 & 0 & 0 & 1 & 1 \\ 0 & 1 & 0 & 0 & 1 & 0 & 1 \\ 0 & 0 & 1 & 0 & 1 & 1 & 0 \\ 0 & 0 & 0 & 1 & 1 & 1 & 1 \end{pmatrix} \mod 2 = \begin{pmatrix} 0 & 0 & 1 & 1 & 0 & 0 & 1 \end{pmatrix}$$

What is the minimum distance between codewords? Every codeword is a
sum of rows of H. Each single row of H has weight at least 3. The sum of
two different rows has weight at least 3 because the two rows differ at two spots
in the first four columns, and at one or more spots in the last three columns.
Obviously the weight of the sum of any three distinct rows is at least 3, from
the first four columns alone, and the sum of all four rows is 7.

Since $d(\mathbf{u}, \mathbf{v}) = w(\mathbf{u} + \mathbf{v})$, and the sum of two codewords is a codeword, the
distance between any two different codewords is at least 3.

The distance function $d(\mathbf{u}, \mathbf{v})$ has the following properties.

Theorem 5.8 *If* $\mathbf{u}$, $\mathbf{v}$, *and* $\mathbf{w}$ *are codewords, then*

1. $d(\mathbf{u}, \mathbf{u}) = 0$

2. $d(\mathbf{u}, \mathbf{v}) = d(\mathbf{v}, \mathbf{u})$

3. $d(\mathbf{u}, \mathbf{v}) \leq d(\mathbf{u}, \mathbf{w}) + d(\mathbf{w}, \mathbf{v})$

These three properties hold for any reasonable notion of distance. The most interesting one is the last, which says that the distance from **u** to **v** can be no greater than the distance from **u** to **w** plus the distance from **w** to **v**. It is often called the **triangle inequality** because it can be thought of as saying that a side of a triangle can be no greater than the sum of the other two sides.

What happens when an error occurs, when one of the bits in a codeword gets changed? If the minimum distance between code words is at least 3, then a single error cannot change a codeword into another codeword. A single error will produce a word that is at a distance of 1 from the original codeword, and at a distance of at least 2 from all the other codewords. The decoder simply chooses the nearest codeword to the word that it receives. This is known as **maximum likelihood decoding**.

How many errors can be corrected if the minimum distance between code-words is d? In Figure 5.9, each center dot is a codeword and the white dots are other words.

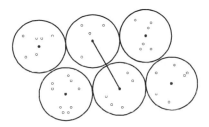

Figure 5.9   Minimum distance

If the circles are disjoint, and of the same radius r, then maximum likelihood decoding assigns to each dot that lies in a circle the codeword that lies within the same circle. If $r < d/2$, then the circles are disjoint. If we let $r = \lfloor (d-1)/2 \rfloor < d/2$, then any dot whose distance from a codeword is at most $\lfloor (d-1)/2 \rfloor$ lies in some circle. This is summarized in the following theorem.

Theorem 5.9 *If d is the minimum distance between codewords, then maximum likelihood decoding will correct any $\lfloor (d-1)/2 \rfloor$ or fewer errors.*

In order to correct any t or fewer errors, it is sufficient to have a minimum distance of $d = 2t + 1$. It's also necessary.

The Hamming (7, 4) code has minimum distance 3, so it can correct single errors. How is this done?

Table 5.10 shows an array of the 128 words of length seven. The codewords are in the leftmost column. Each of the 16 codewords is within a distance of one of exactly 8 words, which are listed in the same row. This accounts for all of the $128 = 16 \cdot 8$ words. To decode a word, find it in the array and choose the codeword that is in the same row.

Codeword

0000000	←	1000000 0100000 0010000 0001000 0000100 0000010 0000001
0001111	←	1001111 0101111 0011111 0000111 0001011 0001101 0001110
0010110	←	1010110 0110110 0000110 0011110 0010010 0010100 0010111
0011001	←	1011001 0111001 0001001 0010001 0011101 0011011 0011000
0100101	←	1100101 0000101 0110101 0101101 0100001 0100111 0100100
0101010	←	1101010 0001010 0111010 0100010 0101110 0101000 0101011
0110011	←	1110011 0010011 0100011 0111011 0110111 0110001 0110010
0111100	←	1111100 0011100 0101100 0110100 0111000 0111110 0111101
1000011	←	0000011 1100011 1010011 1001011 1000111 1000001 1000010
1001100	←	0001100 1101100 1011100 1000100 1001000 1001110 1001101
1010101	←	0010101 1110101 1000101 1011101 1010001 1010111 1010100
1011010	←	0011010 1111010 1001010 1010010 1011110 1011000 1011011
1100110	←	0100110 1000110 1110110 1101110 1100010 1100100 1100111
1101001	←	0101001 1001001 1111001 1100001 1101101 1101011 1101000
1110000	←	0110000 1010000 1100000 1111000 1110100 1110010 1110001
1111111	←	0111111 1011111 1101111 1110111 1111011 1111101 1111110

Table 5.10   Array decoding

Each word is at a distance of at most 1 from exactly one codeword. Because of this we say that the Hamming (7, 4) code is a **perfect code**.

There is an algebraic way to decode the Hamming (7, 4) code. The string

$$(\quad a_1 \quad a_2 \quad a_3 \quad a_4 \quad)$$

is encoded as

$$(\quad a_1 \quad a_2 \quad a_3 \quad a_4 \quad a_2 + a_3 + a_4 \quad a_1 + a_3 + a_4 \quad a_1 + a_2 + a_4 \quad)$$

where the arithmetic is done modulo 2. So if $(b_1 \ b_2 \ b_3 \ b_4 \ b_5 \ b_6 \ b_7)$ is a codeword, then

$$b_2 + b_3 + b_4 + b_5 = 0$$
$$b_1 + b_3 + b_4 + b_6 = 0$$
$$b_1 + b_2 + b_4 + b_7 = 0$$

Conversely, if these three equations hold, then

$$b_5 = b_2 + b_3 + b_4$$
$$b_6 = b_1 + b_3 + b_4$$
$$b_7 = b_1 + b_2 + b_4$$

so $(b_1 \ b_2 \ b_3 \ b_4 \ b_5 \ b_6 \ b_7)$ is a codeword. We can write the three equations as a single equation using a matrix product

$$(\quad b_1 \quad b_2 \quad b_3 \quad b_4 \quad b_5 \quad b_6 \quad b_7 \ ) P = (\ 0 \quad 0 \quad 0 \ )$$

where P is the matrix

$$P = \begin{pmatrix} 0 & 1 & 1 \\ 1 & 0 & 1 \\ 1 & 1 & 0 \\ 1 & 1 & 1 \\ 1 & 0 & 0 \\ 0 & 1 & 0 \\ 0 & 0 & 1 \end{pmatrix}$$

If $\mathbf{u}P = 000$, then $\mathbf{u}$ is a codeword and we decode it by taking the first four bits. Otherwise, $\mathbf{u}P$ tells us where the error occurred. For example, if the codeword 1010101 is corrupted and received as 1010001, then

$$\begin{pmatrix} 1 & 0 & 1 & 0 & 0 & 0 & 1 \end{pmatrix} P = \begin{pmatrix} 1 & 0 & 0 \end{pmatrix}$$

which is the fifth row of the matrix P. That tells us that an error occurred in the fifth bit. We correct the fifth bit of 1010001 to get 1010101, then decode this string as 1010, its first four bits.

The rows of P are the binary representations of the numbers 1, 2, ..., 7. If we could rearrange the rows so that row i is the binary representation of the number i, then decoding a corrupted word would be even easier because $\mathbf{u}P$ would tell us directly at which bit the error occurred.

To do that, we would have to rearrange the columns of the encoding matrix H in the same way. We want

$$P = \begin{pmatrix} 0 & 1 & 1 \\ 1 & 0 & 1 \\ 1 & 1 & 0 \\ 1 & 1 & 1 \\ 1 & 0 & 0 \\ 0 & 1 & 0 \\ 0 & 0 & 1 \end{pmatrix} \quad \text{to become} \quad \begin{pmatrix} 0 & 0 & 1 \\ 0 & 1 & 0 \\ 0 & 1 & 1 \\ 1 & 0 & 0 \\ 1 & 0 & 1 \\ 1 & 1 & 0 \\ 1 & 1 & 1 \end{pmatrix}$$

That is, the rows $R_1, R_2, R_3, R_4, R_5, R_6, R_7$ of P are rearranged in the order $R_7, R_6, R_1, R_5, R_2, R_3, R_4$. So we must also rearrange the columns $C_1, C_2, C_3, C_4, C_5, C_6, C_7$ of H in the order $C_7, C_6, C_1, C_5, C_2, C_3, C_4$. Thus

$$H = \begin{pmatrix} 1 & 0 & 0 & 0 & 0 & 1 & 1 \\ 0 & 1 & 0 & 0 & 1 & 0 & 1 \\ 0 & 0 & 1 & 0 & 1 & 1 & 0 \\ 0 & 0 & 0 & 1 & 1 & 1 & 1 \end{pmatrix} \quad \text{must become} \quad \begin{pmatrix} 1 & 1 & 1 & 0 & 0 & 0 & 0 \\ 1 & 0 & 0 & 1 & 1 & 0 & 0 \\ 0 & 1 & 0 & 1 & 0 & 1 & 0 \\ 1 & 1 & 0 & 1 & 0 & 0 & 1 \end{pmatrix}$$

Note that with this new H, the information bits are in positions 3, 5, 6, and 7, the columns that each contain a single 1, and the parity-check bits are in positions 1, 2, and 4.

Example 5.10 Consider the message word (1101). Using $v = uH \bmod 2$ encoding, the codeword is given by

$$
\begin{pmatrix} 1 & 1 & 0 & 1 \end{pmatrix}
\begin{pmatrix}
1 & 1 & 1 & 0 & 0 & 0 & 0 \\
1 & 0 & 0 & 1 & 1 & 0 & 0 \\
0 & 1 & 0 & 1 & 0 & 1 & 0 \\
1 & 1 & 0 & 1 & 0 & 0 & 1
\end{pmatrix}
\bmod 2 = \begin{pmatrix} 1 & 0 & 1 & 0 & 1 & 0 & 1 \end{pmatrix}
$$

Zap. Some random gamma radiation intercepts the codeword and changes it to (1110101). The decoder on the receiving end tests for an error by computing the product

$$
\begin{pmatrix} 1 & 1 & 1 & 0 & 1 & 0 & 1 \end{pmatrix}
\begin{pmatrix}
0 & 0 & 1 \\
0 & 1 & 0 \\
0 & 1 & 1 \\
1 & 0 & 0 \\
1 & 0 & 1 \\
1 & 1 & 0 \\
1 & 1 & 1
\end{pmatrix}
\bmod 2 = \begin{pmatrix} 0 & 1 & 0 \end{pmatrix}
$$

This result is the binary equivalent for 2. We assume bit two of (1110101) was corrupted with error in the second bit, so we change the word to (1010101). The message bits are located at coordinates 3, 5, 6, and 7, so the corrected codeword is decoded as (1101). This can also be computed using the matrix product

$$
\begin{pmatrix} 1 & 1 & 1 & 0 & 1 & 0 & 1 \end{pmatrix}
\begin{pmatrix}
0 & 0 & 0 & 0 \\
0 & 0 & 0 & 0 \\
1 & 0 & 0 & 0 \\
0 & 0 & 0 & 0 \\
0 & 1 & 0 & 0 \\
0 & 0 & 1 & 0 \\
0 & 0 & 0 & 1
\end{pmatrix}
\bmod 2 = \begin{pmatrix} 1 & 1 & 0 & 1 \end{pmatrix}
$$

Each character in the full set of 256 ASCII characters can be represented by 8-bits, that is, by one **byte**. To use a Hamming (7, 4) code on the ASCII character set, each byte is broken up into two four-bit pieces, or **nibbles**, and each nibble is encoded using the encoding matrix H. This produces codewords that are 7 bits long.

An eighth parity bit is often added to the 7-bit codeword to make the weight of the codeword even. When a word is received, its parity is tested. If the parity is odd, then it is assumed that a single error has occurred. The product $\mathbf{e} = \mathbf{v}P$ is computed using the first 7 bits of the received word. If $\mathbf{e} = 000$, then the single error occurred in the parity bit 8 and the message word consists of bits 3, 5, 6, and 7. If $\mathbf{e} \neq 000$, then the bit in $\mathbf{v}$ corresponding to the binary number $\mathbf{e}$ is changed.

If the parity is even, then it is assumed that either zero or two errors occurred. If $\mathbf{e} = \mathbf{v}P = 000$, then no error occurred and the message word consists of bits

3, 5, 6, and 7. If $\mathbf{e} = \mathbf{v}P \neq 000$, then at least two errors occurred. There is no way to correct these errors, so a message is generated saying that a double error has been detected.

The Hamming (8, 4) code is the Hamming (7, 4) code supplemented with an extra parity bit. This code corrects single errors and detects double errors. The codewords are 8 bits long, so each codeword is exactly one byte. This makes the Hamming (8, 4) code convenient to implement.

The decoder is described in Algorithm 5.1.

Algorithm 5.1   Hamming (8, 4) decoding

Input: An 8-bit word $\mathbf{v} = (v_1 v_2 v_3 v_4 v_5 v_6 v_7 v_8)$

Output: A 4-bit message word $\mathbf{u} = (u_1 u_2 u_3 u_4)$

or a double-error warning

$$\text{Set } P = \begin{pmatrix} 0 & 0 & 1 \\ 0 & 1 & 0 \\ 0 & 1 & 1 \\ 1 & 0 & 0 \\ 1 & 0 & 1 \\ 1 & 1 & 0 \\ 1 & 1 & 1 \end{pmatrix}$$

Set $(e_2 e_1 c_0) = (v_1 v_2 v_3 v_4 v_5 v_6 v_7) P \bmod 2$

If $(e_2 e_1 e_0) = (000)$ then
    Set $(u_1 u_2 u_3 u_4) = (v_3 v_5 v_6 v_7)$
Else
    Set parity $= \sum_{i=1}^{8} v_i \bmod 2$
    If parity $= 0$ then
        Return "Double error detected"
    End If
    Set $n = (e_2 \cdot 2 \mid e_1) 2 + e_0$
    Set $v_n = v_n + 1 \bmod 2$
    Set $(u_1 u_2 u_3 u_4) = (v_3 v_5 v_6 v_7)$
End If
Return $(u_1 u_2 u_3 u_4)$

Problems 5.4

1. Use the Hamming (7, 4) code to encode the message word (1011).

2. Use the Hamming (7, 4) code to decode the word (0110111).

3. Use the Hamming (8, 4) code to decode the word (10110100).

4. Use the Hamming (8, 4) code to decode the word (10110111).

5. Use the Hamming (8, 4) code to decode the word (10110101).

6. Use the Hamming (8, 4) code to decode the word (10110110).

7. Explain how the Hamming (7, 4) code could be adapted for use as a bar code, or why this would be impractical. Explain the advantages and disadvantages. What if the codewords (0000000) and (1111111) were eliminated? How could the remaining 14 codewords be used to design a bar code?

8. Use the Hamming (7, 4) code to encode the message word (1100).

9. Use the Hamming (7, 4) code to decode the word (0000011).

10. Use the Hamming (8, 4) code to decode the word (10110000).

11. Use the Hamming (8, 4) code to decode the word (01001011).

12. Use the Hamming (8, 4) code to decode the word (11001010).

13. Use the Hamming (8, 4) code to decode the word (11111110).

Chapter 6

Chinese Remainder Theorem

Chinese graduate students tell a wonderful story of a Chinese general who used an unusual method to count his troops. He gathered them together on a large parade field, told them to form groups of 11, and asked the leftovers to gather in front of the reviewing stand. He counted the number of leftovers, and then told the troops to reassemble into groups of 12, again with the leftovers up front. After repeating this process with 13, 17, and 19, he did some quick mental calculations to determine the exact number of troops on the parade field.

Some of the details of this story may have been lost, but the techniques were known as early as the first century and appeared in the writings of the Chinese mathematician Sun Tsu Suan-Ching (fourth century A.D.), who posed the question, "There are certain things whose number is unknown. Repeatedly divided by 3, the remainder is 2; by 5 the remainder is 3; and by 7 the remainder is 2. What will be the number?"

These techniques are ancient but timeless. We will see in this chapter how the Chinese remainder theorem can be used to simplify computer calculations that involve exact arithmetic.

6.1   Systems of Linear Equations Modulo n

There are several ways to solve a system of linear equations modulo n, and we illustrate some by examples. The examples illustrate the case of two equations and two unknowns. The methods can be extended to handle the case of k equations and k unknowns.

Example 6.1 Solve the system

$$4x + 7y = 9$$
$$5x + 2y = 3$$

in integers modulo 13.

Solution. Since $4^{-1} \bmod 13 = 10$ and $5^{-1} \bmod 13 = 8$, we can multiply the first equation by 10 and the second by 8 and reduce the equations modulo 13 to get

$$10\,(4x + 7y)\,\mathrm{mod}\,13 = 10 \cdot 9\,\mathrm{mod}\,13$$
$$8\,(5x + 2y)\,\mathrm{mod}\,13 = 8 \cdot 3\,\mathrm{mod}\,13$$

or

$$x + 5y = 12$$
$$x + 3y = 11$$

Now subtract the second equation from the first to get

$$2y = 1$$

Since $2^{-1} \bmod 13 = 7$, we can multiply both sides by 7 to get

$$y = 7$$

Thus
$$x = (12 - 5 \cdot 7)\,\mathrm{mod}\,13 = 3$$

Checking by substituting $x = 3$, $y = 7$ into the original system of equations, we get

$$(4 \cdot 3 + 7 \cdot 7)\,\mathrm{mod}\,13 = 9$$
$$(5 \cdot 3 + 2 \cdot 7)\,\mathrm{mod}\,13 = 3$$

This system of equations can also be phrased as a matrix problem.

Example 6.2 Use elementary row reduction to put the augmented matrix

$$\begin{pmatrix} 4 & 7 & 9 \\ 5 & 2 & 3 \end{pmatrix}$$

in reduced form mod 13.
Solution.

$$\begin{pmatrix} 4 & 7 & 9 \\ 5 & 2 & 3 \end{pmatrix} \xrightarrow{R_2 - R_1} \begin{pmatrix} 4 & 7 & 9 \\ 1 & -5 & -6 \end{pmatrix} \xrightarrow{\mathrm{mod}\,13}$$

$$\begin{pmatrix} 4 & 7 & 9 \\ 1 & 8 & 7 \end{pmatrix} \xrightarrow{R_1 - 4R_2} \begin{pmatrix} 0 & 1 & 7 \\ 1 & 8 & 7 \end{pmatrix} \xrightarrow{R_2 - 8R_1} \begin{pmatrix} 0 & 1 & 7 \\ 1 & 0 & 3 \end{pmatrix}$$

From this it follows that $x = 3$ and $y = 7$.

This system of equations can also be rephrased in matrix form as

$$\begin{pmatrix} 4 & 7 \\ 5 & 2 \end{pmatrix} \begin{pmatrix} x \\ y \end{pmatrix} = \begin{pmatrix} 9 \\ 3 \end{pmatrix}$$

Example 6.3 To solve the matrix equation

$$\begin{pmatrix} 4 & 7 \\ 5 & 2 \end{pmatrix} \begin{pmatrix} x \\ y \end{pmatrix} = \begin{pmatrix} 9 \\ 3 \end{pmatrix}$$

modulo 13, evaluate the matrix inverse

$$\begin{pmatrix} 4 & 7 \\ 5 & 2 \end{pmatrix}^{-1} \bmod 13 = \begin{pmatrix} 11 & 7 \\ 5 & 9 \end{pmatrix}$$

and use matrix multiplication to get

$$\begin{pmatrix} x \\ y \end{pmatrix} = \begin{pmatrix} 11 & 7 \\ 5 & 9 \end{pmatrix} \begin{pmatrix} 9 \\ 3 \end{pmatrix} \bmod 13$$

$$= \begin{pmatrix} 3 \\ 7 \end{pmatrix}$$

To check the result, multiply matrices.

$$\begin{pmatrix} 4 & 7 \\ 5 & 2 \end{pmatrix} \begin{pmatrix} 3 \\ 7 \end{pmatrix} \bmod 13 = \begin{pmatrix} 9 \\ 3 \end{pmatrix}$$

Since the function $f(a/b) = ab^{-1} \bmod p$ that takes rational numbers to integers modulo p preserves sums and products, it follows that systems of linear equations can be solved by first finding a rational solution, then applying the function f. This can be done with any of the three methods shown above.

Example 6.4 We solve the system

$$4x + 7y = 9$$
$$5x + 2y = 3$$

by multiplying the first equation by $\frac{1}{4}$ and the second by $\frac{1}{5}$ to get

$$x + \frac{7}{4}y = \frac{9}{4}$$
$$x + \frac{2}{5}y = \frac{3}{5}$$

Subtract the second equation from the first to get

$$\left(\frac{7}{4} - \frac{2}{5} \right) y = \frac{9}{4} - \frac{3}{5}$$
$$\frac{27}{20}y = \frac{33}{20}$$
$$y = \frac{33}{20} \cdot \frac{20}{27}$$
$$= \frac{11}{9}$$

Substitute back to get

$$x + \frac{7}{4} \cdot \frac{11}{9} = \frac{9}{4}$$
$$x + \frac{77}{36} = \frac{9}{4}$$
$$x = \frac{9}{4} - \frac{77}{36}$$
$$x = \frac{1}{9}$$

Hence

$$x \bmod 13 = \frac{1}{9} \bmod 13 = 3$$
$$y \bmod 13 = \frac{11}{9} \bmod 13 = 7$$

There are systems of linear equations that have a solution over the rational numbers but no solution modulo n. This can happen even with a single equation, such as $2x = 1 \bmod 2$. There will be a unique solution if $\gcd(\det A, n) = 1$, where A is the matrix of coefficients and $\det A$ is the determinant of A (see page 314).

Problems 6.1

1. Solve the system

$$2x + 3y = 7$$
$$3x + y = 2$$

   in integers modulo 11 by doing mod 11 arithmetic by hand. Check your solution.

2. Solve the system

$$5x + 2y = 3$$
$$2x + 3y = 7$$

   in integers modulo 13 by finding a rational solution by hand, then reducing the solution mod 13. Check your solution.

3. Evaluate

$$\begin{pmatrix} 3 & 5 \\ 4 & 8 \end{pmatrix}^{-1} \bmod 11$$

   by finding

$$\begin{pmatrix} 3 & 5 \\ 4 & 8 \end{pmatrix}^{-1}$$

   (with rational entries) and then reducing each entry modulo 11.

4. Evaluate

$$\begin{pmatrix} 3 & 5 \\ 4 & 8 \end{pmatrix}^{-1} \bmod 11$$

by doing row reduction modulo 11 on the augmented matrix

5. Evaluate

$$\begin{pmatrix} 2 & 3 & 1 \\ 0 & 1 & 5 \\ 3 & 7 & 2 \end{pmatrix}^{-1} \bmod 17$$

by finding

$$\begin{pmatrix} 2 & 3 & 1 \\ 0 & 1 & 5 \\ 3 & 7 & 2 \end{pmatrix}^{-1}$$

(with rational entries) and then reducing each entry modulo 17.

6. Evaluate

$$\begin{pmatrix} 2 & 3 & 1 \\ 0 & 1 & 5 \\ 3 & 7 & 2 \end{pmatrix}^{-1} \bmod 17$$

by doing row reduction modulo 17 on the augmented matrix

$$\begin{pmatrix} 2 & 3 & 1 & 1 & 0 & 0 \\ 0 & 1 & 5 & 0 & 1 & 0 \\ 3 & 7 & 2 & 0 & 0 & 1 \end{pmatrix}$$

7. Solve the system

$$x + 2y + 2z = 5$$
$$2x + y + 2z = 7$$
$$2x + 2y + z = 3$$

in integers modulo 13 by doing mod 13 arithmetic by hand. Check your solution.

8. Solve the system

$$5x + 2y + z = 6$$
$$2x + 3y + 4z = 1$$
$$4x + y + 3z = 2$$

in integers modulo 7 by finding a rational solution by hand, then reducing the solution mod 7. Check your solution.

9. Use a computer algebra system to solve the system

$$
\begin{array}{rcrcrcrcr}
w & + & x & + & y & + & z & = & 23 \\
w & + & 2x & + & 3y & + & 4z & = & 19 \\
w & + & 4x & + & 9y & + & 16z & = & 42 \\
w & + & 8x & + & 27y & + & 64z & = & 73
\end{array}
$$

in integers modulo 91 by solving the system in $\mathbb{Q}$ and reducing modulo 91.

10. Use a computer algebra system to solve the system

$$
\begin{array}{rcrcrcrcr}
w & + & x & + & y & + & z & = & 23 \\
w & + & 2x & + & 3y & + & 4z & = & 19 \\
w & + & 4x & + & 9y & + & 16z & = & 42 \\
w & + & 8x & + & 27y & + & 64z & = & 73
\end{array}
$$

in integers modulo 91 by solving the corresponding matrix equation modulo 91.

6.2   Chinese Remainder Theorem

In contrast with the previous section, many applications require the simultaneous use of several moduli.

Simple systems of congruences can be solved by exhaustion.

Example 6.5 To solve the system

$$x \equiv 3 \pmod 5$$
$$x \equiv 5 \pmod 7$$

look at the possible solutions. Solutions to the congruence $x \equiv 3 \pmod 5$ include

$$\{.., 3, 8, 13, 18, 23, 28, 33, 38, ..\}$$

and solutions to $x \equiv 5 \pmod 7$ include

$$\{.., 5, 12, 19, 26, 33, 40, ..\}$$

The intersection includes $\{.., 33, ..\}$. Any two elements of the intersection differ by a multiple of $35 = 5 \cdot 7$.

As was the case in computing greatest common divisors, the method of exhaustion is just that—very exhausting. We need to be able to solve systems of congruences quickly and efficiently with large moduli.

Here is a smarter method than exhaustion to solve the system

$$x \equiv a \pmod m$$
$$x \equiv b \pmod n$$

where m is relatively prime to n. If $x = a + mt$ for some integer t, then x solves the first congruence. So we want to find an integer t such that $a + mt \equiv b \pmod{n}$. As $m \perp n$, we can find an integer m^{-1} so that $mm^{-1} \equiv 1 \pmod{n}$. Set $t = m^{-1}(b - a)$ and $x = a + mt$. Then certainly $x \equiv a \pmod{m}$ and

$$x = a + mt \equiv a + m(m^{-1}(b - a)) \equiv a + (b - a) = b \bmod n$$

We have proved most of the following theorem, and in the process have given a method for solving a system of two congruences.

Theorem 6.6 *If $m \perp n$, then the system*

$$x \equiv a \pmod{m}$$
$$x \equiv b \pmod{n}$$

has a solution. In fact, if m^{-1} is an integer that is the inverse of m modulo n, and $t = m^{-1}(b - a)$, then $x = a + mt$ is a solution. Any two solutions are congruent modulo mn.

Proof. It remains to prove that any two solutions are congruent modulo mn. Suppose x and y are two such solutions. Then $x - y \equiv 0 \bmod m$ and $x - y \equiv 0 \bmod n$, which means that both m and n divide $x - y$. Since $m \perp n$, it follows that mn divides $x - y$; that is, $x \equiv y \bmod mn$. ∎

Example 6.7 We will use the proof of Theorem 6.6 to give an alternative solution to the system

$$x \equiv 3 \pmod{5}$$
$$x \equiv 5 \pmod{7}$$

We have $5^{-1} \bmod 7 = 3$ since $5 \cdot 3 = 15 \equiv 1 \bmod 7$. Thus

$$t = m^{-1}(b - a) \bmod 7 = 3(5 - 3) \bmod 7 = 6$$

and thus

$$x = a + mt = 3 + 5 \cdot 6 = 33$$

We can extend the theorem to more than two congruences.

Definition 6.8 *A collection $m_1, m_2, \ldots, m_r$ of integers is **pairwise relatively prime** if $m_i \perp m_j$ whenever $i \neq j$.*

Theorem 6.9 (Chinese Remainder Theorem) *Let $m_1, m_2, \ldots, m_r$ be pairwise relatively prime positive integers. Then the system*

$$x \equiv a_1 \pmod{m_1}$$
$$x \equiv a_2 \pmod{m_2}$$

$$\vdots$$

$$x \equiv a_r \pmod{m_r}$$

has a unique solution modulo the product $m_1 m_2 \cdots m_r$.

Proof. The proof is by induction on r. The theorem is easy for $r = 1$ and has been proved for $r = 2$. Consider the system

$$x \equiv a_1 \pmod{m_1}$$
$$x \equiv a_2 \pmod{m_2}$$

$$\vdots$$

$$x \equiv a_r \pmod{m_r}$$
$$x \equiv a_{r+1} \pmod{m_{r+1}}$$

where $m_1, m_2, \ldots, m_r, m_{r+1}$ are pairwise relatively prime positive integers. By induction, we can find a solution b to the first r congruences, a solution that is unique modulo $m_1 m_2 \cdots m_r$. Then the system of $r+1$ congruences is equivalent to the system

$$x \equiv b \pmod{m_1 m_2 \cdots m_r}$$
$$x \equiv a_{r+1} \pmod{m_{r+1}}$$

of two congruences, which has a unique solution modulo the product

$$m_1 m_2 \cdots m_r m_{r+1}$$

by Theorem 6.6, the case $r = 2$ of this theorem, because m_{r+1} is relatively prime to $m_1 m_2 \cdots m_r$. ∎

The proof of the Chinese remainder theorem gives an algorithm for solving a system of congruences (see Algorithm 6.1).

Algorithm 6.1   Chinese remainder algorithm

Input: Integer r and vectors $\mathbf{a} = (a_1, \ldots, a_r)$
and $\mathbf{m} = (m_1, \ldots, m_r)$ with $m_i \perp m_j$
Output: An integer k such that $k \equiv a_i \bmod m_i$

Function $CRA(\mathbf{a}, \mathbf{m}, r)$
    If $r = 1$ **Then**
        Return a_1
    Else
        Set $t = m^{-1}(a_r - a_{r-1}) \bmod m_r$
        Set $a_{r-1} = a_{r-1} + t m_{r-1}$
        Set $m_{r-1} = m_{r-1} m_r$
        Return $CRA(\mathbf{a}, \mathbf{m}, r - 1)$
    End If
End Function

Example 6.10 To solve the system

$$x \equiv 4 \pmod{29}$$
$$x \equiv 7 \pmod{30}$$
$$x \equiv 8 \pmod{31}$$

we first solve the last pair. Since $30 = 30^{-1} \bmod 31$, we have $t = 30(8-7) = 30$, so

$$x = a_2 + m_2 t = 7 + 30 \cdot 30 = 907 \bmod 930$$

We now solve the system

$$x \equiv 4 \pmod{29}$$
$$x \equiv 907 \pmod{930}$$

Since

$$29^{-1} \bmod 930 = 449$$

it follows that

$$t = 449\,(907 - 4) \bmod 930 = 897$$

so

$$x = 4 + 29 \cdot 897 = 26\,017$$

Checking, we see that

$$26\,017 \bmod 29 = 4$$
$$26\,017 \bmod 30 = 7$$
$$26\,017 \bmod 31 = 8$$

Problems 6.2

1. Solve the system

$$x \equiv 5 \pmod{9}$$
$$x \equiv 4 \pmod{11}$$

by listing the solutions to each congruence and then finding the common elements.

2. Solve the system

$$x \equiv 5 \pmod{9}$$
$$x \equiv 4 \pmod{11}$$

by hand, using the Chinese remainder algorithm.

3. Solve the system

$$x \equiv 2 \pmod{3}$$
$$x \equiv 3 \pmod{4}$$
$$x \equiv 4 \pmod{5}$$

by listing the solutions to each congruence and then finding the common elements.

4. Solve the system

$$x \equiv 3 \pmod 7$$
$$x \equiv 2 \pmod 9$$
$$x \equiv 1 \pmod{11}$$

by using the Chinese remainder algorithm.

5. Solve the system

$$x \equiv 3 \pmod 7$$
$$x \equiv 2 \pmod 9$$
$$x \equiv 1 \pmod{11}$$

by using a computer algebra system.

6. Solve the system

$$x \equiv 3 \pmod 5$$
$$x \equiv 4 \pmod 6$$
$$x \equiv 3 \pmod 7$$
$$x \equiv 4 \pmod{11}$$

by using the Chinese remainder algorithm.

7. There are n gold coins on a table. If the coins are counted 2 at a time, 1 remains. If the coins are counted $3, 4, 5, 6, 7$ at a time, then $2, 3, 4, 5, 6$ remain, respectively. What is the smallest possible number of gold coins?

8. Answer the question posed by Sun Tsu Suan-Ching (see page 127).

9. Solve the system of congruences

$$2x + 3 \equiv 7 \pmod{11}$$
$$3x + 4 \equiv 5 \pmod{13}$$

by using the Chinese remainder algorithm.

10. Solve the system

$$x^2 + 3 \equiv 0 \pmod 7$$
$$x^2 + 4x + 3 \equiv 0 \pmod{11}$$

of congruences.

6.3 Extended Precision Arithmetic

There are many ways to represent a positive integer. The representation we pick depends on the problem we face. The usual representation is base 10. We write the integer n as

$$n = a_k 10^k + a_{k-1} 10^{k-1} + \cdots + a_2 10^2 + a_1 10 + a_0 = (a_k a_{k-1} \ldots a_2 a_1 a_0)_{10}$$

where $a_k \neq 0$ and each a_i is a digit between 0 and 9. Then we represent n by the string of digits

$$(a_k a_{k-1} \ldots a_2 a_1 a_0)_{10}$$

We can modify this to get the base b representation, for any integer b greater than 1.

Positive integers can also be represented as a product of primes, or as a product of powers of primes

$$n = \prod_{p \text{ prime}} p^{e_p}$$

where each e_p is a nonnegative integer. For many applications, the prime power representation is more natural than the base ten or base b representations.

The Chinese remainder theorem suggests another representation that is particularly useful for extended precision arithmetic, as we shall see in this section. Let $m_1, m_2, \ldots, m_r$ be pairwise relatively prime positive integers with the property that m_i^2 can be computed exactly in machine arithmetic. Then any positive integer n less than the product $m_1 m_2 \cdots m_r$ is determined by the numbers

$$a_1 = n \bmod m_1, \quad a_2 = n \bmod m_2, \quad \ldots, \quad a_r = n \bmod m_r$$

The representation

$$n = [a_1, a_2, \ldots, a_r]$$

is called the **modular representation** of n relative to the basis

$$m_1, m_2, \ldots, m_r$$

In practice, the moduli $m_1, m_2, \ldots, m_r$ are usually distinct positive primes.

Example 6.11 Consider the collection

$$191, \quad 193, \quad 197, \quad 199$$

of primes. Any positive integer less than

$$191 \cdot 193 \cdot 197 \cdot 199 = 1\,445\,140\,189$$

can be given in a modular representation relative to the modular basis

$$(191, 193, 197, 199)$$

In particular,

$$1000\,000\,000 = [18, 29, 26, 125]$$

since

$$1000\,000\,000 \bmod 191 = 18$$
$$1000\,000\,000 \bmod 193 = 29$$
$$1000\,000\,000 \bmod 197 = 26$$
$$1000\,000\,000 \bmod 199 = 125$$

Modular representations make it easy to do large integer arithmetic. The following rules hold.

Theorem 6.12 *Assume $k = [a_1, a_2, \ldots, a_r]$ and $n = [b_1, b_2, \ldots, b_r]$ are modular representations relative to the modular basis $(m_1, m_2, \ldots, m_r)$. If $m = m_1 m_2 \cdots m_r$ then*

     i. $k + n \bmod m = [a_1 + b_1 \bmod m_1, a_2 + b_2 \bmod m_2, \ldots, a_r + b_r \bmod m_r]$,

     ii. $kn \bmod m = [a_1 b \bmod m_1, a_2 b_2 \bmod m_2, \ldots, a_r b_r \bmod m_r]$, *and*

     iii. $k^j \bmod m = [a_1^j \bmod m_1, a_2^j \bmod m_2, \ldots, a_r^j \bmod m_r]$.

Proof. The proof follows from the fact that if $a_i = k \bmod m_i$ and $b_i = n \bmod m_i$, then $a_i + b_i \equiv k + n \bmod m_i$, $a_i b_i \equiv kn \bmod m_i$ and $a_i^j \equiv k^j \bmod m_i$. ∎

Of particular interest is when the numbers are less than m.

Corollary 6.13 *Assume $k = [a_1, a_2, \ldots, a_r]$ and $n = [b_1, b_2, \ldots, b_r]$ are modular representations relative to the modular basis $(m_1, m_2, \ldots, m_r)$. Let $m = m_1 m_2 \cdots m_r$.*

     i. If $k + n < m$, then

$$k + n = [a_1 + b_1 \bmod m_1, a_2 + b_2 \bmod m_2, \ldots, a_r + b_r \bmod m_r]$$

     ii. If $kn < m$, then

$$kn = [a_1 b_1 \bmod m_1, a_2 b_2 \bmod m_2, \ldots, a_r b_r \bmod m_r]$$

     iii. If $k^j < m$, then

$$k^j = \left[a_1^j \bmod m_1, a_2^j \bmod m_2, \ldots, a_r^j \bmod m_r \right]$$

To see why modular representations are useful for extended precision arithmetic, look at the problem of multiplying two integers k and n whose product is roughly 10^{100}. Assuming that 65535 is the largest integer that can be computed in machine arithmetic, it is natural to use base $b = 256$ notation. Assume $k = \sum_{i=0}^{20} a_i b^i$ and $n = \sum_{i=0}^{20} c_i b^i$. To compute the product kn requires multiplying each term in the expansion of k times each term in the expansion of n, or $21^2 = 441$ multiplications plus a large number of adds and carries.

To compute the same product using the modular representation relative to the modular basis (2, 3, 5, 7, 11, 13, 17, 19, 23, 29, 31, 37, 41, 43, 47, 53, 59, 61, 67, 71, 73, 79, 83, 89, 97, 101, 103, 107, 109, 113, 127, 131, 137, 139, 149, 151, 157, 163, 167, 173, 179, 181, 191, 193, 197, 199, 211, 223, 227, 229, 233, 239, 241, 251) requires 54 multiplications, with no adds or carries.

Example 6.14 To compute the product

$$23\,890\,864\,094 \cdot 1\,883\,289\,456$$

we first use base $b = 1000$ notation. We have

$$23\,890\,864\,094 = 23b^3 + 890b^2 + 864b + 94 = (23, 890, 864, 94)_{1000}$$

and

$$1\,883\,289\,456 = 1b^3 + 883b^2 + 289b + 456 = (1, 883, 289, 456)_{1000}$$

which means

$$
\begin{aligned}
&23\,890\,864\,094 \cdot 1\,883\,289\,456 \\
&= 23 \cdot 1b^6 + (23 \cdot 883 + 890 \cdot 1)\,b^5 \\
&\quad + (23 \cdot 289 + 890 \cdot 883 + 864 \cdot 1)\,b^4 \\
&\quad + (23 \cdot 456 + 890 \cdot 289 + 864 \cdot 883 + 94 \cdot 1)\,b^3 \\
&\quad + (890 \cdot 456 + 864 \cdot 289 + 94 \cdot 883)\,b^2 \\
&\quad + (864 \cdot 456 + 94 \cdot 289)\,b + 94 \cdot 456 \\
&= 23b^6 + 21\,199b^5 + 793\,381b^4 + 1030\,704b^3 \\
&\quad + 738\,538b^2 + 421\,150b + 42\,864 \\
&= (23 + 21)\,b^6 + (199 + 793)\,b^5 + (381 + 1030)\,b^4 \\
&\quad + (704 + 738)\,b^3 + (538 + 421)\,b^2 \\
&\quad + (150 + 42)\,b + 864 \\
&= 44b^6 + 992b^5 + 1411b^4 + 1442b^3 + 959b^2 \\
&\quad + 192b + 864 \\
&= 44b^6 + 993b^5 + 412b^4 + 442b^3 + 959b^2 \\
&\quad + 192b + 864 \\
&= 44\,993\,412\,442\,959\,192\,864
\end{aligned}
$$

Notice that 16 multiplications are required, plus sums and carries. A computer algebra system confirms that

$$23\,890\,864\,094 \cdot 1\,883\,289\,456 = 44\,993\,412\,442\,959\,192\,864$$

Example 6.15 A second method of expressing the product $23\,890\,864\,094 \cdot 1\,883\,289\,456$ is to write it as a product of primes.

$$23\,890\,864\,094 = 2 \times 11\,945\,432\,047$$
$$1\,883\,289\,456 = 2^4 3^2 971 \times 13\,469$$

It now easily follows that

$$23\,890\,864\,094 \cdot 1\,883\,289\,456 = 2^5 \times 3^2 \times 971 \times 13\,469$$
$$\times 11\,945\,432\,047$$

This representation actually replaces product calculations with much simpler sum calculations. However, this method also requires prime factorization, a difficult problem for large integers. While prime factorization makes multiplication simple, this method does not lend itself to the calculation of sums.

Example 6.16 Finally, we use modular representations relative to the modular basis (997, 999, 1000, 1001, 1003, 1007, 1009). We have

$$23\,890\,864\,094 = [350, 872, 94, 97, 879, 564, 218]$$
$$1\,883\,289\,456 = [324, 630, 456, 48, 488, 70, 37]$$

and hence

$$23\,890\,864\,094 \times 1\,883\,289\,456 = [350 \cdot 324 \bmod 997, 872 \cdot 630 \bmod 999,$$
$$94 \cdot 456 \bmod 1000, 97 \cdot 48 \bmod 1001,$$
$$879 \cdot 488 \bmod 1003, 564 \cdot 70 \bmod 1007,$$
$$218 \cdot 37 \bmod 1009]$$
$$= [739, 909, 864, 652, 671, 207, 1003]$$

Notice that seven multiplications plus seven calls to the mod function were required.

Problems 6.3

1. Compute the product $37759097376 \cdot 116389305648$ by using base $b = 1000$ arithmetic.

2. Compute the product $37759097376 \cdot 116389305648$ by using prime power factorization.

3. Compute the product $37759097376 \cdot 116389305648$ by using the modular basis $(997, 999, 1000, 1001, 1003, 1007, 1009)$.

4. Compute $37759097376^{-1} \bmod (997 \cdot 1003 \cdot 1007 \cdot 1009)$ by computing each of

$$37759097376^{-1} \bmod 997$$
$$37759097376^{-1} \bmod 1003$$
$$37759097376^{-1} \bmod 1007$$
$$37759097376^{-1} \bmod 1009$$

and then using the Chinese remainder algorithm.

5. Evaluate $2^{40} = 1099\,511\,627\,776$ on a calculator that displays at most 10 digits.

6. Use 10-digit arithmetic to find the repeating decimal expansion of $1/61$.

7. Find the modular representation of 100! relative to the modular basis

$$(2^{97}, 3^{48}, 5^{24}, 7^{16}, 11^9, 13^7, 17^5, 19^5, 23^4, 29^3, 31^3,$$
$$37^2, 41^2, 43^2, 47^2, 53, 59, 61, 67, 71, 73, 79, 83, 89, 97)$$

Explain why this is a modular basis.

6.4 Greatest Common Divisor of Polynomials

In this section we give a second example of calculations that can be simplified by applying the Chinese remainder theorem. Let $\mathbb{Z}[x]$ denote the set of polynomials with integer coefficients. Given two polynomials $a(x)$ and $d(x)$ in $\mathbb{Z}[x]$ we say that $d(x)$ **divides** $a(x)$ and write $d(x)|a(x)$ if $a(x) = d(x)p(x)$ for some polynomial $p(x)$ in $\mathbb{Z}[x]$. Thus, $(x^2+1)|(2x^3+3x^2+2x+3)$ because $2x^3 + 3x^2 + 2x + 3 = (x^2+1)(2x+3)$.

Definition 6.17 *A **greatest common divisor** of two polynomials $a(x)$ and $b(x)$ in $\mathbb{Z}[x]$ is a polynomial $d(x)$ in $\mathbb{Z}[x]$ of largest degree such that $d(x)|a(x)$ and $d(x)|b(x)$.*

The Euclidean algorithm works with polynomials as well as with integers (See Theorem 9.11 on page 206 for details.) Given polynomials $a(x)$ and $b(x)$ we can compute the polynomial quotient $q(x)$ and remainder $r(x)$ by using the long division algorithm. The following two examples show how to calculate the greatest common divisor of two polynomials in $\mathbb{Z}[x]$, first by doing polynomial calculations with coefficients in the set $\mathbb{Q}$ of rational numbers, then by doing polynomial calculations with coefficients modulo a prime.

Example 6.18 Consider the polynomials

$$a(x) = 12x^5 + 28x^4 - 15x^3 + 17x^2 - 3x + 6$$
$$b(x) = 8x^6 + 16x^5 - 18x^4 + 16x^3 + 8x^2 - 9x + 6$$

We calculate $\gcd\left(a\left(x\right),b\left(x\right)\right) \in Z\left[x\right]$ by doing intermediate calculations with rational coefficients. Long division yields

$$\frac{b\left(x\right)}{a\left(x\right)} = \frac{2}{3}x - \frac{2}{9} + \frac{-\frac{16}{9}x^4 + \frac{4}{3}x^3 + \frac{124}{9}x^2 - \frac{41}{3}x + \frac{22}{3}}{12x^5 + 28x^4 - 15x^3 + 17x^2 - 3x + 6}$$

Define $q\left(x\right)$ and $r\left(x\right)$ by

$$q\left(x\right) = \frac{2}{3}x - \frac{2}{9}$$

$$r\left(x\right) = -\frac{16}{9}x^4 + \frac{4}{3}x^3 + \frac{124}{9}x^2 - \frac{41}{3}x + \frac{22}{3}$$

and note that

$$b\left(x\right) = q\left(x\right)a\left(x\right) + r\left(x\right)$$

Another polynomial division leads to

$$\frac{a\left(x\right)}{r\left(x\right)} = -\frac{27}{4}x - \frac{333}{16} + \frac{\frac{423}{4}x^3 + \frac{423}{2}x^2 - \frac{3807}{16}x + \frac{1269}{8}}{r\left(x\right)}$$

Define $q_1\left(x\right)$ and $r_1\left(x\right)$ by

$$q_1\left(x\right) = -\frac{27}{4}x - \frac{333}{16}$$

$$r_1\left(x\right) = \frac{423}{4}x^3 + \frac{423}{2}x^2 - \frac{3807}{16}x + \frac{1269}{8}$$

and note that

$$a\left(x\right) = q_1\left(x\right)r\left(x\right) + r_1\left(x\right)$$

Continue as before to get

$$\frac{r\left(x\right)}{r_1\left(x\right)} = -\frac{64}{3807}x + \frac{176}{3807}$$

Note that

$$r\left(x\right) = r_1\left(x\right)\left(-\frac{64}{3807}x + \frac{176}{3807}\right)$$

This means that the greatest common divisor of $a\left(x\right)$ and $b\left(x\right)$ is given by

$$\gcd\left(a\left(x\right),b\left(x\right)\right) = \frac{423}{4}x^3 + \frac{423}{2}x^2 - \frac{3807}{16}x + \frac{1269}{8}$$

Multiplication by the factor $\frac{16}{423}$ yields the polynomial

$$\frac{16}{423}\left(\frac{423}{4}x^3 + \frac{423}{2}x^2 - \frac{3807}{16}x + \frac{1269}{8}\right) = 4x^3 + 8x^2 - 9x + 6$$

which has integer coefficients. The equations

$$\left(4x^3 + 8x^2 - 9x + 6\right)\left(3x^2 + x + 1\right) = 12x^5 + 28x^4 - 15x^3$$
$$+ 17x^2 - 3x + 6$$
$$\left(4x^3 + 8x^2 - 9x + 6\right)\left(2x^3 + 1\right) = 8x^6 + 16x^3 + 16x^5$$
$$+ 8x^2 - 18x^4 - 9x + 6$$

show directly that

$$4x^3 + 8x^2 - 9x + 6$$

is a common divisor of $a(x)$ and $b(x)$. Since the polynomials $3x^2 + x + 1$ and $2x^3 + 1$ are irreducible, we conclude that

$$4x^3 + 8x^2 - 9x + 6$$

is indeed a greatest common divisor of $a(x)$ and $b(x)$.

There is a modification of the above algorithm that uses integer arithmetic rather than rational arithmetic, but the integers still get very large and messy.

Example 6.19 We will repeat the problem of finding

$$\gcd\left(a\left(x\right),b\left(x\right)\right) \in \mathbb{Z}\left[x\right]$$

by first calculating

$$\gcd\left(a\left(x\right),b\left(x\right)\right) \bmod 5$$

then calculating

$$\gcd\left(a\left(x\right),b\left(x\right)\right) \bmod 7$$

and finally using the Chinese remainder algorithm.
   Modulo 5, we have

$$a\left(x\right) \bmod 5 = 2x^5 + 3x^4 + 2x^2 + 2x + 1$$
$$b\left(x\right) \bmod 5 = 3x^6 + x^5 + 2x^4 + x^3 + 3x^2 + x + 1$$

Note that

$$b\left(x\right) - a\left(x\right)\left(4x + 2\right) \bmod 5 = x^4 + 3x^3 + x^2 + 3x + 4 = r_1\left(x\right)$$
$$a\left(x\right) - \left(2x + 2\right)r_1\left(x\right) \bmod 5 = 2x^3 + 4x^2 + 3x + 3 = r_2\left(x\right)$$
$$r_1\left(x\right) \quad \left(3x + 3\right)r_2\left(x\right) \bmod 5 = 0$$

and hence

$$\gcd\left(a\left(x\right),b\left(x\right)\right) \bmod 5 = 2x^3 + 4x^2 + 3x + 3$$

Repeating these steps using arithmetic modulo 7, we have

$$a\left(x\right) \bmod 7 = 5x^5 + 6x^3 + 3x^2 + 4x + 6$$
$$b\left(x\right) \bmod 7 = x^6 + 2x^5 + 3x^4 + 2x^3 + x^2 + 5x + 6$$

so

$$b\left(x\right) - \left(3x + 6\right)a\left(x\right) \bmod 7 = 6x^4 + 6x^3 + 6x^2 + 5x + 5 = s_1\left(x\right)$$
$$a\left(x\right) - \left(2x + 5\right)s_1\left(x\right) \bmod 7 = 6x^3 + 5x^2 + 4x + 2 = s_2\left(x\right)$$
$$s_1\left(x\right) - \left(x + 6\right)s_2\left(x\right) \bmod 7 = 0$$

which implies

$$\gcd\left(a\left(x\right), b\left(x\right)\right) \bmod 7 = 6x^3 + 5x^2 + 4x + 2$$

Assume that

$$\gcd\left(a\left(x\right), b\left(x\right)\right) = c_3 x^3 + c_2 x^2 + c_1 x + c_0$$

where $c_i \in Z$. Note that

$$c_3 \mid \gcd\left(12, 8\right) = 4$$

and

$$c_0 \mid \gcd\left(6, 6\right) = 6$$

(see problem 10). Assume $c_3 = 1$. Then

$$\gcd\left(a\left(x\right), b\left(x\right)\right) \bmod 5 = \frac{1}{2}\left(2x^3 + 4x^2 + 3x + 3\right) \bmod 5$$
$$= x^3 + 2x^2 + 4x + 4$$
$$\gcd\left(a\left(x\right), b\left(x\right)\right) \bmod 7 = \frac{1}{6}\left(6x^3 + 5x^2 + 4x + 2\right) \bmod 7$$
$$= x^3 + 2x^2 + 3x + 5$$

leads to

$$c_0 \bmod 5 = 4$$
$$c_0 \bmod 7 = 5$$

and hence

$$c_0 \bmod 35 = 19$$

which does not divide 6. Assume $c_3 = 2$. Then

$$\gcd\left(a\left(x\right), b\left(x\right)\right) \bmod 5 = 2x^3 + 4x^2 + 3x + 3$$
$$\gcd\left(a\left(x\right), b\left(x\right)\right) \bmod 7 = \frac{1}{3}\left(6x^3 + 5x^2 + 4x + 2\right) \bmod 7$$
$$= 2x^3 + 4x^2 + 6x + 3$$

and we solve

$$c_1 \bmod 5 = 3$$
$$c_1 \bmod 7 = 6$$

so that
$$c_1 \bmod 35 = 13$$

Trial division yields

$$a\left(x\right) \bmod 2x^3 + 4x^2 + 13x + 3 = \frac{315}{2} + 175x^2 + \frac{1295}{2}r$$

and hence $2x^3 + 4x^2 + 13x + 3$ is not a divisor of $a\left(x\right)$. Assume $c_3 = 4$. Then

$$\gcd\left(u\left(x\right), b\left(x\right)\right) \bmod 5 = 2\left(2x^3 + 4x^2 + 3x + 3\right) \bmod 5$$
$$= 4x^3 + 3x^2 + x + 1$$
$$\gcd\left(a\left(x\right), b\left(x\right)\right) \bmod 7 = \frac{2}{3}\left(6x^3 + 5x^2 + 4x + 2\right) \bmod 7$$
$$= 4x^3 + x^2 + 5x + 6$$

We now solve the Chinese remainder problems

$c_2 \bmod 5$	=	3	$c_1 \bmod 5$	=	1	$c_0 \bmod 5$ = 1
$c_2 \bmod 7$	=	1	$c_1 \bmod 7$	=	5	$c_0 \bmod 7$ = 0
$c_2 \bmod 35$	=	8	$c_1 \bmod 35$	=	−9	$c_0 \bmod 35$ = 6

to get the trial solution $\gcd\left(a\left(x\right), b\left(x\right)\right) = 4x^3 + 8x^2 - 9x + 6$. Trial division yields

$$a\left(x\right) \bmod 4x^3 + 8x^2 - 9x + 6 = 0$$
$$b\left(x\right) \bmod 4x^3 + 8x^2 - 9x + 6 = 0$$

Thus
$$4x^3 + 8x^2 - 9x + 6$$

is indeed a common divisor of the polynomials $a\left(x\right)$ and $b\left(x\right)$. As in the first example, we conclude that

$$4x^3 + 8x^2 - 9x + 6$$

is the greatest common divisor of the polynomials $a\left(x\right)$ and $b\left(x\right)$.

Problems 6.4

1. Find the greatest common divisor of the two polynomials

$$a(x) = 6x^5 - 7x^3 + 8x^2 - 5x + 4$$
$$b(x) = 8x^4 + 6x^3 + 8x^2 + 3x + 2$$

by completely factoring each polynomial.

2. Find the greatest common divisor of the two polynomials

$$a(x) = 2x^3 + 3x^2 + 3x + 1$$
$$b(x) = 2x^2 + 5x + 2$$

by using the Euclidean algorithm with rational arithmetic and long division.

3. Find the greatest common divisor of the two polynomials

$$a(x) = 2x^3 + 3x^2 + 3x + 1$$
$$b(x) = 2x^2 + 5x + 2$$

by using the Chinese remainder theorem with the modular basis $(5, 7)$.

4. Find the greatest common divisor of the two polynomials

$$a(x) = 2x^3 + 3x^2 + 3x + 1$$
$$b(x) = 2x^2 + 5x + 2$$

by using a computer algebra system.

5. Find the greatest common divisor of the two polynomials

$$a(x) = 6x^5 - 7x^3 + 8x^2 - 5x + 4$$

and

$$b(x) = 8x^4 + 6x^3 + 8x^2 + 3x + 2$$

by using the Euclidean algorithm with rational arithmetic and long division.

6. Find the greatest common divisor of the two polynomials

$$a(x) = 6x^5 - 7x^3 + 8x^2 - 5x + 4$$

and

$$b(x) = 8x^4 + 6x^3 + 8x^2 + 3x + 2$$

by using the Chinese remainder theorem with the modular basis $(17, 19)$.

7. Find the greatest common divisor of the two polynomials

$$a(x) = 6x^5 - 7x^3 + 8x^2 - 5x + 4$$
$$b(x) = 8x^4 + 6x^3 + 8x^2 + 3x + 2$$

by using a computer algebra system.

8. Find the greatest common divisor of the two polynomials

$$a(x) = 15x^{10} + 12x^8 - 19x^7 - 3x^5 + 36x^4 + 39x^3 + 21x^2 + 10x - 21$$
$$b(x) = 20x^8 - 15x^7 + 16x^6 + 34x^5 + 16x^4 - 45x^3 - 20x^2 - 21x + 27$$

using the Chinese remainder theorem with the modular basis $(41, 43)$.

9. Show that

$$3x^3 + 2x^2 + 5$$

is a common factor of the two polynomials

$$a(x) = 6x^6 + 4x^5 + 9x^4 + 19x^3 + 2x^2 + 15x + 5$$
$$b(x) = 12x^5 + 11x^4 + 11x^3 + 26x^2 + 5x + 15$$

by using long division. Then using the Euclidean algorithm, show that the polynomials

$$a(x) / (3x^3 + 2x^2 + 5)$$
$$b(x) / (3x^3 + 2x^2 + 5)$$

are relatively prime.

10. Assume that $c(x) \in \mathbb{Z}[x]$ is a common divisor of the two polynomials $a(x), b(x) \in \mathbb{Z}[x]$.

   (a) Show that the leading coefficient of $c(x)$ must divide the greatest common divisor of the leading coefficients of $a(x)$ and $b(x)$.

   (b) Show that the constant term of $c(x)$ must divide the greatest common divisor of the constant terms of $a(x)$ and $b(x)$.

6.5   Hilbert Matrix

David Hilbert's address of 1900 to the International Congress of Mathematicians in Paris is perhaps the most influential speech ever given to mathematicians, given by a mathematician, or given about mathematics. In it, Hilbert outlined 23 major mathematical problems to be studied in the coming century. Some are broad, such as the axiomatization of physics (problem 6) and might never be considered completed. Others, such as problem 3, were much more specific and solved quickly. Some were resolved contrary to Hilbert's expectations, as the continuum hypothesis (problem 1).

David E. Joyce

The **Hilbert matrix**

$$
H_n = \begin{pmatrix}
1 & \frac{1}{2} & \frac{1}{3} & \cdots & \frac{1}{n} \\
\frac{1}{2} & \frac{1}{3} & \frac{1}{4} & \cdots & \frac{1}{n+1} \\
\frac{1}{3} & \frac{1}{4} & \frac{1}{5} & \cdots & \frac{1}{n+2} \\
\vdots & \vdots & \vdots & \ddots & \vdots \\
\frac{1}{n} & \frac{1}{n+1} & \frac{1}{n+2} & \cdots & \frac{1}{2n-1}
\end{pmatrix}
$$

serves as an effective test bed for algorithms in linear algebra. This matrix presents unusual difficulties for matrix inversion algorithms since the determinant is very close to zero for large n. Using usual floating point arithmetic, it is often difficult to realize that the inverse is actually an integer matrix.

This provides an additional example of how the Chinese remainder theorem can be used to replace a difficult problem with several easy problems, followed by an application of the Chinese remainder algorithm.

Example 6.20 We will use the Chinese remainder algorithm to invert H_3. We will invert the Hilbert matrix twice; once modulo 29 and again modulo 31. First, we consider the matrix

$$
H_3 = \begin{pmatrix}
1 & \frac{1}{2} & \frac{1}{3} \\
\frac{1}{2} & \frac{1}{3} & \frac{1}{4} \\
\frac{1}{3} & \frac{1}{4} & \frac{1}{5}
\end{pmatrix}
$$

as a matrix modulo 29 to get

$$
H_3 \bmod 29 = \begin{pmatrix}
1 & \frac{1}{2} & \frac{1}{3} \\
\frac{1}{2} & \frac{1}{3} & \frac{1}{4} \\
\frac{1}{3} & \frac{1}{4} & \frac{1}{5}
\end{pmatrix} \bmod 29 = \begin{pmatrix}
1 & 15 & 10 \\
15 & 10 & 22 \\
10 & 22 & 6
\end{pmatrix}
$$

Inversion modulo 29 can be computed by doing elementary row operations on the matrix $(H_3 : I)$ modulo 29 as follows.

$$
\begin{pmatrix}
1 & 0 & 0 \\
0 & \frac{1}{15} & 0 \\
0 & 0 & \frac{1}{10}
\end{pmatrix}
\begin{pmatrix}
1 & 15 & 10 & 1 & 0 & 0 \\
15 & 10 & 22 & 0 & 1 & 0 \\
10 & 22 & 6 & 0 & 0 & 1
\end{pmatrix} \bmod 29
$$

$$
= \begin{pmatrix}
1 & 15 & 10 & 1 & 0 & 0 \\
1 & 20 & 15 & 0 & 2 & 0 \\
1 & 8 & 18 & 0 & 0 & 3
\end{pmatrix}
$$

$$\begin{pmatrix} 1 & 0 & 0 \\ -1 & 1 & 0 \\ -1 & 0 & 1 \end{pmatrix} \begin{pmatrix} 1 & 15 & 10 & 1 & 0 & 0 \\ 1 & 20 & 15 & 0 & 2 & 0 \\ 1 & 8 & 18 & 0 & 0 & 3 \end{pmatrix} \bmod 29$$

$$= \begin{pmatrix} 1 & 15 & 10 & 1 & 0 & 0 \\ 0 & 5 & 5 & 28 & 2 & 0 \\ 0 & 22 & 8 & 28 & 0 & 3 \end{pmatrix}$$

$$\begin{pmatrix} 1 & 0 & 0 \\ 0 & \frac{1}{5} & 0 \\ 0 & 0 & \frac{1}{22} \end{pmatrix} \begin{pmatrix} 1 & 15 & 10 & 1 & 0 & 0 \\ 0 & 5 & 5 & 28 & 2 & 0 \\ 0 & 22 & 8 & 28 & 0 & 3 \end{pmatrix} \bmod 29$$

$$= \begin{pmatrix} 1 & 15 & 10 & 1 & 0 & 0 \\ 0 & 1 & 1 & 23 & 12 & 0 \\ 0 & 1 & 3 & 25 & 0 & 12 \end{pmatrix}$$

$$\begin{pmatrix} 1 & -15 & 0 \\ 0 & 1 & 0 \\ 0 & -1 & 1 \end{pmatrix} \begin{pmatrix} 1 & 15 & 10 & 1 & 0 & 0 \\ 0 & 1 & 1 & 23 & 12 & 0 \\ 0 & 1 & 3 & 25 & 0 & 12 \end{pmatrix} \bmod 29$$

$$= \begin{pmatrix} 1 & 0 & 24 & 4 & 23 & 0 \\ 0 & 1 & 1 & 23 & 12 & 0 \\ 0 & 0 & 2 & 2 & 17 & 12 \end{pmatrix}$$

$$\begin{pmatrix} 1 & 0 & 0 \\ 0 & 1 & 0 \\ 0 & 0 & \frac{1}{2} \end{pmatrix} \begin{pmatrix} 1 & 0 & 24 & 4 & 23 & 0 \\ 0 & 1 & 1 & 23 & 12 & 0 \\ 0 & 0 & 2 & 2 & 17 & 12 \end{pmatrix} \bmod 29$$

$$= \begin{pmatrix} 1 & 0 & 24 & 4 & 23 & 0 \\ 0 & 1 & 1 & 23 & 12 & 0 \\ 0 & 0 & 1 & 1 & 23 & 6 \end{pmatrix}$$

$$\begin{pmatrix} 1 & 0 & -24 \\ 0 & 1 & -1 \\ 0 & 0 & 1 \end{pmatrix} \begin{pmatrix} 1 & 0 & 24 & 4 & 23 & 0 \\ 0 & 1 & 1 & 23 & 12 & 0 \\ 0 & 0 & 1 & 1 & 23 & 6 \end{pmatrix} \bmod 29$$

$$= \begin{pmatrix} 1 & 0 & 0 & 9 & 22 & 1 \\ 0 & 1 & 0 & 22 & 18 & 23 \\ 0 & 0 & 1 & 1 & 23 & 6 \end{pmatrix}$$

This implies

$$\begin{pmatrix} 1 & 15 & 10 \\ 15 & 10 & 22 \\ 10 & 22 & 6 \end{pmatrix}^{-1} \bmod 29 = \begin{pmatrix} 9 & 22 & 1 \\ 22 & 18 & 23 \\ 1 & 23 & 6 \end{pmatrix}$$

This answer appears to be correct since multiplication yields

$$\begin{pmatrix} 1 & 15 & 10 \\ 15 & 10 & 22 \\ 10 & 22 & 6 \end{pmatrix} \begin{pmatrix} 9 & 22 & 1 \\ 22 & 18 & 23 \\ 1 & 23 & 6 \end{pmatrix} \bmod 29 = \begin{pmatrix} 1 & 0 & 0 \\ 0 & 1 & 0 \\ 0 & 0 & 1 \end{pmatrix}$$

In a similar manner, we have

$$\begin{pmatrix} 1 & \frac{1}{2} & \frac{1}{3} \\ \frac{1}{2} & \frac{1}{3} & \frac{1}{4} \\ \frac{1}{3} & \frac{1}{4} & \frac{1}{5} \end{pmatrix} \bmod 31 = \begin{pmatrix} 1 & 16 & 21 \\ 16 & 21 & 8 \\ 21 & 8 & 25 \end{pmatrix}$$

$$\begin{pmatrix} 1 & 16 & 21 \\ 16 & 21 & 8 \\ 21 & 8 & 25 \end{pmatrix}^{-1} \bmod 31 = \begin{pmatrix} 9 & 26 & 30 \\ 26 & 6 & 6 \\ 30 & 6 & 25 \end{pmatrix}$$

It remains to determine a matrix A that satisfies

$$A \bmod 29 = \begin{pmatrix} 9 & 22 & 1 \\ 22 & 18 & 23 \\ 1 & 23 & 6 \end{pmatrix}$$

$$A \bmod 31 = \begin{pmatrix} 9 & 26 & 30 \\ 26 & 6 & 6 \\ 30 & 6 & 25 \end{pmatrix}$$

Since the matrices are symmetric, this is equivalent to six Chinese remainder problems. The first

$$a_{11} \bmod 29 = 9$$
$$a_{11} \bmod 31 = 9$$

is easy—just take $a_{11} = 9$. Let $c\,(\mathbf{a}, \mathbf{m})$ denote the Chinese remainder function (see page 134). We have

$$a_{12} \bmod 899 = c\,([22, 26], [29, 31]) = 863$$
$$a_{13} \bmod 899 = c\,([1, 30], [29, 31]) = 30$$
$$a_{22} \bmod 899 = c\,([18, 6], [29, 31]) = 192$$
$$a_{23} \bmod 899 = c\,([23, 6], [29, 31]) = 719$$
$$a_{33} \bmod 899 = c\,([6, 25], [29, 31]) = 180$$

Since two of these numbers are close to 899, we assume

$$a_{12} = 863 - 899 = -36$$
$$a_{23} = 719 - 899 = -180$$

The potential solution

$$A = \begin{pmatrix} 9 & -36 & 30 \\ -36 & 192 & -180 \\ 30 & -180 & 180 \end{pmatrix}$$

is correct since multiplication yields

$$\begin{pmatrix} 9 & -36 & 30 \\ -36 & 192 & -180 \\ 30 & -180 & 180 \end{pmatrix} \begin{pmatrix} 1 & \frac{1}{2} & \frac{1}{3} \\ \frac{1}{2} & \frac{1}{3} & \frac{1}{4} \\ \frac{1}{3} & \frac{1}{4} & \frac{1}{5} \end{pmatrix} = \begin{pmatrix} 1 & 0 & 0 \\ 0 & 1 & 0 \\ 0 & 0 & 1 \end{pmatrix}$$

Problems 6.5

1. Compute H_4^{-1} by using elementary row operations and rational arithmetic.

2. Compute H_4^{-1} by using the Chinese remainder theorem with the modular basis $(997, 1009)$.

3. Compute H_4^{-1} by using a computer algebra system.

4. Compare H_4^{-1} and H_5^{-1} and make a conjecture. Test your conjecture by computing H_6^{-1}.

5. Calculate the determinant of H_n for $n = 2, 3, 4, 5, 6$.

6. Verify that the matrix
$$A = \begin{pmatrix} 100 & 31 \\ 29 & 9 \end{pmatrix}$$

   has determinant 1 and hence that A^{-1} has integer entries.

7. Using the matrix A from problem 6, calculate A^{-1} by doing row reduction on the augmented matrix $(\ A\ :\ I\ )$ using rational arithmetic.

8. Using the matrix A from problem 6, calculate A^{-1} by doing row reduction on the augmented matrix $(\ A\ :\ I\ )$ modulo 11 and modulo 13, then using the Chinese remainder theorem.

9. The matrix
$$A = \begin{pmatrix} 127 & -24 \\ -37 & 7 \end{pmatrix}$$

has determinant 1 and hence its inverse also has integer entries. Find A^{-1} by calculating $A^{-1} \bmod 11$ and $A^{-1} \bmod 13$, then using the Chinese remainder algorithm.

Chapter 7

Theorems of Fermat and Euler

In this chapter we develop some facts from number theory that will be useful when we look at public key ciphers. On the one hand, these tools lead to cryptographic methods that make cipher text appear random and thus difficult to break. On the other hand, they provide elegant methods for deciphering.

> *The theory of numbers, one of the oldest branches of mathematics, has engaged the attention of many gifted mathematicians during the past 2300 years. The Greeks, Indians, and Chinese made significant contributions prior to 1000 A.D., and in more modern times the subject has developed steadily since Fermat, one of the fathers of Western mathematics.*
>
> William Judson LeVeque

7.1   Wilson's Theorem

First a simple lemma about square roots of 1 modulo a prime.

Lemma 7.1 *If p is a prime, then the only solutions to $x^2 \equiv 1 \pmod{p}$ are $x \equiv 1 \pmod{p}$ and $x \equiv -1 \pmod{p}$.*

Proof. If $x^2 = 1 \pmod{p}$. Then

$$p|(x^2 - 1) = (x - 1)(x + 1)$$

and because p is a prime it follows that $p|(x - 1)$ or $p|(x + 1)$. But $p|(x - 1)$ means that $x = 1 \pmod{p}$ and $p|(x + 1)$ means that $x \equiv -1 \pmod{p}$. ∎

Although Wilson's theorem is not much use by itself, it leads to other theorems that are used a lot. It is named for Sir John Wilson (1741–1793).

Theorem 7.2 *(Wilson's Theorem) An integer $p > 1$ is prime if and only if*

$$(p-1)! \equiv -1 \pmod{p}$$

Proof. The theorem is trivially true if $p = 2$, so suppose that $p > 2$ is a prime, and consider the product

$$(p-1)! = 1 \cdot 2 \cdot 3 \cdots (p-2)(p-1)$$

If $1 \le a \le p-1$, then the inverse of a is also one of the integers between 1 and $p-1$. If a is equal to its own inverse, then $a^2 \equiv 1 \pmod{p}$, so $a = 1$ or $a = p-1$ by Lemma 7.1. All the other factors in the product $(p-1)!$ have inverses that are different from themselves. It follows that $(p-1)!$ can be written as $1 \cdot (p-1)$ multiplied by $(p-3)/2$ inverse pairs, each of whose product is 1. So $(p-1)! \equiv (p-1) \equiv -1 \pmod{p}$.

Conversely, suppose that $(p-1)! \equiv -1 \pmod{p}$, that is, $(p-1)! = -1 + kp$ for some integer k. Let a be one of the numbers $1, 2, 3, \ldots, p-1$. We will show that if a divides p, then $a = 1$, so p is prime. Certainly a divides $(p-1)!$. If a also divides p, then a divides $kp - (p-1)! = 1$, so $a = 1$. ∎

Example 7.3 We use Wilson's theorem to test the integers 51 and 53 for primality. Since

$$50! \bmod 51 = 0$$
$$52! \bmod 53 = 52 \equiv -1 \pmod{53}$$

it follows from Wilson's theorem that 51 is composite, and 53 is prime.

Problems 7.1

1. Test the odd integers 101 through 110 for primality by applying Wilson's theorem.

2. If n is not prime, what are the possible values of $(n-1)! \bmod n$? For which composites is $(n-1)! \not\equiv 0 \bmod n$?

3. Calculate $x^2 \bmod 11$ for integers $x = 1, 2, 3, \ldots, 9, 10$ and show $x^2 \bmod 11 = 1$ only for $x = 1$ or 10.

4. For any positive integer $n > 1$, show that $x^2 \bmod n = (n-x)^2 \bmod n$.

5. Find all solutions to $x^2 \bmod 15 = 1$ for $x \in \{1, 2, \ldots, 14\}$.

6. Prove or disprove: If $x^2 \bmod p = 1$ has exactly two solutions $x \in \{1, 2, \ldots, p-1\}$, then p is prime.

7. Let p be an odd prime. Show that $2(p-3)! \bmod p = p-1$.

8. Prove that an integer $p > 2$ is prime if and only if $(p-2)! \bmod p = 1$.

9. Illustrate the proof of Wilson's theorem for $p = 17$ by pairing the integers 2, 3, 4, $\ldots$, 15 and using that to find $16! \bmod 17$.

10. Show that $9! + 1 \bmod 19 = 0$ and $18! + 1 \bmod 19 = 0$.

7.2 Powers Modulo n

Some public key ciphers require computing very large powers modulo n. If n and m are positive integers with hundreds of digits, then it would appear to be impossible to calculate

$$y = x^n \bmod m$$

because it seems to require $n - 1$ multiplications to compute x^n. For example, $x^4 = x \cdot x \cdot x \cdot x$ shows three multiplications.

Algorithm 7.1   Crude method

Input: Integers x, n, m
Output: Integer $x^n \bmod m$
Set $p = 1$
For k **from** 1 **to** n **Do**
  $p = p \cdot x \bmod m$
End Loop
Print p

Example 7.4 Consider the problem of computing $3^{43} \bmod 17$. The crude method requires 42 multiplications to give the result

$$3^{43} \bmod 17 = 7$$

We can't compute these powers using logarithms because of the approximations involved—we require exact calculations.

Of course it is silly to compute x^{16} using 15 multiplications as follows

$$x^{16} = x \cdot x \cdot x \cdot x \cdot x \cdot x \cdot x \cdot x \cdot x \cdot x \cdot x \cdot x \cdot x \cdot x \cdot x$$

We can cut that number down to 8 by first computing $y = x \cdot x$ and then computing y^8. Better still, use successive squaring

$$x^{16} = \left(\left(\left(x^2 \right)^2 \right)^2 \right)^2$$

Each squaring involves requires one multiplication, so we can compute x^{16} with 4 multiplications.

Example 7.5 Here is a better way to compute $3^{43} \bmod 17$. First, calculate the sequence

$$a_1 = 3^2 \bmod 17 = 9$$
$$a_2 = a_1^2 \bmod 17 = 13$$
$$a_3 = a_2^2 \bmod 17 = 16$$
$$a_4 = a_3^2 \bmod 17 = 1$$
$$a_5 = a_4^2 \bmod 17 = 1$$

Now note that

$$3^{43} \bmod 17 = 3^{32+8+2+1} \bmod 17$$
$$= 3^{32} \cdot 3^8 \cdot 3^2 \cdot 3^1 \bmod 17$$

$$= \left(\left(\left(\left(3^2\right)^2\right)^2\right)^2\right)^2 \cdot \left(\left(3^2\right)^2\right)^2 \cdot 3^2 \cdot 3^1 \bmod 17$$

$$= a_5 \cdot a_3 \cdot a_1 \cdot 3 \bmod 17$$
$$= 1 \cdot 16 \cdot 9 \cdot 3 \bmod 17$$
$$= 7$$

This method required $5 + 3 = 8$ multiplications compared to the 42 multiplications required in Example 7.4.

The power method in Algorithm 7.2 extends this idea to an arbitrary exponent n by using the binary representation of n.

Algorithm 7.2   Power method

Input Integers x, n, m
Output Integer $y = x^n \bmod m$
Function power(x, n, m)
Set prod $= 1$
While $n > 0$ **do**
    If $n \bmod 2 = 1$ **then**
        Set prod $=$ prod $\cdot x \bmod m$
    End if
    Set $x = x^2 \bmod m$
    Set $n = \lfloor n/2 \rfloor$
End while
Set power $=$ prod
Return

Example 7.6 We will watch Algorithm 7.2 work on the same problem done in Examples 7.4 and 7.5. This algorithm is an implementation of the power method (illustrated in Example 7.5). Initially $x = 3$, $m = 17$, $n = 43$, and prod $= 1$. The steps are shown in Table 7.1 below.

$x^2 \bmod 17 \to x$	$\lfloor n/2 \rfloor \to n$	$n \bmod 2$	prod $\cdot x \bmod 17 \to$ prod
3	43	1	$3 \cdot 1 \bmod 17 = 3$
$3^2 \bmod 17 = 9$	$\lfloor 43/2 \rfloor = 21$	1	$3 \cdot 9 \bmod 17 = 10$
$9^2 \bmod 17 = 13$	$\lfloor 21/2 \rfloor = 10$	0	
$13^2 \bmod 17 = 16$	$\lfloor 10/2 \rfloor = 5$	1	$10 \cdot 16 \bmod 17 = 7$
$16^2 \bmod 17 = 1$	$\lfloor 5/2 \rfloor = 2$	0	
$1^2 \bmod 17 = 1$	$\lfloor 2/2 \rfloor = 1$	1	$7 \cdot 1 \bmod 17 = 7$
$1^2 \bmod 17 = 1$	$\lfloor 1/2 \rfloor = 0$	0	

Table 7.1   Powers modulo m

So we see again that $3^{43} \bmod 17 = 7$. Notice that ten multiplications are required. Why did we count only eight multiplications before (Example 7.5)?

What about larger exponents? Could this method be used, for example, to compute $2^{1000} \bmod 1009$? What about

$$123^{3749378975395793749} \bmod 28348290348294 = 12479863459095$$

or even

$$79^{10^{100}} \bmod 2893320 = 2338441$$

The calculation of $2^{1000} \bmod 1009$ requires 16 multiplications. The power

$$123^{3749378975395793749} \bmod 28348290348294 = 12479863459095$$

requires 86 multiplications, and $2338441 = 79^{10^{100}} \bmod 2893320$ requires only 438 multiplications. This means that these calculations can be made on a personal computer if extended-precision arithmetic is available.

The calculation of $x^n \bmod m$ requires at least $\lceil \log_2 n \rceil$ multiplications but no more than $2\lceil \log_2 n \rceil$ multiplications, where $\log_2$ is the **base 2 logarithm**. In particular, $\log_2 1024 = 10$ because $2^{10} = 1024$. Also,

$$\lceil \log_2 3749378975395793749 \rceil = 62$$

since

$$
\begin{aligned}
2^{61} &= 2305843009213693952 \\
&< 3749378975395793749 \\
&< 4611686018427387904 \\
&= 2^{62}
\end{aligned}
$$

Similarly, $\lceil \log_2 10^{100} \rceil = 333$.

In all three examples in the previous paragraphs, the actual number of multiplications is indeed between $\lceil \log_2 n \rceil$ and $2\lceil \log_2 n \rceil$. In fact, for a randomly chosen exponent n, about half the digits in the binary representation of n should be 1's, so the total number of multiplications would be about $1.5\lceil \log_2 n \rceil$.

Problems 7.2

1. Use Algorithm 7.1 to evaluate the expression $11^4 \bmod 15$

2. Use Algorithm 7.1 to evaluate the expression $9^3 \bmod 23$

3. Use Algorithm 7.1 to evaluate the expression $16^4 \bmod 29$

4. Use Algorithm 7.1 to evaluate the expression $22^5 \bmod 25$

5. Use Algorithm 7.2 to compute the expression $5^{97} \bmod 127$

6. Use Algorithm 7.2 to compute the expression $4^{126} \bmod 127$

7. Use Algorithm 7.2 to compute the expression $4^{63} \bmod 127$

8. Use Algorithm 7.2 to compute the expression $37^{1000} \bmod 127$

9. Use a computer algebra system to compute

$$12^{72387894339363242} \bmod 243682743764$$

10. Count the number of multiplications required to compute $23^{1234} \bmod 137$ by first finding the base 2 representation of 1234.

11. Compare the following variations of Algorithm 7.2. Are these variations equivalent? Explain.

Input Integers x, n, m	**Input** Integers x, n, m
Output Integer $y = x^n \bmod m$	**Output** Integer $y = x^n \bmod m$
Function power(x, n, m)	**Function** power(x, n, m)
Set prod $= 1$	**Set** prod $= 1$
While $n > 0$ **do**	**While** $n > 0$ **do**
If $n \bmod 2 = 1$ **then**	**If** $n \bmod 2 = 1$ **then**
Set prod $=$ prod $\cdot x \bmod m$	**Set** prod $=$ prod $\cdot x$
End if	**End if**
Set $x = x^2 \bmod m$	**Set** $x = x^2$
Set $n = \lfloor n/2 \rfloor$	**Set** $n = \lfloor n/2 \rfloor$
End while	**End while**
Set power $=$ prod	**Set** power $=$ prod $\bmod m$
Return	**Return**
Powers modulo m	**Powers modulo** m

7.3   Fermat's Little Theorem

It is impossible to divide a cube into two cubes, a fourth power into two fourth powers, and in general any power except the square into two powers with the same exponents, ... I have discovered a truly wonderful proof of this, but the margin is too narrow to hold it.

Pierre de Fermat 1637

So wrote Pierre de Fermat (1601–1665) in the margin of his Latin translation of *Arithmetica* by Diophantus of Alexandria. He was claiming that he could prove that, if $n \geq 3$, then the equation

$$x^n + y^n = z^n$$

has no solutions in positive integers x, y, and z. For over 350 years many professional and amateur mathematicians tried to verify Fermat's statement, but no one succeeded in proving it. The statement became known as "Fermat's Last Theorem" because it was the last of Fermat's claims that people figured out how to prove. One of the remarkable achievements of the twentieth century was that Andrew Wiles (1953–) was able to prove it.

> *I grew up in Cambridge in England, and my love of mathematics dates from those early childhood days. I loved doing problems in school. I'd take them home and make up new ones of my own. But the best problem I ever found, I found in my local public library. I was just browsing through the section of math books and I found this one book, which was all about one particular problem—Fermat's Last Theorem. This problem had been unsolved by mathematicians for 300 years. It looked so simple, and yet all the great mathematicians in history couldn't solve it. Here was a problem, that I, a ten year old, could understand and I knew from that moment that I would never let it go. I had to solve it.*
>
> Andrew Wiles

Fermat's last theorem is also known as his *great theorem*. Our next result is sometimes called *Fermat's little theorem* to distinguish it from his great theorem. It is also called simply *Fermat's theorem*.

The Chinese knew as early as 500 B.C. that $2^p - 2$ is divisible by the prime p. Fermat rediscovered this fact in 1640, and stated that he had a proof that if p is any prime and x is any integer not divisible by p, then $x^{p-1} - 1$ is divisible by p. Euler published the first proof of Fermat's little theorem in 1736.

Theorem 7.7 *(Fermat's Little Theorem) If p is a prime and $a \perp p$, then*

$$a^{p-1} \equiv 1 \pmod{p}$$

Proof. Assume that p is a prime and $a \perp p$. If $na \equiv ma \pmod{p}$ then $n \equiv m \pmod{p}$, and hence it follows that no two of the numbers $a, 2a, 3a, \ldots,$ $(p-1)a$ are congruent modulo p, and none is congruent to zero modulo p. It follows that these integers are congruent to $1, 2, 3, \ldots, (p-1)$ in some order so

$$(a)(2a)(3a) \cdots ((p-1)a) \equiv 1 \cdot 2 \cdot 3 \cdots (p-1) \pmod{p}$$

We can rewrite this as

$$a^{p-1}(p-1)! \equiv (p-1)! \pmod{p}$$

But $p \perp (p-1)!$ and hence $a^{p-1} \equiv 1 \pmod{p}$ ∎

Corollary 7.8 *If p is a prime and a is any integer, then $a^p \equiv a \pmod{p}$.*

Proof. Given a prime p and an integer a, either $a \perp p$ or else $p|a$. If $p \perp a$ then

$$a^{p-1} \equiv 1 \bmod p$$

by Fermat's little theorem. Multiplying both sides by a gives

$$a^p \equiv a \pmod{p}$$

On the other hand, if $p|a$ then, $a \equiv 0 \pmod{p}$ so

$$a^p \equiv 0^p \equiv 0 \equiv a \pmod{p}$$

∎

Example 7.9 To verify Fermat's little theorem for $p = 101$ and $a = 5$, we perform the calculations shown in Table 7.2 and note that $5^{100} \bmod 101 = 1$.

$x^2 \bmod 101 \to x$	$\lfloor e/2 \rfloor \to e$	$e \bmod 2$	$\text{prod} \cdot x \bmod 101 \to \text{prod}$
		0	1
$5^2 \bmod 101 = 25$	$\lfloor 100/2 \rfloor = 50$	0	
$25^2 \bmod 101 = 19$	$\lfloor 50/2 \rfloor = 25$	1	$1 \cdot 19 \bmod 101 = 19$
$19^2 \bmod 101 = 58$	$\lfloor 25/2 \rfloor = 12$	0	
$58^2 \bmod 101 = 31$	$\lfloor 12/2 \rfloor = 6$	0	
$31^2 \bmod 101 = 52$	$\lfloor 6/2 \rfloor = 3$	1	$19 \cdot 52 \bmod 101 = 79$
$52^2 \bmod 101 = 78$	$\lfloor 3/2 \rfloor = 1$	1	$79 \cdot 78 \bmod 101 = 1$

Table 7.2   Calculation of $5^{100} \bmod 101 = 1$

What about the converse? If $a^n \equiv a \pmod{n}$ for every integer a, must n be prime? The answer is no, and an example is provided by the integer $n = 561 = 3 \cdot 11 \cdot 17$. We now show why this is the case. The idea is to look modulo 3, 11, and 17.

Note that

$$560 = (3 - 1) \cdot 280 = (11 - 1)56 = (17 - 1)35$$

so $a^2 \equiv 1 \pmod{3}$ and $a^{10} \equiv 1 \pmod{11}$ and $a^{16} \equiv 1 \pmod{17}$ by Fermat's little theorem.

If $a \perp 3$, then

$$
\begin{aligned}
a^{561} \bmod 3 &= a \cdot a^{560} \bmod 3 \\
&= a \cdot (a^2)^{280} \bmod 3 \\
&= a \cdot (1)^{280} \bmod 3 \\
&= a \bmod 3
\end{aligned}
$$

and if $3|a$, then $a^{561} = 0 \equiv a \pmod 3$. Similarly, if $a \perp 11$, then

$$a^{561} \bmod 11 = a \cdot a^{560} \bmod 11$$
$$= a \cdot (a^{10})^{56} \bmod 11$$
$$= a \cdot (1)^{56} \bmod 11$$
$$= a \bmod 11$$

and if $11|a$, then $a^{561} = 0 = a \bmod 11$. Finally, if $a \perp 17$ then

$$a^{561} \bmod 17 = a \cdot a^{560} \bmod 17$$
$$= a \cdot (a^{16})^{35} \bmod 17$$
$$= a \cdot (1)^{35} \bmod 17$$
$$= a \bmod 17$$

and if $17|a$, then $a^{561} = 0 = a \bmod 17$. Thus for any integer a, we have

$$a \equiv a^{561} \pmod 3$$
$$a \equiv a^{561} \pmod{11}$$
$$a \equiv a^{561} \pmod{17}$$

It follows from the Chinese remainder theorem that $a^{561} \equiv a \pmod{561}$ for every integer a.

Thus it is possible for an integer n to be composite even though $a^n \equiv a \pmod n$ for every integer a.

A composite integer n such that $a^n \equiv a \bmod n$ for every a is called a **Carmichael number** after Robert D. Carmichael (1879–1967), who first constructed such a number in about 1909. The number 561 is the smallest Carmichael number. There are many others, some of which are given in the problems.

Problems 7.3

1. Verify Fermat's little theorem for $a = 2$ and the prime $p = 11$, using Algorithm 7.2.

2. Let p be a prime such that $\gcd(p, a) = 1$. Show directly that if $a^p \equiv a \pmod p$, then $a^{p-1} \equiv 1 \pmod p$, thus deriving Fermat's theorem from its corollary. Is it necessary that p be a prime?

3. Let p be a prime such that $\gcd(p, a) = 1$. Show that if $n \bmod (p-1) = 1$, then $a^n \equiv 1 \pmod p$.

In problems 4–10, use Fermat's little theorem to find the answer.

4. $2^{100} \bmod 13$

5. $2^{1000} \bmod 13$

6. $3^{500} \bmod 17$

7. $5^{2000} \bmod 17$

8. $7^{2222} \bmod 23$

9. $11^{1234} \bmod 29$

10. $2^{100} \bmod 31$

11. Verify that $5^{217} \bmod 217 = 5$. Factor 217.

12. Verify that
$$x^{1105} \equiv x \pmod{1105}$$
for every integer x. Show that 1105 can be written as a sum of two squares in four different ways. (Integers smaller than 1105 can be so written in at most three ways.)

13. Verify that 1729 is a Carmichael number. Verify that 1729 is the smallest positive integer that can be written as the sum of two cubes in two different ways. This is the famous **Hardy-Ramanujan number**, so called because of the conversation between those two mathematicians after Hardy took a cab numbered 1729 to visit Ramanujan in the hospital. See http://en.wikipedia.org/wiki/1729_(number).

14. Verify that $n = 6601$ is a Carmichael number.

15. Prove that a product n of distinct primes is a Carmichael number if n is composite and $n - 1$ is divisible by $p - 1$ for each prime p dividing n.

16. Let $k > 0$ and assume $6k + 1$, $12k + 1$, and $18k + 1$ are all primes. Show that
$$n = (6k + 1)(12k + 1)(18k + 1)$$
is a Carmichael number.

17. Use the result of Problem 16 to construct eight Carmichael numbers of the form
$$n = (6k + 1)(12k + 1)(18k + 1)$$
where $1 \le k \le 100$.

7.4 Rabin's Probabilistic Primality Test

A number is prime if its only positive divisors are 1 and itself. Thus to prove that 7 is prime we have to show that none of the numbers 2, 3, 4, 5, 6 divides 7. This search can be shortened a bit by only testing numbers a such that $a < \sqrt{7}$. This works because if $7 = ab$, and $a > \sqrt{7}$, then $b = 7/a < \sqrt{7}$ and 7 is also divisible by b. Thus 7 is prime because 2 does not divide 7.

That's a pretty good test for small primes p, but what if p is a hundred digits long? Testing p in this manner for primality requires testing all integers a in the range

$$2 \leq a \leq \sqrt{10^{100}} = 10^{50}$$

which is, essentially, an impossible task.

A related question is how to generate all the primes up to a certain number n. One way to do that is to go through all the numbers from 2 to n and test each one to see if it's a prime. That's not too efficient. Another approach is the **sieve of Eratosthenes** (third century B.C.). The idea there is that if you remove the composites, then everything that remains will be prime.

Algorithm 7.3   Sieve of Eratosthenes

Input: Positive integer m
Output: All primes up to m

Initialize $p(k) = 0$ **for** k **from** 1 **to** m
Set $p(1) = 1$
Set $q = 2$
For $r = q^2$ **to** m **step** q
    Set $p(r) = 1$
End For
Set $q = 3$
While $q^2 \leq m$ **do**
    For $r = q^2$ **to** m **step** $2q$
        Set $p(r) = 1$
    End For
    Repeat
        Set $q = q + 2$
    Until $p(q) = 0$
End While
For $r = 1$ **to** m
    If $p(r) = 0$ **then print** r
End For
End

Start with the set $\{2, 3, \ldots, n\}$, and cross out the proper multiples of 2, which are 4, 6, 8, …. Then cross out the proper odd multiples of 3, which are 9, 15, 21, …. The number 4 and all its multiples are already crossed out, so we

next cross out the odd multiples of 5, starting with 25, which are 25, 35, 45, ...,
the odd multiples of 7 starting with 49, and so on until we exceed $\sqrt{n}$ at which
point we stop.

The sieve of Eratosthenes is implemented by Algorithm 7.3.

The primes less than 100 are listed in Table 7.3. We start by crossing out
0 and 1, then the multiples of 2 starting with 4, then the odd multiples of 3
starting with 9, then the odd multiples of 5 starting with 25, then the odd
multiples of 7 starting with 49. The remaining numbers are all primes.

~~0~~	~~1~~	2	3	~~4~~	5	~~6~~	7	~~8~~	~~9~~
~~10~~	11	~~12~~	13	~~14~~	~~15~~	~~16~~	17	~~18~~	19
~~20~~	~~21~~	~~22~~	23	~~24~~	~~25~~	~~26~~	~~27~~	~~28~~	29
~~30~~	31	~~32~~	~~33~~	~~34~~	~~35~~	~~36~~	37	~~38~~	~~39~~
~~40~~	41	~~42~~	43	~~44~~	~~45~~	~~46~~	47	~~48~~	~~49~~
~~50~~	~~51~~	~~52~~	53	~~54~~	~~55~~	~~56~~	~~57~~	~~58~~	59
~~60~~	61	~~62~~	~~63~~	~~64~~	~~65~~	~~66~~	67	~~68~~	~~69~~
~~70~~	71	~~72~~	73	~~74~~	~~75~~	~~76~~	77	~~78~~	79
~~80~~	~~81~~	~~82~~	83	~~84~~	~~85~~	~~86~~	~~87~~	~~88~~	89
~~90~~	91	~~92~~	93	~~94~~	~~95~~	~~96~~	97	~~98~~	~~99~~

Table 7.3   Primes less than 100

Figure 7.1 is a picture of the primes less than 10 000, generated by the sieve
of Eratosthenes.

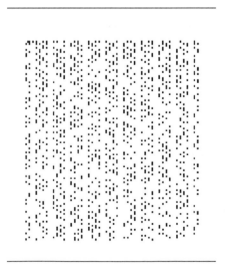

Figure 7.1   Primes < 10 000

Each row of the picture represents 100 integers, starting with zero in the upper
left-hand corner and 9999 in the lower right-hand corner. Each small black

rectangle represents a prime. The first row, which goes from 0 to 99, has black rectangles at positions $2, 3, 5, 7, \ldots$. The second row, which goes from 100 to 199, has black rectangles at positions $101, 103, 107, 109, \ldots$. Notice that there are no primes in the columns corresponding to 15, 25, 35, and so forth.

Prime sieves generate a lot of primes in an efficient manner, but they don't help in testing a particular number for primality. We need other methods to test primality.

As Carmichael numbers are relatively rare, Fermat's little theorem provides a method to test for primality. If Fermat's little theorem fails for an integer n, then n can not be prime. If Fermat's little theorem is satisfied, then further testing is worthwhile. A composite that passes this test is called a **pseudoprime**.

Definition 7.10 *If n is a composite such that $a^n \bmod n = a$, then n is called a* ***pseudoprime*** *to the base a.*

Since $\log_2 n$ is very small compared with $\sqrt{n}$, it is much faster to test the congruence $a^n \bmod n = 0$ than it is to test divisibility. If $a \perp n$ and $a^n \bmod n = a$, then we can divide both sides by a to get $a^{n-1} \bmod n = 1$.

Recall that if p is prime, then the only solutions to $x^2 \equiv 1 \pmod{p}$ are $x \equiv \pm 1$. If n is composite, then $x^2 \bmod n = 1$ always has more than two solutions. In particular, if $n = rs$, where $r \perp s$ and r and s are greater than 2, then $x^2 \bmod n = 1$ has at least four solutions. To see this, consider the system

$$x^2 \equiv 1 \pmod{r}$$
$$x^2 \equiv 1 \pmod{s}$$

of congruences. According to the Chinese remainder theorem, each of the following systems

$$
\begin{array}{ll}
x \equiv 1 \pmod{r} \qquad & x \equiv 1 \pmod{r} \\
x \equiv 1 \pmod{s} & x \equiv -1 \pmod{s}
\end{array}
$$

$$
\begin{array}{ll}
x \equiv -1 \pmod{r} \qquad & x \equiv -1 \pmod{r} \\
x \equiv 1 \pmod{s} & x \equiv -1 \pmod{s}
\end{array}
$$

of congruences has a unique solution modulo rs. Thus the system

$$x^2 \equiv 1 \pmod{r}$$
$$x^2 \equiv 1 \pmod{s}$$

has at least four distinct solutions modulo rs.

Example 7.11 Let $n = 3 \cdot 5 = 15$. Then the system

$$x^2 \bmod 15 = 1$$

has the four solutions

$$x \equiv 1 \pmod{15}$$
$$x \equiv 4 \pmod{15}$$
$$x \equiv 11 \pmod{15}$$
$$x \equiv 14 \pmod{15}$$

This leads to an efficient test for primality, at least in a probabilistic sense.

To test an odd integer $p > 1$ for primality

1. Choose an integer a in the range $1 < a < p$.

2. Test to see if $a \perp p$. If $\gcd(a, p) > 1$, then declare p to be a loser and stop. Otherwise continue with Step 3.

3. Test to see if $a^{p-1} \equiv 1 \pmod{p}$. If not, then declare p to be a loser and stop. Otherwise, continue with Step 4.

4. Set $m = p - 1$.

5. Repeat Steps 6–8 while m is even.

6. Set $m = m/2$.

7. If $a^m \equiv -1 \pmod{p}$, then declare p a winner and stop.

8. If $a^m \not\equiv 1 \pmod{p}$, then declare p to be a loser and stop.

9. If m is odd and $a^m \equiv 1 \pmod{p}$, then declare p to be a winner and stop.

If p is ever declared a loser then it is definitely not a prime. If p is declared a winner, then select another integer a at random and repeat the preceding steps. If p is declared a winner 20 times in a row, then it is very likely that p is a prime. If p is a composite, then the probability of being declared a winner 20 times in a row is less than $(1/4)^{20}$.

A variation of the preceding steps is known as Miller's test (see Algorithm 7.4).

Algorithm 7.4　　　Miller's test

Input Odd integer $p > 1$, integer a in the range $1 < a < p$
Output Message that p is composite or message that p is a potential prime
Factor $p - 1 = 2^t m$ **where** $2 \perp m$
Set $b = a^m \bmod p$
If $b \bmod p = \pm 1$ **then** declare p to be a potential prime and **stop**
For i **from** 1 **to** $t - 1$ **do**
　　Set $b = b^2 \bmod p$
　　If $b = p - 1$ **Then** declare p to be a potential prime and **stop**
　　If $b = 1$ **then** declare p to be composite **and stop**
End For
Declare p to be a composite

The following theorem indicates that Miller's test is an effective filter for generating potential primes. See [25] for a proof.

Theorem 7.12 *(Rabin's Probabilistic Primality Test) Let p be an odd positive integer greater than 1. Randomly select r integers a_i in the range $1 < a_i < p$. If p is composite, the probability that p is declared a potential prime by Miller's test for all r bases a_i is less than $(1/4)^r$.*

In other words, the probability that a composite remains undetected by repeated use of Miller's test is very small. Numbers p that survive the rigors of repeated testing by Miller's test are very likely prime. For the applications that we have in mind, Miller's test will allow us to generate primes p easily that are larger, say, than 10^{50} or even 10^{100}.

A number that survives Rabin's probabilistic primality test is called a **probabilistic prime**. Of course a probabilistic prime either is prime or it isn't. However, we say that such a number is probably a prime because the probability that a composite number would have passed Rabin's test is very small.

Standard algorithms, like the digital signature algorithm (see page 194), specify that a primality test must be good enough so that the probability that it calls a composite a probabilistic prime is less than 10^{-80}. Since

$$(1/4)^{132} \approx 3.373\,5 \times 10^{-80}$$

it follows that Rabin's test is sufficient if r is at least 132.

Example 7.13 We will use Miller's test on the number $n = 105$. We pretend not to notice that $105 = 3 \cdot 5 \cdot 7$. We have $104 = 2^3 \cdot 13$. Selecting the integer $a = 8$, we compute $8^{13} \bmod 105 = 8$, $8^2 \bmod 105 = 64$, and $64^2 \bmod 105 = 1$. This means that 64 is a square root of 1, and hence 105 cannot be a prime.

Similarly, let $n = 49(= 7 \cdot 7)$. Then $48 = 2^4 \cdot 3$ and $18^3 \bmod 49 = 1$, $19^3 \bmod 49 = -1$, $30^3 \bmod 49 = 1$, and $31^3 \bmod 49 = -1$, so 49 is declared a potential prime for each of the bases 18, 19, 30, and 31. However, $5^3 \bmod 49 = 27$, $27^2 \bmod 49 = 43$, $43^2 \bmod 49 = 36$, and $36^2 \bmod 49 = 22$, and hence 49 is declared composite because none of these calculations yielded $-1 \bmod 49$.

Problems 7.4

1. Test 899 for primality by testing for divisibility by integers a in the range $1 < a < \lfloor \sqrt{899} \rfloor$.

2. Use the sieve of Eratosthenes to generate all the primes less than 900. Is 899 on this list?

3. Use Miller's test on $n = 899$.

4. Use Miller's test on $n = 561$ with $a = 13$.

5. Use Fermat's little theorem to show that $p = 205\,193$ is not prime. Factor p.

6. Use Miller's test to determine whether or not $172\,947\,529$ is prime. If p is composite, factor it.

7. Use Miller's test to determine whether or not $187\,736\,503$ is prime. If p is composite, factor it.

8. Use Miller's test on $14\,386\,156\,093$ with $a = 2, 3, 5, 7, 11, 13$.

9. Let $\pi(x)$ denote the number of primes $\leq x$. The prime number theorem states that
$$\lim_{x \to \infty} \frac{\pi(x)}{x / \ln x} = 1$$
Roughly speaking, this means that
$$\pi(x) \sim \frac{x}{\ln x}$$

Use this approximation to determine the average gap between primes for primes $p \approx 10^{100}$. Knowing that large primes are odd, about how many numbers do you expect to test before finding a prime if you start at roughly 10^{100}?

10. Let $p_1 = \text{nextprime}(1000)$, and set $p_{i+1} = \text{nextprime}(1 + p_i)$ for $i = 1, \ldots, 10$. What is the average gap between these primes? How well does this compare with the expected average gap given in problem 9?

11. Let $p_1 = \text{nextprime}(10^{100})$, and set $p_{i+1} = \text{nextprime}(1 + p_i)$ for $i = 1, \ldots, 10$. What is the average gap between these primes? How well does this compare with the expected average gap given in problem 9?

7.5 Exponential Ciphers

Fermat's little theorem leads to a useful exponential cipher. Find a large prime p (using Rabin's test, for example), and select a positive integer e such that $e \perp (p - 1)$. Plaintext must be broken down into integers $1 < x < p$. The ciphertext is then given by
$$y = x^e \bmod p$$

Deciphering requires the calculation of $d = e^{-1} \bmod (p - 1)$. This means that $de = 1 + k(p - 1)$ for some integer k. Then $x \perp p$ and $y = x^e \pmod{p}$ implies that
$$y^d \equiv (x^e)^d \equiv x^{ed} \equiv x^{1+k(p-1)} \equiv x \cdot \left(x^{p-1}\right)^k \equiv x \cdot (1)^k \equiv x \pmod{p}$$

which means that $x = y^d \pmod{p}$.
    We state this as a theorem.

Theorem 7.14 *Let p be a prime. Assume $c \perp (p-1)$ and define d by $d = e^{-1}$ mod $(p-1)$. Then for any integer x we have*

$$x \equiv (x^e)^d \mod p$$

Proof. We have already proved this in the case $x \perp p$. On the other hand, if $p \mid x$, then both sides of the congruence are congruent to 0 modulo p. ∎

The keys are p and e, from which the secondary key d can be computed. Both sender and receiver must have access to these keys.

Example 7.15 Select a prime p such as

$$p = 277\,957\,387\,467\,223\,466\,791$$

by using Miller's test or by using a computer algebra system. Select an encoding key, say

$$e = 1549$$

and compute

$$\gcd(e, p - 1) = 1$$

to verify that e is relatively prime to $p - 1$. Enter the plaintext, say

$$x = 19\,282\,828\,282\,828\,282\,828$$

Compute the ciphertext

$$y = x^e \mod p$$
$$= 128\,676\,138\,005\,901\,478\,327$$

Compute the deciphering key using

$$d = e^{-1} \mod (p - 1)$$
$$= 240\,812\,662\,350\,557\,709\,769$$

Recover the original plaintext.

$$x = 128\,676\,138\,005\,901\,478\,327^d \mod p$$
$$= 19\,282\,828\,282\,828\,282\,828$$

How good are exponential ciphers? One way of measuring the quality of a cipher is to see how well it scrambles messages. That is, what happens to the ciphertext when the plaintext is changed by a small amount? In a good cipher, the ciphertext should be changed completely.

Example 7.16 In Example 7.15, suppose we modify the plaintext by changing one of the 2s to a 3. The modified plaintext is given by

$$x = 19\,282\,828\,282\,828\,282\,828 + 10^{13}$$
$$= 19\,282\,838\,282\,828\,282\,828$$

The ciphertext is now given by

$$x^e \bmod p = 213\,272\,818\,207\,054\,088\,599$$

Notice that the original ciphertext

$$y = 128\,676\,138\,005\,901\,478\,327$$

looks quite different from the new ciphertext

$$z = 213\,272\,818\,207\,054\,088\,599$$

Problems 7.5

1. Test the exponential cipher using $p = 101$, $e = 7$, and $x = 73$.

2. Test the exponential cipher using a computer algebra system to generate a 30-digit prime.

3. Create a scheme for generating a random prime that is exactly 100 decimal digits long.

4. Let nextprime(x) denote the smallest prime $\geq x$. Discuss the strengths and weaknesses of each of the following schemes for generating a random prime p with exactly 50 decimal digits.

   (a) Let $p = $ nextprime(10^{49}).

   (b) Let $p = $ nextprime$(\overbrace{184637\cdots85}^{50 \text{ digits}})$, where the digits inside the brackets are generated by closing your eyes and letting your fingers dance on the top row of your keyboard.

   (c) Let $p = $ nextprime$(7^{61} + 11^{49} \bmod 10^{50})$.

5. Generate a 100-digit prime p, let $x = 55\cdots5$ (string of 99 fives), and let $e = 1009$. Compute $y = x^e \bmod p$ and $z = (x + 10^{43})^e \bmod p$. Compare the numbers y and z. Let y_i be the ith digit of y and z_i the ith digit of z. For how many i does $y_i = z_i$? Is this a surprise? Why or why not?

7.6   Euler's Theorem

Leonhard Euler (1707–1783) was born in Basel, Switzerland. He was one of the most prolific mathematicians of all time, contributing to every mathematical field of his day. In contrast, modern mathematicians tend to be specialists. Mathematicians working in different areas may have a lot of difficulty explaining their research to each other.

Given any positive integer n, consider the numbers $1, 2, \ldots, n$. The number of these that are relatively prime to n is denoted $\varphi(n)$. So $\varphi(12) = 4$ because, of the numbers $1, 2, \ldots, 12$, only 1, 5, 7, and 11 are relatively prime to 12. Small values of this function, called the *Euler phi function* or the *Euler totient function*, are listed in Table 7.4.

$\varphi(1) = 1$	$\varphi(2) = 1$	$\varphi(3) = 2$	$\varphi(4) = 2$	$\varphi(5) = 4$
$\varphi(6) = 2$	$\varphi(7) = 6$	$\varphi(8) = 4$	$\varphi(9) = 6$	$\varphi(10) = 4$
$\varphi(11) = 10$	$\varphi(12) = 4$	$\varphi(13) = 12$	$\varphi(14) = 6$	$\varphi(15) = 8$
$\varphi(16) = 8$	$\varphi(17) = 16$	$\varphi(18) = 6$	$\varphi(19) = 6$	$\varphi(20) = 8$
$\varphi(21) = 12$	$\varphi(22) = 10$	$\varphi(23) = 22$	$\varphi(24) = 8$	$\varphi(25) = 20$
$\varphi(26) = 12$	$\varphi(27) = 18$	$\varphi(28) = 12$	$\varphi(29) = 28$	$\varphi(30) = 8$
$\varphi(31) = 30$	$\varphi(32) = 16$	$\varphi(33) = 20$	$\varphi(34) = 16$	$\varphi(35) = 24$
$\varphi(36) = 12$	$\varphi(37) = 36$	$\varphi(38) = 18$	$\varphi(39) = 24$	$\varphi(40) = 16$
$\varphi(41) = 40$	$\varphi(42) = 12$	$\varphi(43) = 42$	$\varphi(44) = 20$	$\varphi(45) = 24$
$\varphi(46) = 22$	$\varphi(47) = 46$	$\varphi(48) = 16$	$\varphi(49) = 42$	$\varphi(50) = 20$

Table 7.4   Euler phi function

Mathematics may be described as the study of patterns. This table contains many patterns. It appears that $\varphi(p) = p - 1$ for primes p. Is it true that $\varphi(mn) = \varphi(m)\varphi(n)$? How does $\varphi(p^n)$ relate to $\varphi(p)$? Think about these questions for a few minutes before reading further.

Note that $\varphi(1) = 1$ because $1 \le 1$ and $1 \perp 1$. Also, $\varphi(21) = 12$ because the integers $1 \le x \le 21$ such that $x \perp 21$ consist of the 12 integers in the set $\{1, 2, 4, 5, 8, 10, 11, 13, 16, 17, 19, 20\}$.

Definition 7.17 *A reduced residue system modulo n is a set R of $\varphi(n)$ integers such that*

   i. $r \perp n$ for each r in R, and

   ii. if r and s are any two elements of R, then $r \not\equiv s \pmod{n}$.

Example 7.18 Let $n = 21$ and consider the reduced residue system

$$\{1, 2, 4, 5, 8, 10, 11, 13, 16, 17, 19, 20\}$$

modulo 21. Then $5 \perp 21$ and

$$\{5 \cdot 1, 5 \cdot 2, 5 \cdot 4, 5 \cdot 5, 5 \cdot 8, 5 \cdot 10, 5 \cdot 11, 5 \cdot 13, 5 \cdot 16, 5 \cdot 17, 5 \cdot 19, 5 \cdot 20\}$$
$$= \{5, 10, 20, 25, 40, 50, 55, 65, 80, 85, 95, 100\}$$

is also a reduced residue system modulo 21. Each element of the second set is congruent to exactly one element of the original reduced residue system. Indeed,

$$
\begin{array}{llll}
5 \bmod 21 & = & 5 & \qquad 55 \bmod 21 & = & 13 \\
10 \bmod 21 & = & 10 & \qquad 65 \bmod 21 & = & 2 \\
20 \bmod 21 & = & 20 & \qquad 80 \bmod 21 & = & 17 \\
25 \bmod 21 & = & 4 & \qquad 85 \bmod 21 & = & 1 \\
40 \bmod 21 & = & 19 & \qquad 95 \bmod 21 & = & 11 \\
50 \bmod 21 & = & 8 & \qquad 100 \bmod 21 & = & 16
\end{array}
$$

This type of construction always leads to other reduced residue systems.

Theorem 7.19 *Let a be an integer such that $a \perp n$. If R is a reduced residue system modulo n then so is $\{ar \mid r \in R\}$.*

Proof (See problem 6). ∎

Some special properties of the Euler phi function are worth special mention.

Theorem 7.20 *A positive integer p is prime if and only if $\varphi(p) = p - 1$.*

Proof. If p is a prime, then $a \perp p$ for each integer a in the range $1 \leq a \leq p - 1$, and there are $p - 1$ such integers.

Conversely, suppose there are $p - 1$ numbers among the numbers $1, 2, \ldots, p$ that are relatively prime to p. Then $p > 1$ because 1 is relatively prime to itself, and $1 - 1 = 0$. If $p > 1$, then p is not relatively prime to p, so each of the numbers $1, 2, \ldots, p - 1$ must be relatively prime to p. In particular, none of them can divide p, so p is prime. ∎

Theorem 7.21 *If $n = p^k$ is a power of a prime, then $\varphi(n) = p^k - p^{k-1} = p^{k-1}(p-1)$.*

Proof. There are $n = p^k$ integers a in the range $1 \leq a \leq n$. Of these, the integers $p, 2p, 3p, \ldots, p^{k-1}p = p^k$ have a common divisor with p^k that is greater than 1. There are p^{k-1} integers in this list. Hence the number of integers a in the range $1 \leq a \leq p^k$ such that $a \perp p^k$ is $p^k - p^{k-1} = p^{k-1}(p-1)$. ∎

The next theorem describes how to calculate $\varphi(n \cdot m)$ for $n \perp m$. To illustrate the proof, consider $\varphi(8 \cdot 9)$. We place the numbers $\{1, 2, 3, \ldots, 72\}$ into an array and bold the numbers in the array that are relatively prime to 72 (see Table 7.5).

1	2	3	4	5	6	7	8
9	10	**11**	12	**13**	14	**15**	16
17	18	**19**	20	21	22	**23**	24
25	26	27	28	**29**	30	**31**	32
33	34	**35**	36	**37**	38	39	40
41	42	**43**	44	45	46	**47**	48
49	50	51	52	**53**	54	**55**	56
57	58	**59**	60	**61**	62	63	64
65	66	**67**	68	69	70	**71**	72

Table 7.5   Integers relatively prime to 72

Notice that only four of the columns contain bold numbers, corresponding to $\varphi(8) = 4$, and each such column contains exactly six bold numbers, corresponding to $\varphi(9) = 6$. Thus there are exactly $24 = \varphi(8)\varphi(9)$ bold numbers in the array.

Theorem 7.22 *If $n \perp m$, then $\varphi(n \cdot m) = \varphi(n)\varphi(m)$.*

Proof. Observe first that $a \perp mn$ if and only if $a \perp m$ and $a \perp n$. Arrange the numbers $\{1, 2, 3, \ldots, n \cdot m\}$ in a matrix

$$\begin{pmatrix} 1 & 2 & 3 & 4 & \cdots & n \\ n+1 & n+2 & n+3 & n+4 & \cdots & 2n \\ 2n+1 & 2n+2 & 2n+3 & 2n+4 & \cdots & 3n \\ 3n+1 & 3n+2 & 3n+3 & 3n+4 & \cdots & 4n \\ \vdots & \vdots & \vdots & \vdots & \ddots & \vdots \\ (m-1)n+1 & (m-1)n+2 & (m-1)n+3 & (m-1)n+4 & \cdots & mn \end{pmatrix}$$

with m rows and n columns. The number in the ith row and jth column is equal to $(i-1)n + j$. Notice that if $\gcd(n, j) > 1$, then $\gcd(n, (i-1)n + j) > 1$, and hence $\gcd(n \cdot m, (i-1)n + j) > 1$. On the other hand, if $j \perp n$ then $(i-1)n + j \perp n$ for $1 \le i < n$. Thus exactly $\varphi(n)$ columns contain numbers $(i-1)n + j$ that are relatively prime to n.

If $j \perp n$, consider the set

$$R = \{j, n+j, 2n+j, \ldots, (m-1)n+j\}$$

There are exactly m elements in the set R. If

$$rn + j \equiv sn + j \pmod{m}$$

then

$$rn \equiv sn \pmod{m}$$

so

$$r \equiv s \pmod{m}$$

since $n \perp m$, so $r = s$ since both are less than m. Hence R is a complete residue system modulo m, so exactly $\varphi(m)$ elements of R are relatively prime to m. Thus $\varphi(mn) = \varphi(m)\,\varphi(n)$. ∎

We can use Theorems 7.20–7.22 to compute $\varphi(n)$ if we know the prime power factorization of n. In particular, if

$$n = \prod_{i=1}^{r} p_i^{e_i}$$

then

$$\varphi(n) = \prod_{i=1}^{r} p_i^{e_i - 1}(p_i - 1)$$

Example 7.23 Let $1741764024 = 2^3 \cdot 3^2 \cdot 7 \cdot 11^2 \cdot 13^4$. Then

$$\varphi(1741764024) = 2^2 \cdot (2-1) \cdot 3 \cdot (3-1) \cdot (7-1) \cdot 11 \cdot (11-1) \cdot 13^3 \cdot (13-1)$$
$$= 417605760$$

The following theorem of Euler generalizes Fermat's little theorem and leads to an interesting class of public key ciphers.

Theorem 7.24 (Euler's Theorem) *If $a \perp n$, then $a^{\varphi(n)} \equiv 1 \pmod{n}$.*

Proof. Let R be the reduced residue system modulo n consisting of all integers r in the range $1 \le r \le n$ such that $r \perp n$. Assume $a \perp n$. Then $aR = \{ar \mid r \in R\}$ is also a reduced residue system modulo n. It follows that $R = \{ar \bmod n \mid r \in R\}$. Thus

$$a^{\varphi(n)} \prod_{r \in R} r = \prod_{r \in R} ar \equiv \prod_{r \in R} r \pmod{n}$$

Since $n \perp \prod_{r \in R} r$, we can cancel on both sides to get the desired congruence $a^{\varphi(n)} \equiv 1 \pmod{n}$. ∎

Example 7.25 We will show directly that $5^{\varphi(21)} \equiv 1 \pmod{21}$. We have $\varphi(21) = 12$ and the reduced residue system

$$\{1, 2, 4, 5, 8, 10, 11, 13, 16, 17, 19, 20\}$$

modulo 21. Now

$$\{5 \cdot 1, 5 \cdot 2, 5 \cdot 4, 5 \cdot 5, 5 \cdot 8, 5 \cdot 10, 5 \cdot 11, 5 \cdot 13, 5 \cdot 16, 5 \cdot 17, 5 \cdot 19, 5 \cdot 20\}$$
$$= \{5, 10, 20, 25, 40, 50, 55, 65, 80, 85, 95, 100\}$$

and hence

$$1 \cdot 2 \cdot 4 \cdot 5 \cdot 8 \cdot 10 \cdot 11 \cdot 13 \cdot 16 \cdot 17 \cdot 19 \cdot 20 \bmod 21$$
$$= 5 \cdot 10 \cdot 20 \cdot 25 \cdot 40 \cdot 50 \cdot 55 \cdot 65 \cdot 80 \cdot 85 \cdot 95 \cdot 100 \bmod 21$$
$$= 5^{12} \cdot (1 \cdot 2 \cdot 4 \cdot 5 \cdot 8 \cdot 10 \cdot 11 \cdot 13 \cdot 16 \cdot 17 \cdot 19 \cdot 20) \bmod 21$$

and canceling $1 \cdot 2 \cdot 4 \cdot 5 \cdot 8 \cdot 10 \cdot 11 \cdot 13 \cdot 16 \cdot 17 \cdot 19 \cdot 20$ from both sides yields

$$5^{12} \equiv 1 \pmod{21}$$

Problems 7.6

1. Compute $\varphi(24)$ by listing all the integers a in the range $1 \le a \le 24$ such that $a \perp 24$.

2. Compute $\varphi(24)$ by using the prime power factorization of 24 and theorems in this section.

3. Compute $\varphi(27)$ using two different methods.

4. Verify that $4^{\varphi(27)} \bmod 27 = 1$.

5. Compute $\varphi(1001)$ and verify that $5^{\varphi(1001)} \bmod 1001 = 1$.

6. Prove Theorem 7.19.

7. Start with a complete residue system R modulo 10. Note that $7 \perp 10$ and show directly that $\{7r + 4 \mid r \in R\}$ is also a complete residue system modulo 10.

8. Start with a reduced residue system R modulo 15. Note that $7 \perp 15$ and multiply each element of R by 7 to form a new reduced residue system $\{7r \mid r \in R\}$. Show directly that each element of $\{7r \mid r \in R\}$ is congruent modulo 15 to exactly one element of R.

9. Show that if n is a product of distinct primes $a^{\varphi(n)+1} \equiv a \pmod{n}$ for any a.

10. Verify the formula in problem 9 for $n = 30$.

Chapter 8

Public Key Ciphers

The ciphers we have discussed so far require a key for each pair of users. A secure method—such as the use of couriers with locked briefcases—is required for the exchange of these keys. Such methods of key exchange can be expensive, unreliable, and time consuming. If there are many pairs of users in a system, each using a different key, the cost becomes prohibitive. In this chapter we describe cryptosystems that allow keys to be exchanged securely in the open.

8.1   The Rivest-Shamir-Adleman Cipher System

In the mid to late 1970s, several schemes were proposed that allowed keys to be safely exchanged over public channels such as telephone lines and satellite links. Since any potential adversary can listen in on these channels, such schemes are called **public key ciphers**. The most widely known public key cipher is the **RSA cipher** proposed by Ron Rivest, Adi Shamir, and Leonard Adleman. The algorithm is well known, and now it is legal to develop your own version.

BEDFORD, Mass., September 6, 2000—RSA$^{®}$ Security Inc. (NAS-DAQ: RSAS) today announced it has released the RSA public key encryption algorithm into the public domain, allowing anyone to create products that incorporate their own implementation of the algorithm. This means that RSA Security has waived its rights to enforce the patent for any development activities that include the RSA algorithm occurring after September 6, 2000.

RSA Security press release

In a public key system, each participant has a public key which anyone can look up. If you want to send a message to Bob, you use his public key. To decipher a message enciphered with Bob's public key, you have to use a private key that is known only to Bob. The trick is to make it very difficult to compute the private key from the public key.

In the RSA scheme, the public key is a pair (e, m) of positive integers. To decipher a message enciphered with the key (e, m), you need to know the private key (d, m). Although theoretically possible, it is very difficult to compute d from (e, m).

The integer m is constructed by finding two large primes p and q and setting $m = pq$. Here *large* might mean 50 digits, for low risk applications, to hundreds of digits for national security applications. This is a good example of how probabilistic primes are used. It is still hard to prove that a given 200-digit integer is prime, but it is very easy to generate a 200-digit probabilistic prime.

The integer e is chosen relatively prime to $\varphi(m) = (p-1)(q-1)$, and is usually chosen to be moderately sized, perhaps 3 or 4 digits long. Selecting e to be prime makes it more likely that it is relatively prime to $\varphi(m)$.

If x is the plaintext, in the form of an integer less than m, then the corresponding ciphertext is given by

$$y = x^e \bmod m$$

Deciphering uses the private key (d, m), where d is the inverse of e modulo $\varphi(m)$. The plaintext is recovered by calculating

$$x = y^d \bmod m$$

To see why this is true, write $ed = 1 + k\varphi(m)$ and suppose $1 \le x < m$. If x is relatively prime to m, which it normally will be because the probability that x is divisible by p or q is about $1/p + 1/q$, then

$$(x^e)^d \bmod m = x^{ed} \bmod m$$
$$= x^{1+k\varphi(n)} \bmod m$$
$$= x^1 \left(x^{\varphi(n)}\right)^k \bmod m$$
$$= x (1)^k \bmod m$$
$$= x$$

If, miraculously, p divides x, then $\varphi(m) = \varphi(p)\varphi(q) = (p-1)(q-1)$ so

$$x^{ed} \bmod q = x^{1+(p-1)(q-1)k} \bmod q$$
$$= x \left(x^{q-1}\right)^{k(p-1)} \bmod q$$
$$= x (1)^{k(p-1)} \bmod q$$
$$\equiv x \pmod q$$

and

$$x \equiv 0 \pmod p$$
$$\equiv x^{ed} \pmod p$$

By the Chinese remainder theorem it follows that

$$x^{ed} \bmod m = x.$$

The case when q divides x is the same, with the roles of p and q interchanged. Thus for every x in the range $1 \leq x < m$ we have

$$(x^e)^d \equiv x \bmod m$$

We state this result as a theorem.

Theorem 8.1 *Let p and q be distinct primes and $m = pq$. Let e be an integer that is relatively prime to $\varphi(m)$ and let d be its inverse modulo $\varphi(m)$. If x is any integer such that $1 \leq x < m$, then*

$$(x^e)^d \bmod m = x.$$

Theorem 8.1 is the basis for a scheme, which allows anyone to send encrypted messages to Sarah. The first step is to construct keys.

Step 1: Construct keys

    Public key (Published in a directory)   Integers m and e

    Private key (Known only to Sarah)   Primes p and q with $m = pq$
                                           $\varphi(m) = (p-1)(q-1)$
                                           $d = e^{-1} \bmod \varphi(m)$

Step 2: Agree on a Protocol Sarah and Kyle, or anyone else who wishes to send messages to Sarah, must agree on a method for converting text into a sequence of positive integers that are less than m.

Step 3: Kyle Sends a Message to Sarah

    1. Kyle converts a message into a sequence of positive integers that are each less than m.

    2. For each integer x, Kyle calculates $y = x^e \bmod m$ and sends y to Sarah.

Step 4: Sarah Decrypts the Message

    1. For each y, Sarah calculates $x = y^d \bmod m$.

    2. Sarah converts each x to text, using the same method as Kyle.

A method for converting text into integers will be described in a later section. The following example assumes that the plaintext has already been converted into a positive integer.

Example 8.2 First, Sarah generates primes p, q, and an exponent e for enciphering. Then she computes $m = pq$, the Euler phi function $\varphi(m)$, and the exponent d for deciphering. In particular, suppose

$$p = 2847\,893\,757\,848\,938\,511$$
$$q = 92\,734\,928\,626\,327\,511$$
$$m = pq$$
$$\quad = 264\,099\,224\,369\,484\,956\,639\,974\,579\,586\,676\,121$$
$$e = 1009$$

$$\varphi = (p-1)(q-1)$$
$$\quad = 264\,099\,224\,369\,484\,953\,699\,345\,893\,111\,410\,100$$
$$d = e^{-1} \bmod \varphi$$
$$\quad = 5758\,357\,716\,678\,561\,924\,068\,988\,749\,703\,689$$

are the numbers generated by Sarah. Suppose that Kyle decides to send the plaintext message

$$x = 33\,333\,333\,333\,333\,333\,333\,333\,333\,333\,333\,333$$

to Sarah. In order to do this, Kyle looks up Sarah's numbers e and m and computes the ciphertext

$$y = x^e \bmod m$$
$$\quad = 54\,423\,731\,721\,403\,481\,610\,392\,517\,373\,097\,210$$

and sends it to Sarah, who calculates

$$x = y^d \bmod m = 33\,333\,333\,333\,333\,333\,333\,333\,333\,333\,333\,333$$

and wonders why Kyle would bother sending such a dull message.

What makes the RSA cipher secure? Can't you construct d from m and e, in which case anyone could decipher an RSA cipher?

Yes, it is possible to deduce d from m and e; but it is not easy to do. To calculate $\varphi(m)$ is just as hard as factoring m. On the one hand, if $m = pq$, where p and q are known, then we can calculate $\varphi(m)$ by

$$\varphi(m) = \varphi(p)\varphi(q) = (p-1)(q-1)$$

On the other hand, if m and $\varphi(m)$ are known, but not the factorization $m = pq$ with $q < p$, then

$$\varphi(m) = (p-1)(q-1)$$
$$= pq - (p+q) + 1$$
$$= m - (p+q) + 1$$

implies that

$$p + q = m - \varphi(m) + 1$$

and

$$(p + q)^2 = p^2 + 2pq + q^2$$
$$= p^2 - 2pq + q^2 + 4pq$$
$$= (p - q)^2 + 4m$$

implies that

$$p - q = \sqrt{(p + q)^2 - 4m}$$

Thus $p + q$ and $p - q$ are both easily computable. The primes p and q can then be computed by

$$p = \frac{(p + q) + (p - q)}{2}$$

and

$$q = \frac{(p + q) - (p - q)}{2}$$

Factoring is much harder than testing for primality. The security of the RSA cipher is based on the premise that deciphering can only be accomplished by factoring m.

Changing a single bit in x can completely change the value of $y = x^e \bmod m$. For example,

$$5555555555555555^{1234} \bmod 2784892789 = 1172\,041\,861$$

whereas

$$5555555555455555^{1234} \bmod 2784892789 = 356\,849\,431$$

Public channels such as phone lines are noisy and the likelihood of at least one error is significant. By combining RSA with something like the Hamming (8, 4) code, the probability of errors can be greatly reduced. (More powerful codes will be introduced in the next chapter.)

Text is broken up into blocks and each block is enciphered using RSA, then the ciphertext is converted to hex. Hamming (8, 4) is applied to each nibble (see page 124). At the receiving end, Hamming (8, 4) is used to correct single errors and detect double errors. If double errors are detected, then the corresponding block must be retransmitted. The corrected ciphertext is then deciphered using RSA.

Problems 8.1

1. Let $p = 5$ and $q = 7$ so that $m = 35$, and let $e = 11$. Find $d = e^{-1} \bmod \varphi(m)$. Then let $x = 22$ and compute $y = x^e \bmod 35$ and $z = y^d \bmod m$.

2. Show that $a^2 \bmod 24 = 1$ for each $a \perp 24$. Conclude that $a^{-1} \bmod 24 = a \bmod 24$ for each $a \perp 24$.

3. Let $p = 29$ and $q = 31$ so that $m = 899$, and let $e = 101$. Find

$$d = e^{-1} \bmod \varphi(m)$$

Then let $x = 555$ and compute $y = x^e \bmod m$ and $z = y^d \bmod m$.

4. Assume

$$m = 25\,972\,641\,171\,898\,723$$

is a product of two primes p and q and that

$$\varphi(m) = 25\,972\,640\,809\,676\,568$$

Use these two equations to find the primes p and q.

5. Illustrate the RSA algorithm using m from Problem 4 with

$$e = 997$$
$$x = 99\,999\,999\,999\,999$$

6. Assume that $y = x^e \bmod m$ from problem 5 is transmitted over a noisy channel and received as $w = y + 10^{12}$. How similar are y and w? What message is decrypted? How similar are x and the decrypted message? (This explains why most ciphers are wrapped in some type of error correction scheme.)

7. Assume $m = 21\,936\,520\,921\,056\,942\,428\,185\,744\,321\,881\,874\,204\,790\,829\,920$ $570\,235\,226\,904\,516\,467\,385\,564\,406\,736\,567\,597\,367\,535\,979\,699\,930\,859$ 170 $667\,289\,061\,009\,756\,151\,158\,068\,196\,185\,554\,149$ is a product of two primes p and q and that $\varphi(m) = 21\,936\,520\,921\,056\,942\,428\,185\,744\,321\,881\,874\,204$ $790\,829\,920\,570\,235\,226\,904\,516\,467\,385\,548\,925\,092\,800\,907\,044\,879\,409\,345$ $000\,125\,494\,759\,694\,716\,131\,857\,830\,775\,913\,843\,528\,947\,136$. Use these two equations to find the primes p and q.

8. Illustrate the RSA algorithm using m from problem 7 with

$$e = 997$$
$$x = 99\,999\,999\,999\,999$$

9. Let $m = pq$ where p and q are distinct primes. Show that if x is chosen at random in the range $1 \leq x \leq m$, then the probability that $\gcd(x, m) \neq 1$ is equal to $1/p + 1/q - 1/m$.

10. Verify the formula in problem 9 for $m = 6$ and $m = 15$. Show that if x is chosen at random in the range $1 \leq x < m$, then the probability that $\gcd(x, m) \neq 1$ is less than $1/p + 1/q$.

8.2 Electronic Signatures

The RSA cipher can be used for electronic signatures. More precisely, here is how to use RSA to guarantee that a given message came from a given source.

As before, the original message is broken into blocks of fixed length and each block is translated into a number x.

Suppose Kyle's public key is (e_k, m_k) and his private key is (d_k, m_k), that Sarah's public key is (e_s, m_s) and her private key is (d_s, m_s), and that $m_k < m_s$. Kyle enciphers the plaintext x with his private key (d_k, m_k) and then with Sarah's public key (e_s, m_s):

$$y = (x^{d_k} \bmod m_k)^{e_s} \bmod m_s$$

Sarah reads the ciphertext y and then computes the plaintext x by

$$x = (y^{d_s} \bmod m_s)^{e_k} \bmod m_k$$

using her private key (d_s, m_s) and Kyle's public key (e_k, m_k).

How does Sarah know that the message came from Kyle? Because only Kyle knows d_k and so only he can encipher the plaintext x with it.

To see that this procedure works, let $z = x^{d_k} \bmod m_k$. Then

$$(z^{e_s} \bmod m_s)^{d_s} \bmod m_s = z^{e_s d_s} \bmod m_s = z \bmod m_s = z$$

assuming $z < m_s$. But indeed $z = x^{d_k} \bmod m_k < m_k < m_s$. Thus

$$\left(\left((x^{d_k} \bmod m_k)^{e_s} \bmod m_s \right)^{d_s} \bmod m_s \right)^{e_k} \bmod m_k$$

$$= \left((z^{e_s} \bmod m_s)^{d_s} \bmod m_s \right)^{e_k} \bmod m_k$$

$$= z^{e_k} \bmod m_k$$

$$= \left(x^{d_k} \bmod m_k \right)^{e_k} \bmod m_k$$

$$= x^{d_k e_k} \bmod m_k$$

$$= x$$

A message from Sarah to Kyle goes as follows. Given the plaintext x, Sarah sends the ciphertext y, where

$$y = (x^{e_k} \bmod m_k)^{d_s} \bmod m_s$$

Kyle then deciphers y by using

$$x = (y^{e_s} \bmod m_s)^{d_k} \bmod m_k$$

Each person uses his or her own private key and the other person's public key. Enciphering is done by using the smaller modulus first, then the larger modulus. Deciphering uses the larger modulus first, then the smaller.

What if the larger modulus were applied first, followed by the smaller modulus for enciphering? It is likely that $x^{e_s} \bmod m_s > m_k$, and hence some information might get lost.

If the deciphered text makes sense, it could only have been enciphered by the person who used his or her own private key. There is a possible flaw in this reasoning. If Sean intercepts the message $y = (x^{e_k} \bmod m_k)^{d_s} \bmod m_s$, he could strip off Sarah's signature by computing $z = y^{e_s} \bmod m_s$ and then attach his own signature by calculating $z^{d_h} \bmod m_h$ and forwarding this message to Kyle. It would appear to Kyle that Sean had sent the original message, although Sean would have no idea what he had actually sent.

There are several ways to minimize the effects of this problem. One would be for Sarah to include within the message her name and a time stamp.

Another possibility is always to encipher by using the private key first, say $y = x^{d_s} \bmod m_s$. If $y > m_k$, then break y into blocks $y_i < m_k$ and calculate $y_i^{e_k} \bmod m_k$ for each block and send these blocks to Kyle.

Finally, each person could publish two keys; one for sending and one for receiving. The moduli for the send-keys would be smaller, say $10^{50} < m_s < 10^{51}$, with the moduli for the receive-keys larger, say $10^{51} < m_r < 10^{52}$. That way each person would be sending a message to someone else with a larger moduli, so enciphering would use the formula $y = (x^{d_s} \bmod m_s)^{e_r} \bmod m_r$, where (d_s, m_s) is the private sender-key and (e_r, m_r) is the public receiver-key.

Problems 8.2

1. Send the plaintext $x = 12$ from Kyle to Sarah assuming Kyle's private key is $(3, 15)$ and Sarah's public key is $(5, 21)$.

2. Sean uses the primes $p_s = 38\,490\,587$ and $q_s = 983\,299$ with exponent $e_s = 7349$. Find Sean's public and private keys.

3. Brendon uses the primes $p_b = 2748\,401$ and $q_b = 157\,849\,763$ with exponent $e_b = 9587$. Find Brendon's public and private keys.

4. Brendon wishes to send the message $1234\,567\,890$ to Sean, using Sean's public key and Brendon's private key. What message should Sean receive?

5. Show how Sean decrypts the message from Problem 4.

   Use the following table of primes and exponents to answer Problems 6–9.

	Brendon	Janet	Kyle	Preston	Sarah
p	8329	2333	9967	7451	3469
q	3499	6469	1931	6689	7643
e	5501	7853	4663	7907	4637

6. Compute private keys for Brendon, Janet, Kyle, Preston, and Sarah.

7. Brendon encrypts the plaintext message $99\,999\,999$ using his private key and Janet's public key. What is the encrypted message?

8. Preston receives the encrypted message 34 568 007 from Kyle, who used his own private key and Preston's public key. Decrypt the message.

9. The plaintext message 2000 000 is encrypted by first applying Preston's public key, then Sarah's private key. Show that decryption fails. Explain what went wrong.

For problems 10–13, use the following table of primes and exponents.

	Brendon	Janet	Kyle
p	41 381 059 613	688 223 072 857	10 410 338 677
q	23 841 287 213	10 214 358 881	20 604 499 379
e	387 420 499	177 264 463	96 889 010 447

	Preston	Sarah
p	862 891 037 453	16 321 363 081
q	510 246 412 969	11 162 261 477
e	304 839 373	98 329 493

10. Compute private keys for Brendon, Janet, Kyle, Preston, and Sarah.

11. Brendon encrypts the plaintext message 99 999 999 using his private key and Janet's public key. What is the encrypted message?

12. Preston receives the encrypted message 182 654 067 428 694 778 930 238 from Kyle, who used his own private key and Preston's public key. Decrypt the message.

13. The plaintext message 101010101010101010101 is encrypted by first applying Preston's public key, then Sarah's private key. Show that decryption fails. Explain what went wrong.

8.3 A System for Exchanging Messages

In practice, the primes p and q are chosen so that $n = pq$ has about 200 digits. We will go through an example where n has about 50 digits. Kyle constructs two primes whose product is greater than 10^{50}. He then chooses a positive integer e and tests to make sure that it is relatively prime to $\varphi(n)$. Kyle then sends an e-mail message:

TO: Brendon, Sean, Sarah
FROM: Kyle
$n_k = 126\,284\,756\,413\,553\,050\,978\,360\,366\,248\,406\,608\,817\,836\,065\,776\,277$
$e_k = 1009$

Kyle constructed n_k by computing $n_k = p_k\, q_k$, where

$$p_k = 8273\,488\,995\,874\,738\,949\,376\,607$$
$$q_k = 15\,263\,784\,900\,967\,433\,245\,564\,811$$

but he would never tell Sarah (or anyone else) how n_k was defined. Then Kyle computes d_k by first computing

$$\varphi_k = (p_k - 1)(q_k - 1)$$
$$= 126\,284\,756\,413\,553\,050\,978\,360\,342\,711\,132\,711\,975\,663\,870\,834\,860$$

then setting

$$d_k = 1009^{-1} \bmod \varphi_k$$
$$= 54\,694\,190\,835\,205\,830\,800\,340\,406\,109\,777\,002\,114\,336\,086\,773\,869$$

Sarah also sends an e-mail message that contains her public key.

TO: Brendon, Sean, Kyle
FROM: Sarah
$n_s = 7649\,659\,803\,155\,869\,454\,870\,134\,280\,817\,974\,298\,450\,529\,391\,711\,731$
$e_s = 997$

Sarah remembers that her primes are

$$p_s = 77839584736125364985746393$$
$$q_s = 98274673857627364665373867$$

and computes

$$\varphi_s = (p_s - 1)(q_s - 1)$$
$$= 7649\,659\,803\,155\,869\,454\,870\,134\,104\,703\,715\,704\,697\,799\,740\,591\,472$$

and

$$d_s = 997^{-1} \bmod \varphi_s$$
$$= 5723\,817\,666\,152\,736\,823\,804\,533\,643\,038\,086\,174\,227\,240\,327\,463\,629$$

Rather than restricting messages to strings of the capital letters A–Z, we allow a full set of 256 ASCII characters.

Example 8.3 Kyle decides to send the message

```
The #@%&* computer is acting up again!
```

to Sarah. He first writes down the ASCII value of each character and produces the vectors

$$A = (84, 104, 101, 32, 35, 64, 37, 38, 42, 32, 99, 111,$$
$$109, 112, 117, 116, 101, 114, 32, 106)$$
$$B = (115, 32, 97, 99, 116, 105, 110, 103, 32, 119, 112,$$
$$32, 97, 103, 97, 105, 110, 33, 0, 0)$$

He then calculates

$$x_1 = 600\,167\,620\,106\,771\,636\,211\,628\,585\,849\,721\,838\,540\,876\,507\,220$$
$$x_2 = 2912\,276\,570\,798\,527\,424\,777\,621\,159\,210\,842\,274\,734\,195$$

using the sums

$$x_1 = \sum_{i=0}^{19} a_i c^i$$

$$x_2 = \sum_{i=0}^{19} b_i c^i$$

where $c = 256$. Kyle tests the moduli to see which of n_k or n_s is larger. Since

$$n_k - n_s = -7649\,659\,803\,155\,869\,454\,870\,134$$
$$272\,544\,485\,302\,575\,790\,442\,335\,124$$

Kyle concludes that $n_k < n_s$. Thus Kyle enciphers by first using his private key (d_k, n_k), then Sarah's public key (e_s, n_s). Kyle enciphers the message by calculating

$z_1 = x_1^{d_k} \bmod n_k$
$= 125\,819\,736\,344\,175\,718\,268\,932\,263\,205\,870\,616\,867\,460\,585\,594\,806$

$z_2 = x_2^{d_k} \bmod n_k$
$= 110\,431\,267\,001\,056\,501\,628\,738\,304\,699\,976\,888\,214\,174\,820\,174\,836$

$y_1 = z_1^{e_s} \bmod n_s$
$= 6490\,122\,138\,798\,845\,932\,205\,029\,063\,697\,475\,651\,026\,315\,779\,005\,336$

$y_2 = z_2^{e_s} \bmod n_s$
$5206\,997\,138\,452\,155\,799\,019\,530\,511\,244\,204\,675\,487\,260\,820\,178\,238$

Kyle then sends Sarah the following e-mail message:

TO. Sarah
FROM: Kyle
6490 122 138 798 845 932 205 029 063 697 475 651 026 315 779 005 336
5206 997 138 452 155 799 019 530 511 244 204 675 487 260 820 178 238

Sarah then deciphers the message by calculating some inverse operations:

$$w_1 = y_1^{d_s} \bmod n_s$$
$$w_1 = 125\,819\,736\,344\,175\,718\,268\,932\,263\,205\,870\,616\,867\,460\,585\,594\,806$$
$$w_2 = y_2^{d_s} \bmod n_s$$
$$w_2 = 110\,431\,267\,001\,056\,501\,628\,738\,304\,699\,976\,888\,214\,174\,820\,174\,836$$
$$x_1 = w_1^{e_k} \bmod n_k$$
$$= 600\,167\,620\,106\,771\,636\,211\,628\,585\,849\,721\,838\,540\,876\,507\,220$$
$$x_2 = w_2^{e_k} \bmod n_k$$
$$= 2912\,276\,570\,798\,527\,424\,777\,621\,159\,210\,842\,274\,734\,195$$

Sarah now calculates the sequence $\{a_i\}$ of ASCII values by performing the following steps on x_1 and x_2:

$$a_0 = x \bmod 256 = 84$$
$$x = \frac{x - a_0}{256} = 2344\,404\,766\,042\,076\,703\,951\,674\,163\,475\,475\,931\,800\,298\,856$$
$$a_1 = x \bmod 256 = 104$$
$$x = \frac{x - a_1}{256} = 9157\,831\,117\,351\,862\,124\,811\,227\,201\,076\,077\,858\,594\,917$$
$$a_2 = x \bmod 256 = 101$$

$$\vdots$$

$$a_{19} = x \bmod 256 = 105$$
$$x = 2912\,276\,570\,798\,527\,424\,777\,621\,159\,210\,842\,274\,734\,195$$
$$a_{20} = x \bmod 256 = 115$$

$$\vdots$$

$$a_{37} = x \bmod 256 = 33$$
$$a_{38} = 0$$
$$a_{39} = 0$$

Translated back into text, Sarah reads the message,

The #@%&* computer is acting up again!

Problems 8.3

1. Convert the plaintext "2^4=16" to a positive integer, using ASCII values and base 256 arithmetic.

2. Convert to plaintext the three numbers

   653 481 561 706 256 160 607 300 716 034 424 225 952 184 750 930

   556 034 090 672 982 451 998 688 663 006 597 581 940 164 228 207

   199 505 372 523

In problems 3-8, use the keys

	Sarah
e	997
d	5723 817 666 152 736 823 804 533 643 038 086 174 227 240 327 463 629
n	7649 659 803 155 869 454 870 134 280 817 974 298 450 529 391 711 731

	Kyle
e	1009
d	54 694 190 835 205 830 800 340 406 109 777 002 114 336 086 773 869
n	126 284 756 413 553 050 978 360 366 248 406 608 817 836 065 776 277

3. Encrypt the message

   Ron Rivest is a professor at MIT

   from Sarah to Kyle.

4. Kyle received the encrypted message

   580 735 046 350 033 514 934 592 803 992 352 023 532 099 155 378 947

   5291 232 071 949 948 925 471 032 807 295 136 508 630 705 523 321 750

   from Sarah. Read the message.

5. Sarah received the encrypted message

   1695 609 128 579 432 034 342 230 443 933 539 315 666 505 917 090 109

   3526 148 554 529 184 887 055 643 524 345 287 218 828 172 175 185 624

   2837 766 027 562 936 344 846 748 331 694 487 638 574 287 462 071 682

   from Kyle. Read the message.

6. Encrypt the message

   RSA is used to transfer keys for fast encryption schemes

   from Sarah to Kyle.

7. Kyle received the encrypted message

$$1491\,543\,853\,684\,791\,090\,861\,904\,364\,888\,457\,101\,214\,455\,628\,031\,970$$
$$6920\,136\,844\,206\,351\,338\,082\,229\,186\,328\,578\,558\,592\,587\,709\,058\,521$$
$$464\,188\,094\,287\,035\,035\,931\,921\,710\,687\,540\,480\,117\,216\,797\,612\,537$$

from Sarah. Read the message.

8. Sarah received the encrypted message

$$2027\,208\,815\,204\,265\,182\,328\,141\,518\,312\,028\,960\,081\,474\,989\,240\,004$$
$$3151\,590\,210\,265\,147\,454\,601\,370\,266\,850\,881\,158\,943\,783\,670\,753\,786$$
$$2698\,042\,079\,593\,028\,266\,457\,969\,044\,162\,648\,597\,639\,273\,126\,469\,652$$

from Kyle. Read the message.

8.4   Knapsack Ciphers

Modern cryptology is based on the construction of functions that are easy to calculate in one direction, but difficult to compute in the other direction (without special information). Such a function is called a **trap door function**. In particular, we seek one-to-one functions $y = f(x)$ for which $y = f^{-1}(x)$ is difficult to determine unless the user has a particular key.

In the case of the RSA algorithm, $y = x^e \bmod n$ is easy to compute, but the inverse function is difficult unless a factorization of n is known or $d = e^{-1} \bmod \varphi(n)$ is known.

In this section we present another class of functions that are easy to calculate, but for which the inverse problem is more difficult.

In the sequence $1, 2, 2^2, 2^3, 2^4, \ldots$, each term is greater than the sum of the preceding terms because $1 + 2 + 2^2 + \cdots + 2^{k-1} = 2^k - 1$.

Definition 8.4 *A sequence $a_1, a_2, \ldots, a_n$ of positive integers is **superincreasing** if each term is greater than the sum of the preceding terms. That is,*

$$a_k > \sum_{i<k} a_i$$

for $k = 1, 2, \ldots, n - 1$.

In the superincreasing sequence $1, 2, 2^2, 2^3, 2^4, \ldots$, each term is twice the preceding one. We can relax that condition a little bit and still get a superincreasing sequence.

Theorem 8.5 *A sequence $a_1, a_2, \ldots, a_n$ is super increasing if $a_{i+1} > 2a_i$ for $i = 1, \ldots, n - 1$.*

Proof. The theorem holds for $n = 2$ because $a_2 \geq 2a_1 > a_1 = \sum_{i=1}^{1} a_i$. Suppose that the theorem holds for $n = k$. Then

$$a_{k+1} > 2a_k > a_k + \sum_{i<k} a_i = \sum_{i<k+1} a_i \qquad \blacksquare$$

Example 8.6 The sequence 1, 3, 8, 21, 45, 97, 207, 415, 843, 1691, 3389, 6787, 13581, 27166, 54335, 108671 is a superincreasing sequence because each term is more than twice as large as the preceding term.

In the **knapsack problem**, you are given a set K of positive integers and you want to write a given positive integer as a sum of some of the numbers in K. This is known to be a very difficult problem to solve if K is large.

A **knapsack cipher** works on binary plaintext. We will illustrate with a small example. Suppose we base our knapsack cipher on a superincreasing sequence $a_1, a_2, a_3, \ldots, a_{16}$. Then plaintext is broken up into blocks of length 16. The text "Hi" is converted to ASCII hex code 4869, which is 0100 1000 0110 1001 in binary. The ciphertext is the sum of the corresponding terms in the sequence

$$y = a_2 + a_5 + a_{10} + a_{11} + a_{13} + a_{16}$$

Consider the particular superincreasing sequence

$$1, 3, 8, 21, 45, 97, 207, 415, 843, 1691, 3389, 6787, 13581, 27166, 54335, 108671$$

Note that each term is more than twice the preceding term, so the sequence is indeed superincreasing. Then the ciphertext is

$$y = a_2 + a_5 + a_{10} + a_{11} + a_{13} + a_{16} = 127380$$

Deciphering is easy– much like conversion from decimal to binary. We need to find coefficients x_i in $\{0, 1\}$ such that

$$\sum_{i=1}^{16} x_i a_i = y$$

The key observation is that if

$$\sum_{i=1}^{k} x_i a_i = w$$

then $x_k = 0$ if $a_k > w$, for otherwise the sum would be larger than w, and that $x_k = 1$ if $a_k \leq w$, because otherwise the sum would be less than a_i, as the sequence is superincreasing.

So, $x_{16} = 1$, because $a_{16} = 108671 \leq 127380$. Subtracting a_{16} from both sides of the equation gives

$$\sum_{i=1}^{15} x_i a_i = 127380 - 108671 = 18709$$

so $x_{15} = 0$ because $a_{15} = 54335 > 18709$. Similarly $x_{14} = 0$ because $a_{14} = 27166 > 18709$. But $x_{13} = 1$ because $a_{13} = 13581 \le 18709$. We then subtract a_{13} from 18709 and continue. We show the interesting steps, when $x_k = 1$ and we subtract a_k from w, in the table:

k	a_k	w
16	108671	127380
13	13581	18709
11	3389	5128
10	1691	1739
5	45	48
2	3	0

Thus $x_k = 1$ exactly when $k = 2, 5, 10, 11, 13, 16$ and the binary plaintext is

$$x_1 x_2 \cdots x_{16} = 0100\ 1000\ 0110\ 1001$$

which is 4869 in hex, so the plaintext is "Hi".

Algorithm 8.1   To solve easy knapsack problem

Input: superincreasing sequence $a_1, \ldots, a_n$ and sum y
Output: Binary sequence $x_1, \ldots, x_n$

Set $x_i = 0$ for $i = 1 \ldots n$
For i from n down to 1
    If $a_i < y$ **Then**
        Set $x_i = 1$
        Set $y = y - a_i$
    End If
Next i
End

This is an easy algorithm. In fact, it is too easy. Anyone with a general knowledge of superincreasing sequences could do the same thing, and hence deduce the plaintext.

One way to disguise the superincreasing sequence is to select a modulus $m > 2 \cdot a_n$ and a multiplier k such that $k \perp m$, and define a new sequence b_i by

$$b_i = k a_i \bmod m$$

The sequence $b_1, b_2, \ldots, b_n$ is no longer superincreasing, even if it were arranged in order, but if

$$y = \sum_{i=1}^{n} x_i a_i$$

then set

$$z = ky \bmod m$$

In this variation, the public key is $(b_1, b_2, \ldots, b_n)$ and the private key is (m, k). Given the binary plaintext $x_1, x_2, \ldots, x_n$, the ciphertext is the number

$$z = \sum_{i=1}^{n} x_i b_i \bmod m$$

To recover the plaintext $x_1, x_2, \ldots, x_n$ if z and (m, k) are known, first compute $k^{-1} \bmod m$, and then solve

$$y = ykk^{-1} \bmod m = \sum_{i=1}^{n} k^{-1} x_i b_i = \sum_{i=1}^{n} k^{-1} x_i k a_i = \sum_{i=1}^{n} x_i a_i$$

for $x_1, x_2, \ldots, x_n$, using the easy method for a superincreasing sequence.

The name "knapsack cipher" comes from the problem of guessing the contents of a knapsack without looking inside. All you know is the weight of the filled knapsack and the weights of the various objects that might be inside.

The general knapsack problem is provably difficult and therefore looks like an attractive basis for ciphers. A key that is 100 bits long may require 2^{100} trials to solve a knapsack problem. However, the version we described is not so secure. Knowing that modular arithmetic is used, the multiplier k can be deduced with moderate effort. The search continues for a secure knapsack cipher that is easy for users to encipher and decipher.

Problems 8.4

1. Show that $2, 3, 6, 12, 24, 48, 96, 200$ is a superincreasing sequence.

2. Write 313 as a sum of (some of) the integers $2, 3, 6, 12, 24, 48, 96, 200$.

3. Given the sequence $2, 3, 6, 12, 24, 48, 96, 200$, let $m = 453$ and $k = 61$. What is the public key? Encrypt the plaintext 10010110. Use the private key $(453, 61)$ to decrypt the message.

4. Write $361\,675$ as a sum of (some of) the numbers

   $$86\,977, 75\,580, 29\,992, 71\,381, 21\,594, 54\,585, 97\,773, 85\,775,$$
   $$38\,985, 2390, 103\,154, 62\,346, 94\,700, 45\,438, 79\,479, 60\,584$$

   using the fact that these numbers were created by starting with a superincreasing sequence and multiplying each number by $98\,374$ and reducing the results modulo $109\,771$.

5. Show that $1, 3, 5, 10, 20, 40, 80, 160, 320, 640, 1280, 2560, 5120, 10240, 20480, 40960, 81920, 163840$ is a superincreasing sequence.

6. Write $315\,428$ as a sum of (some of) the integers $1, 3, 5, 10, 20, 40, 80, 160,$ $320, 640, 1280, 2560, 5120, 10\,240, 20\,480, 40\,960, 81\,920, 163\,840$.

7. Given the sequence $1, 3, 5, 10, 20, 40, 80, 160, 320, 640, 1280, 2560, 5120,$ $10\,240, 20\,480, 40\,960, 81\,920, 163\,840$, let $m = 327\,819$ and $k = 997$. What is the public key? Encrypt the plaintext 101100101001011010. Calculate the private key and decrypt the message.

8. Write $2410\,809$ as a sum modulo $4272\,954$ of (some of) the numbers

875 372	51 953	3553 441	311 718	1498 808
475 406	3524 975	254 786	1384 944	2821 841
3996 844	1198 524	1625 582	1552 373	3156 699
393 606	1714 537	31 492	3564 472	2959 896

using the fact that these numbers were created by starting with a superincreasing sequence and multiplying each number by $2574\,163$ and reducing the results modulo $4272\,954$.

8.5 Digital Signature Standard

The digital signature algorithm (DSA) allows a digital signature to be used on an electronic document instead of a traditional signature. It is intended to assure that signatures cannot be forged and that anyone can verify a validly signed document.

Digital Signature Algorithm

DSA uses the following parameters:

1. A prime p between 2^{L-1} and 2^L, where L is a multiple of 64 and $512 \leq L \leq 1024$.

2. A prime q between 2^{159} and 2^{160} that divides $p - 1$.

3. An integer $g > 1$ of the form

$$g = h^{(p-1)/q} \bmod p$$

The order of g modulo p is q.

4. A random integer x in the range

$$0 < x < q$$

5. The integer

$$y = g^x \bmod p$$

6. A random integer k in the range

$$0 < k < q$$

The public keys p, q, and g can be shared by a group of users. An individual user has a private key x and public key y. A new choice of k is required for each signed document.

Digital Signature

Let m be the 160-bit output of the secure hash algorithm for the message M (see page 196). The number m will almost certainly be different for different messages. A digital signature for the message M is the pair of integers (r, s) computed by the formulas

$$r = \left(g^k \bmod p\right) \bmod q$$
$$s = k^{-1}\left(m + xr\right) \bmod q$$

The point of incorporating m into the digital signature is to assure that this signature applies to the message M and wasn't just copied from some other message.

Verification

Knowing p, q, g, the user's public key y, the electronic document M, and the digital signature (r, s), the signature is verified by calculating

$$t = ms^{-1} \bmod q$$
$$u = rs^{-1} \bmod q$$
$$v = \left(g^t y^u \bmod p\right) \bmod q$$

If $v = r$, then the signature is verified. To see this, note that

$$s^{-1} \bmod q = k\left(m + xr\right)^{-1} \bmod q$$

implies

$$s^{-1}\left(m + xr\right) \bmod q - k \bmod q$$

so

$$
\begin{aligned}
v &= \left(g^t y^u \bmod p\right) \bmod q \\
&= \left(g^{ms^{-1} \bmod q} y^{rs^{-1} \bmod q} \bmod p\right) \bmod q \\
&= \left(g^{ms^{-1} \bmod q} g^{xrs^{-1} \bmod q} \bmod p\right) \bmod q \\
&= \left(g^{(m+rx)s^{-1} \bmod q} \bmod p\right) \bmod q \\
&= \left(g^{k \bmod q} \bmod p\right) \bmod q \\
&= r
\end{aligned}
$$

Example 8.7 We look at an example with smaller numbers. Let $q = 101$ and $p = 1248\,563$. Then $p - 1 \bmod q = 0$. Let $g = 2^{\frac{p-1}{q}} \bmod p = 588\,634$. Then $g^q \bmod p = 1$ and hence g has order q modulo p. Let $x = 23$. Then $y = g^x \bmod p = 702\,644$. Suppose $m = 53$ and let $k = 42$. The digital signature is given by

$$r = \left(g^k \bmod p\right) \bmod q = 52$$
$$s = k^{-1}\left(m + xr\right) \bmod q = 61$$

Now compute

$$t = ms^{-1} \bmod q = 82$$
$$u = rs^{-1} \bmod q = 29$$
$$v = \left(g^t y^u \bmod p\right) \bmod q = 52$$

Secure Hash Algorithm

The **secure hash algorithm** produces a 160-bit output called a **message digest**. A more suggestive name is **digital fingerprint**. The message digest detects even slight changes in a lengthy document, so it is almost impossible for different readable documents to have the same message digest.[1]

The algorithm uses bitwise logical operations, the circular shift operators, and the modular sum

$x \wedge y$	and
$x \vee y$	or
$x \oplus y$	xor
$\tilde{}\,x$	complement
$S^n(x)$	left circular shift
$x + y \bmod 2^{32}$	sum

where x and y are 32-bit words, and S^n shifts x left by n bits, filling in the vacated bits on the right with those that fell off the left. Thus,

$$S^4\left(x_0 x_1 x_2 \cdots x_{31}\right) = \left(x_4 x_5 \cdots x_{31} x_0 x_1 x_2 x_3\right)$$

Problems 8.5

1. Verify that

$$q = 1091\,381\,946\,026\,415\,813\,817\,094\,712\,889\,364\,130\,135\,176\,179\,079$$

is a prime in the range $2^{159} < q < 2^{160}$.

---

[1]This algorithm, called SH-1, is described in the *Federal Information Processing Standard Publication 180-1*, which is available at http://www.itl.nist.gov/fipspubs/fip180-1.htm. A public competition for a new algorithm, to be called SH-3, was announced by the National Institute of Standards and Technology with entries due by October 31, 2008.

2. Verify that

$$p = 1001\,233\,193\,091\,254\,238\,180\,294\,974\,169\,661\,662\,027\,929\,267$$
$$542\,507\,108\,664\,330\,1413485\,006\,681\,346\,209\,714\,994\,069\,015$$
$$138\,048\,278\,778\,614\,844\,442\,236\,115\,774458\,221\,400\,826\,017$$
$$032\,280\,284\,335\,829\,147$$

is a prime in range $2^{511} < p < 2^{512}$.

3. Verify that $p - 1 \bmod q = 0$, where p and q are given in problems 1 and 2.

4. Find the smallest positive integer h that satisfies $h^{(p-1)/q} \bmod p \neq 1$.

5. If $q = h^{(p-1)/q} \bmod p$, where h is given in Problem 4, verify that $g^q \bmod p = 1$.

6. Let q be a prime in the range $2^{159} < q < 2^{160}$ and consider the following algorithm for finding a prime p.

---
p := 2^352*q + 1
while isprime(p) = false do
    p := p + 2*q
end do

Verify that the prime p given by this algorithm satisfies $q \mid p$ 1.

7. Let $n \perp m$ and define $\pi_{n,m}(x)$ to be the number of primes of the form $p = nk \mid m$ that are less than x. The generalized prime number theorem states that

$$\lim_{x \to \infty} \frac{\pi_{n,m}(x)}{x/\ln x} = \frac{1}{\varphi(n)}$$

where $\varphi(n)$ is the Euler phi function. Using the algorithm given in Problem 6, how many times do you expect the algorithm to repeat the loop before finding a prime?

Class Project

1. Get an e-mail address for each student in class.

2. Generate two primes p and q such that $m = pq > 10^{50}$.

3. Publish your public key (e, m), where e is a three- or four-digit positive integer with $e \perp \varphi(n)$ by sending e-mail to your classmates.

4. Share a secret (50 to 100 characters) with each student in class by breaking the original secret message into 20-character blocks and enciphering each block.

5. Decipher each incoming message.

6. Hand in a list of deciphered messages to your instructor.

Chapter 9

Finite Fields

Finite fields are called **Galois fields** in honor of the French mathematician Evariste Galois (1811–1832), the first to study them in full generality. Recently there has been a renewed interest in finite fields because of their applications to coding theory and cryptography.

In all the history of science there is no completer example of the triumph of crass stupidity over untamable genius than is afforded by the all too brief life of Evariste Galois.

E. T. Bell

At the age of 20, Galois engaged in a pistol duel at 25 paces. He was shot in the intestines and died a day later. Much of his mathematical legacy can be traced to the ideas that he frantically put on paper the night before the duel.

9.1 The Galois Field GF_p

The simplest example of a Galois field is $\mathbb{Z}_p = \{0, 1, 2, \ldots, p-1\}$ where p is a prime and addition and multiplication are done modulo p. Here is a list of its properties:

Theorem 9.1 *Let p be a prime. Then for all a, b, and c in $\mathbb{Z}_p$ we have*

1.	$a + (b + c) = (a + b) + c$	*Associative law*
2.	$a + b = b + a$	*Commutative law*
3.	$a + 0 = a$	*Additive identity*
4.	$a + m = 0$ *for some m*	*Negatives*

199

5. $a(bc) = (ab)c$ *Associative law*

6. $ab = ba$ *Commutative law*

7. $a1 = a$ *Multiplicative identity*

8. $a(b + c) = ab + ac$ *Distributive law*

9.   *If* $a \neq 0$, *then* $am = 1$ *for some* m *Inverses*

The first eight properties follow from the corresponding properties of $\mathbb{Z}$. The existence of inverses comes from the fact that if p does not divide a, then there is an integer b such that $ab \equiv 1 \pmod{p}$. The element m in property 4 is $-a$. We write the element m in property 9 as a^{-1}.

We can think of GF_p as the set of congruence classes modulo p. The integers that are congruent to a modulo p form the **congruence class** of a, denoted by $[a]$. Note that

$$[a] = \{\ldots, \ a - 3p, \ a - 2p, \ a - p, \ a, \ a + p, \ a + 2p, \ a + 3p, \ \ldots\}$$

The sum of the classes $[a]$ and $[b]$ is $[a + b]$, their product is $[ab]$. That is,

$$[a] + [b] = [a + b]$$
$$[a][b] = [ab]$$

(See Problem 10.)

The idea of a **field** is obtained by abstracting the properties of addition and multiplication in $\mathbb{Z}_p$.

Definition 9.2 *A **field** is a set F with an addition and a multiplication satisfying*

1. $a + (b + c) = (a + b) + c$ *for all* a, b, c *in* F *(Associative law)*

2. $a + b = b + a$ *for all* a, b *in* F *(Commutative law)*

3. *There is an element* 0 *in* F *such that* $a + 0 = a$ *for all* a *in* F *(Additive identity)*

4. $a + x = 0$ *for some* x *(Negatives. We write* $x = -a$.*)*

5. $a(bc) = (ab)c$ *for all* a, b, c *in* F *(Associative law)*

6. $ab = ba$ *for all* a, b *in* F *(Commutative law)*

7. $a1 = a$ *(Multiplicative identity)*

8. $a(b + c) = ab + ac$ *(Distributive law)*

9. *If* $a \neq 0$, *then* $ax = 1$ *for some* x *(Inverses. We write* $x = a^{-1}$.*)*

If we take F to be $\mathbb{Z}_p$ with addition and multiplication modulo p, then we see that $\mathbb{Z}_p$ is a field. This field is also denoted by GF_p, and called the **Galois field** of order p.

The rational numbers $\mathbb{Q}$ form a field as do the real numbers $\mathbb{R}$, and the complex numbers $\mathbb{C}$. These three fields are infinite. We are interested here in finite fields.

Example 9.3 Here are the addition and multiplication tables for GF_7.

+	0	1	2	3	4	5	6
0	0	1	2	3	4	5	6
1	1	2	3	4	5	6	0
2	2	3	4	5	6	0	1
3	3	4	5	6	0	1	2
4	4	5	6	0	1	2	3
5	5	6	0	1	2	3	4
6	6	0	1	2	3	4	5

·	0	1	2	3	4	5	6
0	0	0	0	0	0	0	0
1	0	1	2	3	4	5	6
2	0	2	4	6	1	3	5
3	0	3	6	2	5	1	4
4	0	4	1	5	2	6	3
5	0	5	3	1	6	4	2
6	0	6	5	4	3	2	1

These are called Cayley tables. They are symmetric about the main diagonal because of the commutative laws. Except for the row of zeros in the multiplication table, each nonzero element appears exactly once in each row. This shows that the equations $a + x = b$ and $cx = d$ have unique solutions in GF_7 when $a \neq 0$.

Example 9.4 We can think of GF_5 as the set of congruences classes modulo 5. There are five congruence classes:

$$[0] = \{\ldots, \quad 10, -5, 0, 5, 10, \ldots\}$$
$$[1] = \{\ldots, -9, -4, 1, 6, 11 \ldots\}$$
$$[2] = \{\ldots, -8, -3, 2, 7, 12 \ldots\}$$
$$[3] = \{\ldots, -7, -2, 3, 8, 13 \ldots\}$$
$$[4] = \{\ldots, -6, -1, 4, 9, 14 \ldots\}$$

Note that

$$[1] = [6] = [-4]$$

since each is equal to the set of all integers x such that $x \bmod 5 = 1$.

Example 9.5 The addition and multiplication tables of GF_5 can be written as

+	[0]	[1]	[2]	[3]	[4]
[0]	[0]	[1]	[2]	[3]	[4]
[1]	[1]	[2]	[3]	[4]	[0]
[2]	[2]	[3]	[4]	[5]	[6]
[3]	[3]	[4]	[5]	[6]	[7]
[4]	[4]	[5]	[6]	[7]	[8]

·	[0]	[1]	[2]	[3]	[4]
[0]	[0]	[0]	[0]	[0]	[0]
[1]	[0]	[1]	[2]	[3]	[4]
[2]	[0]	[2]	[4]	[6]	[8]
[3]	[0]	[3]	[6]	[9]	[12]
[4]	[0]	[4]	[8]	[12]	[16]

Since

$$[0] = [5] \qquad\qquad [3] = [8]$$
$$[1] = [6] = [16] \qquad [4] = [9]$$
$$[2] = [7] = [12]$$

these tables can be rewritten as

+	[0]	[1]	[2]	[3]	[4]
[0]	[0]	[1]	[2]	[3]	[4]
[1]	[1]	[2]	[3]	[4]	[0]
[2]	[2]	[3]	[4]	[0]	[1]
[3]	[3]	[4]	[0]	[1]	[2]
[4]	[4]	[0]	[1]	[2]	[3]

·	[0]	[1]	[2]	[3]	[4]
[0]	[0]	[0]	[0]	[0]	[0]
[1]	[0]	[1]	[2]	[3]	[4]
[2]	[0]	[2]	[4]	[1]	[3]
[3]	[0]	[3]	[1]	[4]	[2]
[4]	[0]	[4]	[3]	[2]	[1]

Systems of linear equations can be solved over the finite field GF_p by first solving the system over the rational number field $\mathbb{Q}$ and then reducing the solution modulo p.

Example 9.6 To solve the system

$$26x + 10y + 23z = 22$$
$$18x + 35y + 7z = 17$$
$$7x + 15y + 26z = 30$$

over the field GF_{41},

1. Find the solution

$$x = \frac{87}{449}, \quad y = \frac{481}{2245}, \quad z = \frac{486}{449}$$

to the system of equations over the field Q of rational numbers.

2. Evaluate the solution modulo 41 to get

$$x = 23, \quad y = 38, \quad z = 3$$

3. Substitute the solution back into the system of equations and evaluate modulo 41 to verify that

$$26 \cdot 23 + 10 \cdot 38 + 23 \cdot 3 \bmod 41 = 22$$
$$18 \cdot 23 + 35 \cdot 38 + 7 \cdot 3 \bmod 41 = 17$$
$$7 \cdot 23 + 15 \cdot 38 + 26 \cdot 3 \bmod 41 = 30$$

Let F be a finite field. When necessary, we will denote the 1 of F by 1_F to distinguish it from the integer 1. We can form the sum $1_F + 1_F$, which we denote by $2 \cdot 1_F$, the sum $1_F + 1_F + 1_F$, which we denote by $3 \cdot 1_F$, and so on. As F is finite, we will eventually come across positive integers $m < n$ so that $m \cdot 1_F = n \cdot 1_F$, from which it follows that $(n - m) \cdot 1_F = 0$.

The smallest positive integer c such that $c \cdot 1_F = 0$ is called the **characteristic** of the field F.

The characteristic c of a field is a prime number. Why is this? Suppose $c = ab$. Then $ab \cdot 1_F = 0$, so $(a \cdot 1_F)(b \cdot 1_F) = 0$. It follows that either $a \cdot 1_F = 0$ or $b \cdot 1_F = 0$ (see Problem 2). But c is the smallest positive integer such that $c \cdot 1_F = 0$, so either $a = c$ or $b = c$.

If F is a field of characteristic p, then we can think of GF_p as being contained in F under the correspondence $a \longleftrightarrow a \cdot 1_F$.

For $a \neq 0$ in F, consider the sequence $a, a^2, a^3, \ldots$. As F is finite, there must be positive integers $m < n$ such that $a^m = a^n$. Repeatedly multiplying this equation by the inverse of a, we get $1 = a^{n-m}$. So some power of any nonzero element of F is equal to 1. The least positive integer k such that $a^k = 1$ is called the **order** of a.

Problems 9.1

1. Show that if F is a field, then $a0 = 0$ for any a in F. (Look at $a0 + a0$.)

2. Show that if F is a field, and $ab = 0$, then $a = 0$ or $b = 0$. (Show that if $a \neq 0$, then $b = 0$.) Another way to state this is that if a and b are nonzero, then so is ab.

3. Solve the following system of equations over GF_5:

$$3x + 7y + 6z = 2$$
$$4x + 5y + 3z = 2$$
$$2x + 4y + 5z = 4$$

4. Construct the Cayley tables for GF_{11}.

5. Solve the system

$$3x + 9y + 4z = 8$$
$$2x + 3y + 5z = 2$$
$$5x + 4y + 9z = 10$$

   over GF_{11}.

6. Construct the Cayley tables for GF_{13}.

7. Construct the Cayley tables for GF_{17}.

8. Explain how to locate a^{-1} in the multiplicative Cayley table for GF_p.

9. Explain why the set of integers modulo 10 under addition and multiplication is not a field.

10. Explain why the operations on congruence classes mod p, $[a]+[b] = [a+b]$ and $[a][b] = [ab]$, are well defined. That is, why does the result not depend on the particular choice of a and b from the congruence class?

11. Solve the equation $x^2 + 7x + 10 = 0$ over GF_{11}.

12. Factor the polynomial $3x^3 + x^2 + 11x + 1$ over the field GF_{17}.

9.2   The Ring $GF_p[x]$ of Polynomials

To construct finite fields other than GF_p, we look at polynomials with coefficients in the field GF_p. First, consider polynomials with coefficients in an arbitrary field F. Given two polynomials

$$f(x) = a_0 + a_1 x + a_2 x^2 + \cdots + a_n x^n$$
$$g(x) = b_0 + b_1 x + b_2 x^2 + \cdots + b_m x^m$$

we add them by adding their coefficients. The coefficient of x^i in the sum $f(x) + g(x)$ is $a_i + b_i$, where we set $a_i = 0$ if $i > n$ and $b_i = 0$ if $i > m$.

Multiplication of polynomials is more complicated. The coefficient d_k of x^k in the product $f(x) \cdot g(x)$ is given by

$$d_k = \sum_{i+j=k} a_i b_j$$

that is,

$$d_0 = a_0 b_0$$
$$d_1 = a_0 b_1 + a_1 b_0$$

$$\vdots$$

$$d_{m+n-1} = a_{n-1} b_m + a_n b_{m-1}$$
$$d_{m+n} = a_n b_m$$

This is the product you would get if you simply multiplied $f(x)$ and $g(x)$ using the usual laws.

Example 9.7 The product in $GF_7[x]$ of the two polynomials $2 + 3x + 5x^2$ and $1 + 5x + 3x^2 + 2x^3$ is

$$(2 + 3x + 5x^2)(1 + 5x + 3x^2 + 2x^3)$$
$$= (2 \cdot 1) + (2 \cdot 5 + 3 \cdot 1)x + (2 \cdot 3 + 3 \cdot 5 + 5 \cdot 1)x^2$$
$$+ (2 \cdot 2 + 3 \cdot 3 + 5 \cdot 5)x^3 + (3 \cdot 2 + 5 \cdot 3)x^4 + (5 \cdot 2)x^5$$
$$= 2 + 6x + 5x^2 + 3x^3 + 3x^5$$

The set $F[x]$ of polynomials satisfies almost all of the properties of a field. However, a polynomial of degree greater than zero does not have an inverse. The more general notion is that of **ring**. In a ring you can add, subtract, and multiply. In a field, you can also divide.

Definition 9.8 *A **ring** is a set R with an addition and a multiplication satisfying*

1. $a + (b + c) = (a + b) + c$ for all a, b, c in R

2. $a + b = b + a$ for all a, b in R

3. there is an element 0 in R such that $a + 0 = a$ for all a in R

4. for each a in R there is an element $-a$ in R such that $a + (-a) = 0$

5. $a(bc) = (ab)c$ for all a, b, c in R

6. $a(b + c) = ab + ac$ for all a, b, c in R

   A ring R is a **ring with identity** if

7. there is an element 1 in R such that $1a = a1 = a$ for all a in R

   A ring R is **commutative** if

8. $ab = ba$ for all a, b in R

   A commutative ring with identity is a **field** if

9. for each a in R with $a \neq 0$ there exists an element a^{-1} in R such that $aa^{-1} = 1$

The integers are a commutative ring with identity. The 2×2 matrices over the integers with the usual matrix addition and multiplication are a noncommutative ring with identity. The even integers are a commutative ring without an identity.

One important ring is $\mathbb{Z}_n = \{0, 1, 2, \ldots, n-1\}$ with addition and multiplication modulo n. This ring is a field when n is prime. We will look at $\mathbb{Z}_6$ to see what happens when n is composite.

The multiplication table

×	0	1	2	3	4	5
0	0	0	0	0	0	0
1	0	1	2	3	4	5
2	0	2	4	0	2	4
3	0	3	0	3	0	3
4	0	4	2	0	4	2
5	0	5	4	3	2	1

is the Cayley table for multiplication modulo 6. None of the integers 2, 3, or 4 has an inverse modulo 6, so $\mathbb{Z}_6$ is not a field. Also, equations such as $2x = 1$, $2x = 3$, $2x = 5$ have no solutions.

The set $F[x]$ of polynomials with coefficients in a field F is a commutative ring with identity. It behaves in many ways like the ring $\mathbb{Z}$ of integers.

Definition 9.9 *The **degree** of a nonzero polynomial*

$$a_0 x^0 + a_1 x^1 + a_2 x^2 + \cdots$$

*is the largest m such that $a_m \neq 0$. If $a_m = 1$ for this m, we say that the polynomial is **monic**.*

We write $\deg(p(x))$ for the degree of the polynomial $p(x)$. The polynomial $2x^3 + 3x^2 + 1$ has degree 3 in $\mathrm{GF}_3[x]$ and degree 2 in $\mathrm{GF}_2[x]$. It is not monic in either ring. The polynomial $x^4 + 2x^3 + 3x^2 + 1$ is monic in any ring.

Theorem 9.10 *If F is a field, and $p(x)$ and $q(x)$ are nonzero polynomials in $F[x]$, then $p(x)q(x) \neq 0$ and*

$$\deg(p(x)q(x)) = \deg(p(x)) + \deg(q(x))$$

Proof. Write $p(x) = a_0 + a_1 x + \cdots + a_m x^m$ and $q(x) = b_0 + b_1 x^1 + \cdots + b_n x^n$ with a_m and b_n nonzero. The term of the largest degree in $p(x)q(x)$ is $a_m b_n x^{m+n}$, which is not zero because F is a field. ∎

The division algorithm for polynomials over a field is just like the division algorithm for integers.

Theorem 9.11 (Division Algorithm) *Let F be any field. Given polynomials $a(x)$ and $b(x)$ in $F[x]$ with $b(x) \neq 0$, there are polynomials $q(x)$ and $r(x)$ in $F[x]$, where either $r(x) = 0$ or $\deg(r(x)) < \deg(b(x))$, such that*

$$a(x) = b(x) q(x) + r(x)$$

Proof. If $a(x) = 0$, then take $q(x) = r(x) = 0$. Otherwise write

$$a(x) = a_0 + a_1 x + \cdots + a_m x^m$$
$$b(x) = b_0 + b_1 x + \cdots + b_n x^n$$

with $a_m \neq 0$ and $b_n \neq 0$. If $m < n$, take $q(x) = 0$ and $r(x) = a(x)$. If $m \geq n$ consider the polynomial

$$b(x)\left(a_m b_n^{-1} x^{m-n}\right) = b_0 a_m b_n^{-1} x^{m-n} + b_1 a_m b_n^{-1} x^{m-n+1} + \cdots + b_n a_m b_n^{-1} x^m$$

This is a polynomial of degree m whose last term is $a_m x^m$ which is the same as the last term of $a(x)$. If we subtract it from $a(x)$, we get the polynomial

$$f_1(x) = a(x) - b(x)\left(a_m b_n^{-1} x^{m-n}\right)$$

which is either equal to 0 or has degree at most $m - 1$. So, if we let $q_1(x) = a_m b_n^{-1} x^{m-n}$ then

$$a(x) = b(x) q_1(x) + f_1(x)$$

with $f_1(x) = 0$ or $\deg(f_1(x)) < \deg(a(x))$. If $\deg(f_1(x)) \geq \deg(b(x))$, write $f_1(x) = q_2(x)b(x) + f_2(x)$ with $f_2(x) = 0$ or $\deg(f_2(x)) < \deg(f_1(x))$. Eventually we get

$$a(x) = q_1(x)b(x) + q_2(x)b(x) + \cdots + q_k(x)b(x) + f_k(x)$$

with $f_k(x) = 0$ or $\deg(f_k(x)) < \deg(b(x))$. Taking

$$q(x) = q_1(x) + q_2(x) + \cdots + q_k(x)$$
$$r(x) = f_k(x)$$

we have shown that

$$a(x) = q(x)b(x) + r(x)$$

with $r(x) = 0$ or $\deg(r(x)) < \deg(b(x))$. ∎

The polynomial $q(x)$ is called the **quotient** and the polynomial $r(x)$ is called the **remainder**. It's not hard to see that the quotient and remainder are uniquely determined (see Problem 14). Computer algebra systems will compute the remainder, which it is natural to call $a(x) \bmod b(x)$ as we do for integers.

Definition 9.12 *Let $a(x)$, $b(x)$, and $m(x)$ be polynomials in $\mathbb{Q}[x]$. Then $a(x)$ and $b(x)$ are **congruent modulo** $m(x)$ if $a(x) - b(x) = m(x)k(x)$ for some polynomial $k(x)$ in $\mathbb{Q}[x]$. In this case we write*

$$a(x) \equiv b(x) \bmod m(x)$$

Note that operations can be carried out in $GF_p[x]$ by doing the operations in $\mathbb{Q}[x]$, then reducing the coefficients modulo p.

Example 9.13 To find the quotient and remainder when $x^3 + x + 1$ is divided by $x^2 + x + 1$, we begin by doing long division. This produces

$$\frac{x^3 + x + 1}{x^2 + x + 1} = x - 1 + \frac{x + 2}{x^2 + x + 1}$$

which means that

$$x^3 + x + 1 = (x^2 + x + 1)(x - 1) + (x + 2)$$

In this case we write

$$x^3 + x + 1 \bmod x^2 + x + 1 = x + 2$$

and say that the polynomials $x+2$ and x^3+x+1 are congruent modulo x^2+x+1, that is,

$$x^3 + x + 1 \equiv x + 2 \ \left(\bmod x^2 + x + 1\right)$$

Note that $\deg(x + 2) < \deg(x^2 + x + 1)$.

Example 9.14 To evaluate

$$3x^4 + 5x + 2 \bmod 2x^3 + x^2 + 5$$

in $GF_{11}[x]$, first use long division to rewrite the quotient in the form

$$\frac{3x^4 + 5x + 2}{2x^3 + x^2 + 5} = \frac{3}{2}x - \frac{3}{4} + \frac{\frac{23}{4} - \frac{5}{2}x + \frac{3}{4}x^2}{2x^3 + x^2 + 5}$$

Thus

$$3x^4 + 5x + 2 = \left(\tfrac{3}{2}x - \tfrac{3}{4}\right)\left(2x^3 + x^2 + 5\right) + \tfrac{23}{4} - \tfrac{5}{2}x + \tfrac{3}{4}x^2$$

Reducing the polynomials with rational coefficients modulo 11, we have

$$\tfrac{3}{2}x - \tfrac{3}{4} \bmod 11 = 7x + 2$$
$$\tfrac{23}{4} - \tfrac{5}{2}x + \tfrac{3}{4}x^2 \bmod 11 = 9x^2 + 3x + 3$$

Thus

$$3x^4 + 5x + 2 \bmod 2x^3 + x^2 + 5 = 9x^2 + 3x + 3$$

in $GF_{11}[x]$. As a check, note that

$$\left(3x^4 + 5x + 2\right) - \left(9x^2 + 3x + 3\right) \bmod 11 = 3x^4 + 2x + 10 + 2x^2$$
$$\left(7x + 2\right)\left(2x^3 + x^2 + 5\right) \bmod 11 = 3x^4 + 2x + 10 + 2x^2$$

and hence $2x^3 + x^2 + 5$ divides the difference

$$\left(3x^4 + 5x + 2\right) - \left(9x^2 + 3x + 3\right)$$

modulo 11.

Computer algebra systems have built-in functions that produce a polynomial of smallest degree that is congruent to a given polynomial.

The division algorithm can be used to compute the greatest common divisor of two polynomials $p_1(x)$ and $p_2(x)$, and to write it as a linear combination of $p_1(x)$ and $p_2(x)$.

Theorem 9.15 (Extended Euclidean algorithm) *Let F be any field. Given polynomials a and b in $F[x]$, there are polynomials s and t in $F[x]$ such that*

$$d = sa + tb$$

is the greatest common divisor of a and b.

Proof. If $b = 0$, take $s = 1$ and $t = 0$. Otherwise, writing $p_1 = a$ and $p_2 = b$ we apply the division algorithm repeatedly to get a sequence of polynomials p_1, p_2, p_3, ... :

$$p_1 = p_2 q_1 + p_3 \text{ with } \deg p_3 < \deg p_2$$
$$p_2 = p_3 q_2 + p_4 \text{ with } \deg p_4 < \deg p_3$$
$$p_3 = p_4 q_3 + p_5 \text{ with } \deg p_5 < \deg p_4$$

$$\vdots$$

$$p_{k-3} = p_{k-2} q_{k-3} + p_{k-1} \text{ with } \deg p_{k-1} < \deg p_k 2$$
$$p_{k-2} = p_{k-1} q_{k-2} + p_k \text{ with } \deg p_k < \deg p_{k-1}$$
$$p_{k-1} = p_k q_{k-1}$$

We stop when the remainder is zero. Then $d = p_k$ is the greatest common divisor of $p_1 = a$ and $p_2 = b$. Moreover, d may be written as a linear combination of a and b by repeated substitution, from bottom to top, in the preceding equations. ∎

The procedure is exactly like the extended Euclidean algorithm for the integers (see page 47). See page 141 for examples with rational coefficients.

Every nonzero element u in a finite field K of characteristic p satisfies a monic polynomial in $GF_p[x]$ because some power of u is equal to 1 (See page 203.) Of course 0 satisfies the polynomial x.

Definition 9.16 *Let K be a finite field containing a field F, and 0 an element of K. The* **minimal polynomial** *of 0 over F is the monic polynomial $m(x)$ in $F[x]$ of smallest degree such that $m(0) = 0$.*

Let θ be an element of a finite field K containing a field F. Let $F[\theta]$ denote the set of elements $f(\theta)$ in K where $f(x)$ ranges over $F[x]$. It's clear that $F[\theta]$ is a ring because $F[x]$ is a ring. In fact, $F[\theta]$ is a field because some power of any nonzero element of K is equal to 1. We will show that the order of $F[\theta]$ is q^n where q is the order of F and n is the degree of the minimal polynomial of θ over F.

If $m(x)$ is the minimal polynomial of θ over F, and $f(x)$ is any polynomial in $F[x]$, then we can write $f(x) = s(x)m(x)+r(x)$ where $r(x) = 0$ or $\deg r(x) < \deg m(x) = n$. So $f(\theta) = r(\theta)$. In particular, if $f(\theta) = 0$, then $r(\theta) = 0$, so $r(x) = 0$ because $m(x)$ is the minimal polynomial of θ over F and $\deg r(x) < \deg m(x)$. That is, $m(x)$ divides any polynomial that has θ as a root.

Different polynomials $r(x)$ of degree less than n give different values of $f(\theta)$ because if $r(\theta) = r'(\theta)$, then $r(x) - r'(x)$ is a polynomial of degree less than n satisfied by θ, so must be zero because $m(x)$ is the minimal polynomial of θ. So the number of elements in $F[\theta]$ is the same as the number of polynomials in $F[x]$ of degree less than n. But such a polynomial looks like

$$a_0 + a_1 x + a_2 x^2 + \cdots + a_{n-1}x^{n-1}$$

and there are q choices for each of the coefficients a_i giving a total of q^n polynomials. We can use this result to show that each finite field has order a power of its characteristic, hence has order p^n for some prime p and positive integer n. More generally we show

Theorem 9.17 *If K is a finite field containing a field F, then the order of K is a power of the order of F.*

Proof. Let K have order k and F have order q. If $K = F$ we are done. Otherwise choose an element θ in K but not in F and let $m_\theta(x)$ be its minimal polynomial over F. If $n = \deg m_\theta$, then the field $F[\theta]$ has order q^n. So k is a power of q^n by complete induction (see problem 10) on $k - q$ (we have replaced q by q^n, thus decreasing $k - q$), whence k is a power of q. ∎

Given polynomials $a(x)$ and $b(x) \neq 0$, the division algorithm for polynomials yields the quotient $q(x)$ and remainder $r(x)$. The remainder is given by

$$r(x) = a(x) \bmod b(x)$$

The **cyclic redundancy-check** codes (CRC codes) are based on polynomial remainders. Given a fixed polynomial $g(x)$ of degree n, called the generator polynomial, the codeword for the plaintext polynomial $p(x)$ is $c(x) = x^n p(x) + (x^n p(x) \bmod g(x)) \bmod 2$. The codeword $c(x)$ is transmitted, and received as $d(x)$. If no errors occur in transmission, then $(d(x) \bmod g(x)) \bmod 2 = 0$. If $(d(x) \bmod g(x)) \bmod 2 \neq 0$, then at least one error must have occurred during transmission, and the codeword would need to be retransmitted. In the case of a barcode, the item would be rescanned.

Example 9.18 Given the generator polynomial $g(x) = x^4 + x^3 + 1$ and the plaintext polynomial $p(x) = x^7 + x^5 + x^2 + 1$, the codeword is given by

$$
\begin{aligned}
c(x) &= x^n p(x) + (x^n p(x) \bmod g(x)) \bmod 2 \\
&= x^4 \left(x^7 + x^5 + x^2 + 1\right) + \left(x^4 \left(x^7 + x^5 + x^2 + 1\right) \bmod x^4 + x^3 + 1\right) \bmod 2 \\
&= x^{11} + x^9 + x^6 + x^4 + x^3 + x^2 + x
\end{aligned}
$$

If an error x^k occurs, then the received message is

$$d(x) = \left(x^{11} + x^9 + x^6 + x^4 + x^3 + x^2 + x\right) + x^k$$

Since

$$(d(x) \bmod g(x)) \bmod 2 = x^k$$

is not zero, at least one error is detected. If no errors occur, then

$$\left(x^{11} + x^9 + x^6 + x^4 + x^3 + x^2 + x \bmod x^4 + x^3 + 1\right) \bmod 2 = 0$$

and the plaintext polynomial is the quotient

$$\frac{x^{11} + x^9 + x^6 + x^4 + x^3 + x^2 + x}{x^4 + x^3 + 1} = x^7 + x^5 + x^2 + 1$$

The critical factor in designing a good CRC code is the generator polynomial $g(x)$. One of the most popular CRC codes is CRC-16, which uses the generator polynomial $g(x) = x^{16} + x^{15} + x^2 + 1$. This code

1. detects all single errors.

2. detects all double errors.

3. detects any odd number of errors.

4. detects any error burst of length at most 16; that is, an error of the form $x^k + x^{k+1} + \cdots + x^{k+n-1}$ where $n \leq 16$.

Problems 9.2

1. Compute the sum $(5x^3 + 4x^2 + 3) + (4x^4 + 6x^2 + 3x + 5)$ in $GF_7[x]$.

2. Compute the sum $\left(10x^4 + 8x^3 + 6x + 7\right) + \left(7x^4 + 6x^2 + 8x + 9\right)$ in $GF_{11}[x]$.

3. Compute the product $(5x^3 + 4x^2 + 3)(4x^4 + 6x^2 + 3x + 5)$ in $GF_7[x]$.

4. Compute $\left(10x^4 + 8x^3 + 6x + 7\right)\left(7x^4 + 6x^2 + 8x + 9\right)$ in $GF_{11}[x]$.

5. Compute $4x^4 + x^2 + 3x + 1 \bmod 5x^3 + 4x^2 + 3$ in $GF_7[x]$.

6. Compute $2x^8 + 6x^3 + 4 \bmod 3x^5 + 7x^4 + 2x + 5$ in $GF_{11}[x]$.

7. Factor $x^5 + x^3 + 6x^2 + x + 2$ in $GF_7[x]$.

8. Factor $x^6 + x^5 + x^4 + 4x^3 + 3x^2 + 3x + 3$ in $GF_7[x]$.

9. Use the generator polynomial $g(x) = x^4 + x^3 + 1$ to generate the CRC codeword corresponding to the plaintext $p(x) = x^{11} + x^9 + x^2$.

10. The principle of **complete induction**: Let $P(n)$ be a statement that depends on the positive integer n. If $P(k)$ is true whenever $P(j)$ is true for all $j < k$, then $P(n)$ is true for each positive integer n.

Let $P'(n) = P(1) \, \& \, P(2) \, \& \cdots \& \, P(n)$. Use ordinary induction on the statement $P'(n)$ to prove the principle of complete induction.

11. Show by complete induction that every positive integer is either 1, a prime, or a product of primes.

12. The polynomial

$$
\begin{aligned}
r(x) \quad = \quad & x^{31} + x^{29} + x^{27} + x^{25} + x^{23} + x^{21} + x^{19} + x^{17} + x^{15} + x^{14} + x^{13} \\
& + x^{12} + x^{11} + x^{10} + x^9 + x^8 + x^7 + x^6 + x^5 + x^4 + x^2 + x
\end{aligned}
$$

is received after being encoded by the CRC-16 generator polynomial $g(x) = x^{16} + x^{15} + x^2 + 1$. Test to see whether an error has occurred. If no error is detected, find the plaintext polynomial.

13. The polynomial

$$
\begin{aligned}
r(x) \quad = \quad & x^{31} + x^{29} + x^{27} + x^{25} + x^{23} + x^{21} + x^{20} + x^{18} + x^{17} + x^{15} + x^{14} \\
& + x^{13} + x^{12} + x^{11} + x^{10} + x^9 + x^8 + x^7 + x^6 + x^5 + x^4 + x^2 + x
\end{aligned}
$$

is received after being encoded by the CRC-16 generator polynomial $g(x) = x^{16} + x^{15} + x^2 + 1$. Test to see whether an error has occurred. If no error is detected, find the plaintext polynomial.

14. Suppose $a(x) = q(x)b(x) + r(x)$ and also $a(x) = q'(x)b(x) + r'(x)$ as in the division algorithm. Show that $r(x) = r'(x)$ and $q(x) = q'(x)$. (Start by subtracting one equation from the other, and use the fact that the degrees of the remainders are small.)

9.3   The Galois Field GF_4

As we observed in Section 9.1, there is a finite field GF_p with p elements for each prime p. For $p = 2$, addition and multiplication in GF_2 are given by the tables

+	0	1
0	0	1
1	1	0

×	0	1
0	0	0
1	0	1

For handling data, fields with 2^n elements are convenient because information is often represented as strings of 0s and 1s. The immediate candidate for a field of size 4 is $\mathbb{Z}_4$, but its multiplication table reveals that it is not a field.

×	0	1	2	3
0	0	0	0	0
1	0	1	2	3
2	0	2	0	2
3	0	3	2	1

Note that $2 \times 2 = 0$, so 2 has no inverse.

To construct a field with four elements, we consider $GF_2[x]$ modulo the polynomial $x^2 + x + 1$.

Definition 9.19 *Let F be a field and $p(x)$ a polynomial in $F[x]$ of degree at least 1. We say that $p(x)$ is **reducible** if we can write $p(x) = a(x)b(x)$ where $\deg a(x) \geq 1$ and $\deg b(x) \geq 1$. If $p(x)$ is not reducible, it is said to be **irreducible**. So $p(x)$ is irreducible if whenever $p(x) = a(x)b(x)$, then either $\deg a(x) = 0$ or $\deg b(x) = 0$.*

A polynomial $p(x)$ in $F[x]$ of degree 2 or 3 is reducible exactly when it has a root in F. Indeed, if $p(x)$ is reducible, say $p(x) = a(x)b(x)$ with $\deg a(x) \geq 1$ and $\deg b(x) \geq 1$, then, because $\deg a(x) + \deg b(x) = \deg p(x) \leq 3$, either $a(x)$ or $b(x)$ has degree 1. Thus $p(x)$ has a linear factor, so it has a root in F. On the other hand, suppose that $p(x)$ has a root α in F. By the division algorithm, there are polynomials $q(x)$ and $r(x)$ in $F[x]$ such that

$$p(x) = (x - \alpha) q(x) + r(x)$$

where either $r(x) = 0$ or $\deg r(x) < \deg(x - \alpha) = 1$. So $r(x) = r$ is a constant. Since α is a root of $p(x)$, $0 = p(\alpha) = (\alpha - \alpha)q(\alpha) + r = r$ so $p(x) = (x - \alpha)q(x)$ is reducible.

Consider the polynomial $p(x) = x^2 + x + 1$ in $GF_2[x]$. As $p(0) = 0^2 + 0 + 1 = 1 \neq 0$ and $p(1) = 1^2 + 1 + 1 - 1 \neq 0$, this polynomial has no roots in GF_2, so it is irreducible.

For any polynomial $f(x) \in GF_2[x]$, the remainder $f(x) \bmod (x^2 + x + 1)$ has degree less than 2. Thus each polynomial in $GF_2[x]$ is congruent modulo $x^2 + x + 1$ to a polynomial of degree at most 1—that is, any polynomial in $GF_2[x]$ is congruent, modulo $x^2 + x + 1$, to one of the four polynomials

$$x, \quad x + 1, \quad 0, \quad 1$$

The Galois Field GF_4 as a Set of Linear Polynomials

We use the irreducible polynomial $x^2 + x + 1$ in $GF_2[x]$ to construct a four-element field. The basic idea is to introduce a symbol θ and require that $\theta^2 + \theta + 1 = 0$; that is, that θ be a root of the irreducible polynomial $x^2 + x + 1$. Because $\theta^2 + \theta + 1 = 0$, we can write $\theta^2 = \theta + 1$ (remember that the coefficients are in GF_2 so $-1 = 1$).

Definition 9.20 *The **Galois field** $GF_{2^2} = GF_4$ consists of the expressions $a\theta + b$, where a and b are in GF_2, with the straightforward definition of addition and multiplication defined by setting $\theta^2 = \theta + 1$.*

For example,

$$(\theta + 1) + \theta = 2\theta + 1 = 1$$
$$(\theta + 1)\theta = \theta^2 + \theta = \theta + 1 + \theta = 2\theta + 1 = 1$$

Similar calculations lead to the addition and multiplication tables for GF_4.

+	0	1	θ	$\theta+1$
0	0	1	θ	$\theta+1$
1	1	0	$\theta+1$	θ
θ	θ	$\theta+1$	0	1
$\theta+1$	$\theta+1$	θ	1	0

$\cdot$	0	1	θ	$\theta+1$
0	0	0	0	0
1	0	1	θ	$\theta+1$
θ	0	θ	$\theta+1$	1
$\theta+1$	0	$\theta+1$	1	θ

The multiplication table can be constructed by calculating the matrix product

$$
\begin{pmatrix} 0 \\ 1 \\ \theta \\ \theta+1 \end{pmatrix}
\begin{pmatrix} 0 & 1 & \theta & 1+\theta \end{pmatrix}
=
\begin{pmatrix}
0 & 0 & 0 & 0 \\
0 & 1 & \theta & 1+\theta \\
0 & \theta & \theta^2 & \theta(1+\theta) \\
0 & 1+\theta & \theta(1+\theta) & (1+\theta)^2
\end{pmatrix}
$$

then setting $\theta^2 = \theta + 1$ throughout to get

$$
\begin{pmatrix}
0 & 0 & 0 & 0 \\
0 & 1 & \theta & 1+\theta \\
0 & \theta & 1+\theta & 1 \\
0 & 1+\theta & 1 & \theta
\end{pmatrix}
$$

Notice that setting $\theta^2 = \theta + 1$ amounts to reducing modulo $\theta^2 + \theta + 1$.

We can represent the element $a\theta + b$ in GF_4 by the binary string ab. So θ is represented by 10 and $\theta + 1$ by 11. The addition and multiplication tables become

+	00	01	10	11
00	00	01	10	11
01	01	00	11	10
10	10	11	00	01
11	11	10	01	00

$\cdot$	00	01	10	11
00	00	00	00	00
01	00	01	10	11
10	00	10	11	01
11	00	11	01	10

If, for shorthand, we let $0 = 00$, $1 = 01$, $2 = 10$, and $3 = 11$, then the addition and multiplication tables become

+	0	1	2	3
0	0	1	2	3
1	1	0	3	2
2	2	3	0	1
3	3	2	1	0

$\cdot$	0	1	2	3
0	0	0	0	0
1	0	1	2	3
2	0	2	3	1
3	0	3	1	2

Notice that these tables do *not* give addition and multiplication modulo 4. No matter how the elements are relabeled, the diagonal elements of the addition table would all be the same, which is not the case for $\mathbb{Z}_4$. All nine of the field properties listed in Section 9.1 are satisfied by the addition and multiplication of these tables. This field is the Galois field GF_4.

The Galois Field GF_4 as a Set of Congruence Classes

The Galois field GF_4 can also be defined in terms of congruence classes.

Definition 9.21 *Define a **congruence relation** on $GF_2[x]$ by*

$$f(x) \equiv g(x)$$

if

$$f(x) - g(x) = k(x)(x^2 + x + 1)$$

for some $k(x) \in GF_2[x]$. Let $[f(x)]$ denote the congruence class containing $f(x)$. The Galois field GF_4 is the set of congruence classes with addition and multiplication defined by

$$[f(x)] + [g(x)] = [f(x) + g(x)]$$
$$[f(x)][g(x)] = [f(x)g(x)]$$

Every congruence class $[f(x)]$ contains a unique representative of degree at most 1, namely $f(x) \bmod (x^2 + x + 1)$. The connection with the notation $a\theta \mid b$ is that the element of θ of GF_4 is the congruence class $[x]$ of the polynomial x.

The Galois Field GF_4 as Linear Combinations of a Fixed Root

A third way to define the Galois field GF_4 is to represent the elements of GF_4 by linear polynomials over GF_2, and reduce modulo $x^2 + x + 1$ after multiplying two of these linear polynomials. This amounts to replacing the congruence class $[f(x)]$ by the linear polynomial $f(x) \bmod x^2 + x + 1$.

Definition 9.22 *Let GF_4 be the set $\{0, 1, x, x+1\}$ with addition and multiplication defined by*

+	0	1	x	$x+1$
0	0	1	x	$x+1$
1	1	0	$x+1$	x
x	x	$x+1$	0	1
$x+1$	$x+1$	x	1	0

$\cdot$	0	1	x	$x+1$
0	0	0	0	0
1	0	1	x	$x+1$
x	0	x	$x+1$	1
$x+1$	0	$x+1$	1	x

The Galois Field GF_4 as Powers of a Fixed Root

A fourth way to define the Galois field GF_4 is to let θ be a root of the irreducible polynomial $x^2 + x + 1$ and write each of the nonzero elements of GF_4 as a power of θ:

$$\theta$$
$$\theta^2 = \theta + 1$$
$$\theta^3 = \theta(\theta + 1) = \theta^2 + \theta = \theta + \theta + 1 = 1$$

Definition 9.23 *Let θ be a root of the polynomial $x^2 + x + 1$ and define GF_4 to be the set $\{0, 1, \theta, \theta^2\}$ with addition and multiplication given by*

+	0	1	θ	θ^2
0	0	1	θ	θ^2
1	1	0	θ^2	θ
θ	θ	θ^2	0	1
θ^2	θ^2	θ	1	0

$\cdot$	0	1	θ	θ^2
0	0	0	0	0
1	0	1	θ	θ^2
θ	0	θ	θ^2	1
θ^2	0	θ^2	1	θ

Problems 9.3

1. Verify the addition and multiplication tables for GF_4 by evaluating sums and products of the four expressions 0, 1, θ, and $\theta + 1$ where $\theta^2 = \theta + 1$.

2. Prove that addition and multiplication are commutative in GF_4.

3. Verify that the associative laws for addition and multiplication hold in GF_4.

4. Verify that the distributive law holds in GF_4.

5. Let $h(x) = x^2 + x + 1$. Show that if

$$f_1(x) \equiv f_2(x) \pmod{h(x)}$$

and

$$g_1(x) \equiv g_2(x) \pmod{h(x)}$$

then

$$f_1(x) + g_1(x) \equiv f_2(x) + g_2(x) \pmod{h(x)}$$

and

$$f_1(x)\, g_1(x) \equiv f_2(x)\, g_2(x) \pmod{h(x)}$$

In other words, show that the addition and multiplication in GF_4

$$[f(x)] + [g(x)] = [f(x) + g(x)]$$
$$[f(x)] \cdot [g(x)] = [f(x)\, g(x)]$$

is well defined.

6. If $\theta \in GF_4$ is a root of $x^2 + x + 1$, then $\beta = \theta + 1$ is also a root of $x^2 + x + 1$.

7. Use θ and β in Problem 6 and define a function $f : GF_4 \to GF_4$ by $f(0) = 0$, $f(1) = 1$, $f(\theta) = \beta$, and $f(\theta + 1) = \beta + 1$. Verify that $f(s + t) = f(s) + f(t)$ and $f(st) = f(s) f(t)$ for all $s, t \in GF_4$. (We say that f is the automorphism of GF_4 that interchanges the roots of the polynomial $x^2 + x + 1$.)

8. Let F be any field and let $f(x)$ and $g(x)$ be nonzero polynomials in $F[x]$. Prove the following.

    (a) If $f(x) g(x) \in F$, then $f(x)$ and $g(x)$ are both elements of F, that is, they are polynomials of degree 0.

    (b) If $f(x) = (x - a) q(x) + r(x)$ for some $q(x), r(x) \in F[x]$ with $a \in F$ and $r(x) = 0$ or $\deg(r(x)) = 0$, then $r(x) = f(a) \in F$, that is,

    $$f(x) = (x - a) q(x) + f(a)$$

    (c) The polynomial $x - a$ divides $f(x)$ in $F[x]$ if and only if a is a root of $f(x)$.

9. For the congruence relation on $GF_2[x]$ given in Definition 9.21 on page 215, prove that $f(x) \bmod (x^2 + x + 1)$ is the only polynomial of degree at most 1 in the equivalence class $[f(x)]$.

10. Show that the multiplication table for GF_4 is essentially different from the multiplication table for $\mathbb{Z}_4$. That is, show that no relabeling of $\mathbb{Z}_4$ would create a multiplication table for GF_4.

9.4 The Galois Fields GF_8 and GF_{16}

The elements of a field of size 2^n can be represented in a natural way as binary numbers and easily manipulated by computers and networks. We will look at GF_{2^n} for $n = 3$ and $n = 4$ in this section, and then develop the theory for GF_{p^n} with p an arbitrary prime in the following section.

The Field GF_8

The elements of the field GF_8 can be thought of as polynomials of the form $ax^2 + bx + c$ in $GF_2[x]$. There are two choices for each of the three coefficients, so there are $2 \cdot 2 \cdot 2 = 8$ of these polynomials. Addition is polynomial addition. How do we multiply?

To define the multiplication, we need an irreducible cubic polynomial. The polynomials $x^3 + x^2 + 1$ and $x^3 + x + 1$ are irreducible over GF_2 because neither 0 nor 1 is a root of either polynomial, and any reducible cubic has a linear factor, so has a root in GF_2. We choose the simpler polynomial $x^3 + x + 1$.

We note in passing that $x^8 + x = x(x+1)(x^3+x+1)(x^3+x^2+1)$ in $GF_2[x]$, and the factors on the right are a complete list of all the irreducible polynomials of degrees 1 and 3 over GF_2.

To multiply in GF_8, we take the polynomial product and reduce modulo $x^3 + x + 1$. So we can form the multiplication table by taking the matrix product

$$
\begin{pmatrix}
1 \\
x \\
x+1 \\
x^2 \\
x^2+1 \\
x^2+x \\
x^2+x+1
\end{pmatrix}
\begin{pmatrix}
1 & x & 1+x & x^2 & x^2+1 & x+x^2 & x^2+x+1
\end{pmatrix}
$$

$$
= \left(
\begin{array}{ccc}
1 & x & x+1 \\
x & x^2 & x^2+x \\
x+1 & x^2+x & x^2+2x+1 \\
x^2 & x^3 & x^3+x^2 \\
x^2+1 & x^3+x & x^3+x+x^2+1 \\
x^2+x & x^3+x^2 & x^3+2x^2+x \\
x^2+x+1 & x^3+x^2+x & x^3+2x^2+2x+1
\end{array}
\right.
$$

$$
\begin{array}{cc}
x^2 & x^2+1 \\
x^3 & x^3+x \\
x^3+x^2 & x^3+x+x^2+1 \\
x^4 & x^4+x^2 \\
x^4+x^2 & x^4+2x^2+1 \\
x^4+x^3 & x^4+x^3+x^2+x \\
x^4+x^3+x^2 & x^4+x^3+2x^2+x+1
\end{array}
$$

$$
\left.
\begin{array}{cc}
x^2+x & x^2+x+1 \\
x^3+x^2 & x^3+x^2+x \\
x^3+2x^2+x & x^3+2x^2+2x+1 \\
x^4+x^3 & x^4+x^3+x^2 \\
x^4+x^3+x^2+x & x^4+x^3+2x^2+x+1 \\
x^4+2x^3+x^2 & x^4+2x^3+2x^2+x \\
x^4+2x^3+2x^2+x & x^4+2x^3+3x^2+2x+1
\end{array}
\right)
$$

This is a 7×7 matrix whose entries are the polynomial products of all pairs of quadratic polynomials with coefficients in $\{0, 1\}$.

Now reduce each polynomial in the matrix modulo the irreducible polynomial $x^3 + x + 1$ to get

$$
\begin{pmatrix}
1 & x & 1+x & x^2 \\
x & x^2 & x+x^2 & -1-x \\
1+x & x+x^2 & 1+2x+x^2 & x^2-1-x \\
x^2 & -1-x & x^2-1-x & -x-x^2 \\
x^2+1 & -1 & x^2 & -x \\
x+x^2 & x^2-1-x & 2x^2-1 & -2x-x^2-1 \\
x^2+x+1 & x^2-1 & x+2x^2 & -2x-1 \\
\end{pmatrix}
$$

$$
\begin{pmatrix}
x^2+1 & x+x^2 & x^2+x+1 \\
-1 & x^2-1-x & x^2-1 \\
x^2 & 2x^2-1 & x+2x^2 \\
-x & -2x-x^2-1 & -2x-1 \\
1+x^2-x & -1-x & x^2-x \\
-1-x & -3x-2 & x^2-2-2x \\
x^2-x & x^2-2-2x & -1+2x^2-x \\
\end{pmatrix}
$$

Here we have left the coefficients as integers. To get the multiplication table for GF_8, we reduce this matrix modulo 2 to get

$$
\begin{pmatrix}
1 & x & x+1 & x^2 \\
x & x^2 & x^2+x & x+1 \\
x+1 & x^2+x & x^2+1 & x^2+x+1 \\
x^2 & x+1 & x^2+x+1 & x^2+x \\
x^2+1 & 1 & x^2 & x \\
x^2+x & x^2+x+1 & 1 & x^2+1 \\
x^2+x+1 & x^2+1 & x & 1 \\
\end{pmatrix}
$$

$$
\begin{pmatrix}
x^2+1 & x^2+x & x^2+x+1 \\
1 & x^2+x+1 & x^2+1 \\
x^2 & 1 & x \\
x & x^2+1 & 1 \\
x^2+x+1 & x+1 & x^2+x \\
x+1 & x & x^2 \\
x^2+x & x^2 & x+1 \\
\end{pmatrix}
$$

The multiplication table of GF_8 can be represented in binary form by using the string of coefficients of each polynomial. The addition table is just mod 2 vector addition (see page 104), so we can compute the sums directly. Addition and multiplication by zero have been omitted from these tables because they are trivial: $0 + a = a + 0 = a$ and $0a = a0 = 0$.

+	001	010	011	100	101	110	111
001	000	011	010	101	100	111	110
010	011	000	001	110	111	100	101
011	010	001	000	111	110	101	100
100	101	110	111	000	001	010	011
101	100	111	110	001	000	011	010
110	111	100	101	010	011	000	001
111	110	101	100	011	010	001	000

·	001	010	011	100	101	110	111
001	001	010	011	100	101	110	111
010	010	100	110	011	001	111	101
011	011	110	101	111	100	001	010
100	100	011	111	110	010	101	001
101	101	001	100	010	111	011	110
110	110	111	001	101	011	010	100
111	111	101	010	001	110	100	011

We can also write these tables in decimal (or hex) notation.

+	1	2	3	4	5	6	7
1	0	3	2	5	4	7	6
2	3	0	1	6	7	4	5
3	2	1	0	7	6	5	4
4	5	6	7	0	1	2	3
5	4	7	6	1	0	3	2
6	7	4	5	2	3	0	1
7	6	5	4	3	2	1	0

·	1	2	3	4	5	6	7
1	1	2	3	4	5	6	7
2	2	4	6	3	1	7	5
3	3	6	5	7	4	1	2
4	4	3	7	6	2	5	1
5	5	1	4	2	7	3	6
6	6	7	1	5	3	2	4
7	7	5	2	1	6	4	3

Suppose that θ is a root in GF_8 of the polynomial x^3+x+1. Then $\theta^3+\theta+1 = 0$, so that $\theta^3 = \theta + 1$. Thus

$$\theta^4 = \theta\,(\theta + 1) = \theta^2 + \theta$$
$$\theta^5 = \theta\left(\theta^2 + \theta\right) = \theta^3 + \theta^2 = \theta + 1 + \theta^2$$
$$\theta^6 = \theta\left(\theta + 1 + \theta^2\right) = \theta^2 + \theta + \theta^3 = \theta^2 + \theta + \theta + 1 = \theta^2 + 1$$
$$\theta^7 = \theta\left(\theta^2 + 1\right) = \theta^3 + \theta = \theta + 1 + \theta = 1$$

So each nonzero element of GF_8 is a power of θ. This leads to a representation for GF_8 that makes both multiplication and addition easy.

Power Notation	Polynomial Notation	Binary Notation	Decimal (Hex) Notation
0	0	000	0
θ^0 $(=\theta^7)$	1	001	1
θ^1	θ	010	2
θ^2	θ^2	100	4
θ^3	$\theta+1$	011	3
θ^4	$\theta^2+\theta$	110	6
θ^5	$\theta^2+\theta+1$	111	7
θ^6	θ^2+1	101	5

Table 9.1 $GF_{2^3} = GF_8$

To add, convert field elements to polynomial or binary notation and use polynomial or vector addition modulo 2. To multiply, convert to powers of θ and add exponents, with exponents added modulo 7. For example, $3 \cdot 5 = \theta^3 \cdot \theta^6 = \theta^9 = \theta^2 = 100 = 4$.

Notice that 2 is a root of the polynomial $q(x) = x^3 + x + 1$ because

$$2^3 + 2 + 1 = 3 + 2 + 1 = (011) + (010) + (001) = (000) = 0$$

In fact, each field element is a root of the polynomial $x^8 + x$ because $0^8 + 0 = 0$ and $(\theta^i)^8 = (\theta^i)^7\theta^i = (\theta^7)^i\theta^i = 1^i\theta^i = \theta^i$. Also,

$$x^8 + x = x(x+1)(x^3 + x^2 + 1)(x^3 + x + 1)$$

so every field element is a root of exactly one of the factors on the right.

Suppose θ is a element of GF_8 such that $\theta^3 + \theta + 1 = 0$ and $u = \theta + 1$, then

$$\begin{aligned}
u^3 + u^2 + 1 &= (\theta+1)^3 + (\theta+1)^2 + 1 \\
&= \theta^3 + 4\theta^2 + 5\theta + 3 \\
&= \theta^3 + \theta + 1 \\
&= 0
\end{aligned}$$

Thus $u = \theta + 1$ is a root of the polynomial $x^3 + x^2 + 1$. This polynomial is irreducible over GF_2 since it is of degree 3 and has no roots in GF_2. Thus it is the minimal polynomial of $\theta + 1$. The following table shows the minimal polynomial for each element of the Galois field GF_8.

Power Notation	Polynomial Notation	Binary Notation	Decimal Notation	Minimal Polynomial
0	0	000	0	x
2^0	1	001	1	$x + 1$
2^1	θ	010	2	$x^3 + x + 1$
2^2	θ^2	100	4	$x^3 + x + 1$
2^3	$\theta + 1$	011	3	$x^3 + x^2 + 1$
2^4	$\theta^2 + \theta$	110	6	$x^3 + x + 1$
2^5	$\theta^2 + \theta + 1$	111	7	$x^3 + x^2 + 1$
2^6	$\theta^2 + 1$	101	5	$x^3 + x^2 + 1$

Table 9.2   Minimal polynomials of elements in GF_8

The Field GF_{16}

The field GF_{2^4} (also written GF_{16}) will be especially important to us. To construct this field, we need an irreducible polynomial of degree 4 over GF_2. One way to construct one is to generate all of the polynomials of degrees 1 through 3, then make a list of all polynomials of degree 4 that can be written as products of those polynomials, then pick a polynomial of degree 4 that is not on the list. This is essentially the sieve of Eratosthenes applied to polynomials.

Another way is to factor the polynomial $x^{16} - x = x^{16} + x$ (remember that $-1 = 1$ modulo 2). Factoring $x^{16} + x$ modulo 2 in a computer algebra system yields

$$x^{16} + x = x\,(x + 1)\left(x^2 + x + 1\right)\left(x^4 + x^3 + 1\right)$$
$$\cdot \left(x^4 + x + 1\right)\left(x^4 + x^3 + x^2 + x + 1\right)$$

There are three irreducible polynomials of degree 4 in this factorization. We pick $x^4 + x + 1$ because of its simple form.

The method we used to construct GF_8 also works for GF_{16}, but the matrices are a bit large to display here.

Before giving a multiplication table for GF_{16}, we construct a table of equivalent representations for elements in GF_{16}. Let θ be a root of $x^4 + x + 1$, so $\theta^4 + \theta + 1 = 0$. Rewrite this as $\theta^4 = \theta + 1$. Thus θ^4 can be replaced by $\theta + 1$. Multiplying both sides of the equation $\theta^4 = \theta + 1$ by θ gives $\theta^5 = \theta^2 + \theta$. Similar calculations can be used to calculate the second column of Table 9.3. (The rest of this table is left as problem 8.)

Power	Polynomial	Binary	Hex	Decimal	Minimal Polynomial
0	0	0000	0	0	x
θ^0	1	0001	1	1	$x + 1$
θ^1	θ	0010	2	2	$x^4 + x + 1$
θ^2	θ^2	0100	4	4	$x^4 + x + 1$
θ^3	θ^3	1000	8	8	$x^4 + x^3 + x^2 + x + 1$
θ^4	$\theta + 1$	0011	3	3	$x^4 + x + 1$
θ^5	$\theta^2 + \theta$	0110	6	6	$x^2 + x + 1$
θ^6	$\theta^3 + \theta^2$	1100	C	12	$x^4 + x^3 + x^2 + x + 1$
$\vdots$	$\vdots$	$\vdots$	$\vdots$	$\vdots$	$\vdots$

Table 9.3   Representations of elements in GF_{16}

The binary representation is the string of coefficients in the polynomial representation. That is, the binary representation of $c_3\theta^3 + c_2\theta^2 + c_1\theta + c_0$ is $(c_3c_2c_1c_0)_2$.

The minimal polynomials of elements of GF_{16} can be computed using the identities $(a + b)^2 = a^2 + b^2$ for $a, b \in GF_{16}$, and $x^2 - x$ for $x \in GF_2$.

If u in GF_{16} is a root of a polynomial with coefficients in GF_2, then u^2 is a root of the same polynomial. To see this, consider a polynomial $c_3x^3 + c_2x^2 + c_1x + c_0$ such that $c_3u^3 + c_2u^2 + c_1u + c_0 = 0$. Then

$$0 = \left(c_3u^3 + c_2u^2 + c_1u + c_0\right)^2$$
$$= c_3^2\left(u^3\right)^2 + c_2^2\left(u^2\right)^2 + c_1^2u^2 + c_0^2$$
$$= c_3\left(u^2\right)^3 + c_2\left(u^2\right)^2 + c_1u^2 + c_0$$

Thus u^2 is also a root of $c_3x^3 + c_2x^2 + c_1x + c_0$. Repeating this argument on u^2 shows that $\left(u^2\right)^2 = u^4$ is also a root of this polynomial, and so on.

Now let θ be a root of the defining polynomial $x^4 + x + 1$ for GF_{16}. We have seen that θ, θ^2, $\left(\theta^2\right)^2 = \theta^4$, and $\left(\theta^4\right)^2 = \theta^8$ are all roots of $x^4 + x + 1$. This means that $x - \theta$, $x - \theta^2$, $x - \theta^4$, and $x - \theta^8$ are all factors of $x^4 + x + 1$. And, sure enough, the product of these four polynomials, computed in $GF_{16}[x]$, is

$$(x - \theta)\,(x - \theta^2)\,(x - \theta^4)\,(x - \theta^8) = x^4 + x + 1$$

the minimal polynomial for each of θ, θ^2, θ^4, and θ^8.

Similarly, the minimal polynomial for θ^3 is

$$\left(x + \theta^3\right)\left(x + \left(\theta^3\right)^2\right)\left(x + \left(\theta^3\right)^4\right)\left(x + \left(\theta^3\right)^8\right) = x^4 + x^3 + x^2 + x + 1$$

The powers of $\theta^2 + \theta$ are

$$\left(\theta^2 + \theta\right)^2 = \theta^2 + \theta + 1$$
$$\left(\theta^2 + \theta + 1\right)^2 = \theta^2 + \theta$$

so repeated squaring yields two distinct powers. Hence the minimal polynomial of $\theta^2 + \theta$ is

$$\left(x + \left(\theta^2 + \theta\right)\right)\left(x + \left(\theta^2 + \theta\right)^2\right) = x^2 + x + 1$$

In the following examples, we compute the product

$$\left(\alpha^2 + \alpha + 1\right)\left(\alpha^3 + \alpha^2 + 1\right)$$

in three different ways.

Example 9.24 First, we expand the product to get

$$\left(\theta^2 + \theta + 1\right)\left(\theta^3 + \theta^2 + 1\right) = \theta^5 + 2\theta^4 + 2\theta^2 + 2\theta^3 + \theta + 1$$

Then reduce the product modulo the irreducible polynomial $q\left(\theta\right) = \theta^4 + \theta + 1$ to produce

$$\theta^5 + 2\theta^4 + 2\theta^2 + 2\theta^3 + \theta + 1 \bmod q\left(\theta\right) = -1 + \theta^2 + 2\theta^3 - 2\theta$$

Then reduce the coefficients modulo 2 to get the reduced product

$$-1 + \theta^2 + 2\theta^3 - 2\theta \bmod 2 = \theta^2 + 1$$

It follows that in GF_{16},

$$\left(\theta^2 + \theta + 1\right)\left(\theta^3 + \theta^2 + 1\right) = \theta^2 + 1$$

Example 9.25 On the other hand, we have $\theta^{15} = 1 = \theta^0$, so multiplication can be computed by adding exponents of θ modulo 15. Since $\theta^2 + \theta + 1 = \theta^{10}$ and $\theta^3 + \theta^2 + 1 = \theta^{13}$, it follows that

$$\left(\theta^2 + \theta + 1\right)\left(\theta^3 + \theta^2 + 1\right) = \theta^{10}\theta^{13}$$
$$= \theta^{(10+13)\bmod 15}$$
$$= \theta^8 = \theta^2 + 1$$

Example 9.26 Expanding using the distributive law gives

$$\left(\theta^2 + \theta + 1\right)\left(\theta^3 + \theta^2 + 1\right) = \theta^2\left(\theta^3 + \theta^2 + 1\right) + \theta\left(\theta^3 + \theta^2 + 1\right)$$
$$+ \left(\theta^3 + \theta^2 + 1\right)$$

Multiplication by θ yields

$$\theta\left(\theta^3 + \theta^2 + 1\right) = \theta^4 + \theta^3 + \theta$$
$$= \left(\theta + 1\right) + \left(\theta^3 + \theta\right) = \theta^3 + 1$$

since $\theta^4 = \theta + 1$ and $\theta + \theta = 0$; hence

$$\theta^2 (\theta^3 + \theta^2 + 1) = \theta (\theta^3 + 1) = \theta^4 + \theta$$
$$= (\theta + 1) + \theta = 1$$

Therefore,

$$(\theta^2 + \theta + 1) (\theta^3 + \theta^2 + 1) - (1) + (\theta^3 + 1) + (\theta^3 + \theta^2 + 1)$$
$$= \theta^2 + 1 = 5$$

Problems 9.4

1. Find $(\theta^2 + 1)(\theta^2 + \theta)$ in GF_8 by first converting to powers of θ and then expressing the result as a polynomial in θ.

2. Calculate $(\theta^3 + \theta^2)(\theta^3 + \theta + 1)$ in GF_{16} by using polynomial arithmetic.

3. Verify the products $E \cdot E = B$ and $8 \cdot A = F$ (written in hex notation) in GF_{16}.

4. Solve the system

$$3u + Dv = 7$$
$$Eu + 5v = 9$$

(written in hex notation) for elements u and v of GF_{16}.

5. Find an irreducible polynomial of degree 5 in $GF_2[x]$.

6. Use your polynomial $q(x)$ from Problem 5 to construct an element of order 31 in GF_{32}.

7. Factor $x^5 + x + 1$ over GF_2.

8. Complete the representations of elements in GF_{16} begun in Table 9.3. page 223.

9. Verify that $x^3 + x^2 + 1$ and $x^3 + x + 1$ are the only irreducible polynomials of degree 3 in $GF_2[x]$.

10. Verify that $x^4 + x^3 + 1$, $x^4 + x + 1$, and $x^4 + x^3 + x^2 + x + 1$ are the only irreducible polynomials of degree 4 in $GF_2[x]$.

9.5 The Galois Field GF_{p^n}

We have seen how the ring $GF_2[x]$ can be used to construct a field with 2^n elements. It turns out that any finite field has p^n elements for some prime p, a fact we will prove later. Moreover, there is essentially only one field with p^n elements.

In this section we construct a field with p^n elements for an arbitrary prime p. We do this by constructing a field K containing GF_p in which the polynomial $x^{p^n} - x$ factors into linear factors

$$x^{p^n} - x = \prod_{i=1}^{p^n} (x - \theta_i)$$

where $\theta_1, ..., \theta_{p^n}$ are in K. It turns out that the elements $\theta_1, ..., \theta_{p^n}$ are all different and form a field with p^n elements. This field is called the **splitting field** of $x^{p^n} - x$ over GF_p.

We will show how to construct a field of order p^m from an irreducible polynomial of degree m over GF_p. Later we will show how to construct such an irreducible polynomial. The idea of how to construct a field from a field and an irreducible polynomial is due to Kronecker.

Theorem 9.27 *Let F be a field and $h(x)$ an irreducible polynomial of degree m in $F[x]$. Define K to be the set of congruence classes of polynomials in $F[x]$ modulo $h(x)$. Let θ be the congruence class of the polynomial x, so if $f(x) \in F[x]$, then the congruence class of $f(x)$ is $f(\theta)$.*

Define multiplication in K by setting $f(\theta) g(\theta) = r(\theta)$ where

$$r(x) = (f(x) g(x)) \bmod h(x).$$

This turns the set K into a field. If F is a finite field of order q, then the order of K is q^m.

Proof. Note that each element of K can be written uniquely as $r(\theta)$ where $r(x)$ is a polynomial in $F[x]$ of degree less than m. Indeed, if $f(\theta)$ is an element of K, then $r(x) = f(x) \bmod h(x)$. Notice also that $f(\theta) = 0$ exactly when $h(x)$ divides $f(x)$.

We need to show that nonzero elements of K have inverses. Let $r(\theta)$ be a nonzero element of K where $\deg r(x) < m$. As $h(x)$ is irreducible and does not divide $r(x)$, there exist polynomials $s(x)$ and $t(x)$ with coefficients in F so that

$$s(x) r(x) + t(x) h(x) = 1$$

by the extended Euclidean algorithm for polynomials. (See page 209.) As $h(\theta) = 0$, we have

$$s(\theta) r(\theta) = s(\theta) r(\theta) + t(\theta) q(\theta) = 1$$

so $r(\theta)$ has an inverse in K.

All of the other field properties follow from the corresponding properties of $F[x]$.

Finally, if the order of F is q, then there are exactly q^m polynomials

$$a_{m-1} x^{m-1} + a_{m-2} x^{m-2} + \cdots + a_2 x^2 + a_1 x + a_0$$

of degree less than m in $F[x]$, because there are exactly q choices for each of the m coefficients. So K has order q^m.   ∎

As a corollary, we will show that every nonconstant polynomial over a field has a root in some extension field.

Corollary 9.28 *Let $f(x)$ be a nonconstant polynomial with coefficients in a field F. Then there is a field K containing F that also contains a root of $f(x)$.*

Proof. If $f(x)$ is not irreducible, then $f(x)$ has a factor $f_1(x)$ in $F[x]$ such that $1 \leq \deg f_1(x) < \deg f(x)$. Either $f_1(x)$ is irreducible, or $f_1(x)$ has a factor $f_2(x)$ such that $1 \leq \deg f_2(x) < \deg f_1(x)$. At each step the degree of the polynomial $f_i(x)$ gets smaller, so you will eventually find an irreducible factor $q(x)$ of $f(x)$.

Use $q(x)$ to construct the field K as congruence classes modulo $q(x)$ of polynomials over F. Let θ be the element of K that is the congruence class of the polynomial x. Then $f(\theta) = 0$ because $q(x)$ is a factor of $f(x)$ in $F[x]$. ∎

Theorem 9.29 *There is a field K containing GF_p so that $x^{p^n} - x$ factors into linear factors over K, that is,*

$$x^{p^n} - x = (x - \theta_1)(x - \theta_2)\cdots(x - \theta_{p^n})$$

*with each θ_i in K. Moreover, the set of roots $\theta_1, \theta_2, \ldots, \theta_{p^n}$ of $x^{p^n} - x$ is itself a field, called a **splitting field** of $x^{p^n} - x$.*

Proof. We can construct K by repeating the construction of the corollary, starting with the polynomial $x^{p^n} - x$ over the field GF_p. At each stage we will have written the polynomial $x^{p^n} - x = x\left(x^{p^n - 1} - 1\right)$ as a product of linear factors times a polynomial $g(x)$ of degree greater than 1 over some field F. We apply the corollary to get a field F' containing F in which $g(x)$ has a root θ, so we can write $g(x) = (x - \theta)h(x)$ where the degree of $h(x)$ is smaller than the degree of $g(x)$. Thus we have written $x^{p^n} - x$ as a product of linear factors times $h(x)$, and we then apply the corollary to the polynomial $h(x)$ over the field F'.

To see that the set of roots $\theta_1, \theta_2, \ldots, \theta_{p^n}$ of $x^{p^n} - x$ is itself a field, note that if θ is a root of $x^{p^n} - x$, then $\theta^{p^n} - \theta = 0$ —that is, $\theta^{p^n} = \theta$.. Thus if θ_i, θ_j are roots of $x^{p^n} - x$, then

$$(\theta_i + \theta_j)^{p^n} - (\theta_i + \theta_j) = \theta_i^{p^n} + \theta_j^{p^n} - (\theta_i + \theta_j) = 0$$

and

$$(\theta_i \theta_j)^{p^n} - \theta_i \theta_j = \theta_i^{p^n} \theta_j^{p^n} - \theta_i \theta_j = 0$$

Also, if $\theta \neq 0$, then

$$(\theta^{-1})^{p^n} = \left(\theta^{p^n}\right)^{-1} = \theta^{-1}$$

So the set of roots of $x^{p^n} - x$ is a field. If there are no multiple roots, then this field has exactly p^n elements, as desired. To see that there are no multiple roots, let $m = p^n$. As $x^m - x = x\left(x^{m-1} - 1\right)$ it is clear that 0 is not a multiple root. Suppose r is a nonzero root of $x^m - x$. Then $r^m = r$ so $r^{m-1} = 1$. It follows that

$$x^m - x = (x - r)\left(x^{m-1} + rx^{m-2} + r^2 x^{m-3} + \cdots + r^{m-2}x\right)$$

The issue is whether r is a root of the second factor on the right. But if we plug r into that polynomial we get $(m - 1)\, r^{m-1} = (m - 1) \cdot 1_K = -1_K \neq 0$ because m is divisible by p.   ∎

As another application of the division algorithm, we show that a polynomial of degree n has at most n distinct roots. The key fact is

Theorem 9.30 *An element θ of a field is a root of a polynomial $f(x)$ if and only if $f(x)$ is divisible by the polynomial $x - \theta$.*

Proof. Indeed, the division algorithm tells us that we can write

$$f(x) = q(x)(x - \theta) + r(x)$$

where either $r(x) = 0$ or $\deg r(x) < \deg(x - \theta) = 1$. So $r(x)$ is a constant r, and

$$f(\theta) = q(\theta)(\theta - \theta) + r = r$$

is equal to 0 exactly when $f(x) = q(x)(x - \theta)$.   ∎

Corollary 9.31 *Let F be a field and $s(x)$ a polynomial of degree n in $F[x]$. Then $s(x)$ has at most n distinct roots in F.*

Proof. Suppose θ is a root of $s(x)$. From the above theorem, $s(x) = q(x)(x - \theta)$ for some polynomial $q(x)$. If β is a root of $s(x)$ different from θ, then $0 = s(\beta) = q(\beta)(\beta - \theta)$ and $\beta - \theta \neq 0$ implies that $q(\beta) = 0$. But, by $\deg q(x) = n - 1$, so by induction on the degree, the polynomial $q(x)$ has at most $n - 1$ distinct roots in F. Thus $s(x)$ has at most n distinct roots: θ and at most $n - 1$ more.∎

Problems 9.5

1. Find an irreducible polynomial of degree 2 in $GF_5[x]$.

2. Find an element of order 24 in GF_{25}.

3. Show that if $a^n = 1$ and $a^{n/p} \neq 1$ for each prime divisor p of n, then a is an element of order n.

4. Assume that $q(x)$ is an irreducible polynomial in $F[x]$ for some field F, and assume that $q(x) \mid r(x)p(x)$, where $r(x)$, $p(x) \in F[x]$ with $\deg(r(x)) < \deg(q(x))$. Prove that $q(x) \mid p(x)$.

5. A polynomial $q(x)$ of degree 2 or 3 is irreducible in $F[x]$ if and only if $q(x)$ has no roots in F. Use this fact to verify that the polynomials $x^2 + 1$, $x^2 + x + 2$ and $x^2 + 2x + 2$ are irreducible in $GF_3[x]$.

6. Use the technique from Problem 5 to find an irreducible polynomial of degree 3 in $GF_3[x]$.

7. Find an element of order 26 in GF_{27}, using the irreducible polynomial from Problem 6.

8. Compute $(2x^2 + x + 2)(x^2 + 2)$ in GF_{27}, using the irreducible polynomial from Problem 6.

9. Factor $x^{27} - x$ over GF_3.

10. Do the following.

   (a) Construct the Cayley tables for GF_3.

   (b) Find all polynomials of degree 2 that are irreducible in $GF_3[x]$. (*Hint:* The only monic irreducible polynomials of degree 1 are x, $x + 1$ and $x + 2$, and reducible polynomials of degree 2 must be products of these.)

   (c) Factor $x^9 - x$ modulo 3 into irreducible factors in $GF_3[x]$.

11. Analyze the Galois field GF_9.

   (a) Complete the following table for powers $y^3, ..., y^8$ where $y = x + 1$.

Power	Product	Expand	mod $x^2 + 1$	mod 3
y	$x + 1$	$x + 1$	$x + 1$	$x + 1$
y^2	$(x + 1)(x + 1)$	$x^2 + 2x + 1$	$2x$	$2x$

   Powers of y

   (b) Create addition and multiplication tables using the representation of elements from the far right column in the preceding table, and the base 3 shorthand notation $ax + b = (ab)_3$ (e.g., $2x + 0 = 20$).

   (c) Convert the tables to decimal (or hex).

12. Prove that $x^{p^m} - x \mid x^{p^n} - x$ in $\mathbb{Q}[x]$ if and only if $m \mid n$.

9.6   The Multiplicative Group of GF_{p^n}

In Chapter 4, we looked at permutation ciphers. The permutations of the set $\{0, 1, ..., n - 1\}$, with the product of two permutations being their composition, is called the **symmetric group** of degree n. (See Problem 2 at the end of this section.) In order to show that there are irreducible polynomials of every degree in $GF_p[x]$, we need some facts about a different kind of group—a commutative group. The general definition of a group is

Definition 9.32 *A **group** is a set G with one binary operation, often indicated simply by juxtaposition, such that*

1. $a(bc) = (ab)c$ for all a, b, c in G

2. there is an element 1 in G such that $a1 = 1a = a$ for all a in G

3. for each a in G, there is an element a^{-1} in G such that $aa^{-1} = a^{-1}a = 1$.

   *The group is **commutative** if*

4. $ab = ba$ for all a, b in G

   *The group is **cyclic** if*

5. *there is an element $g \in G$ such that every element of G can be written as g^m for some integer m, positive, negative, or zero. Here the convention is that $g^{-m} = \left(g^{-1}\right)^m$ and $g^0 = 1$.*

From the definition of a field (page 200) we see that the nonzero elements of a field F form a commutative group under multiplication. This group is called the **multiplicative group** of the field F, and is denoted F^*.

Let a be an element of a finite group G, and consider the list of positive powers of a

$$a = a^1, a^2, a^3, a^4, \ldots$$

As G is finite, this list must have duplicates, so there are positive integers $j < k$ such that $a^j = a^k$. Multiplying both sides of this equation by $\left(a^{-1}\right)^j$ gives

$$1 = a^j \left(a^{-1}\right)^j = a^k \left(a^{-1}\right)^j = a^{k-j}$$

so there is a positive integer $t = k - j$ such that $a^t = 1$. The smallest such positive integer is called the **order** of a, and is denoted $o(a)$. Note that the list of positive powers of a repeats after $a^{o(a)} = 1$, that is, it looks like

$$a, a^2, a^3, \ldots, a^{o(a)-1}, 1, \ a, a^2, a^3, \ldots, a^{o(a)-1}, 1, \ a, a^2, \ldots$$

From this it is clear that $a^n = 1$ exactly when $o(a)$ divides n. This is a very useful fact.

The subset $\langle a \rangle = \left\{a, a^2, a^3, \ldots, a^{o(a)-1}, 1\right\}$ of G is itself a group (see Problem 4). A subset of G that is also a group is called a **subgroup**. If G is a finite group, the number of elements of G is called the **order** of G, denoted $o(G)$. Note that $o(a) = o(\langle a \rangle)$, so the two uses of the word "order" are consistent.

Let S be a subgroup of a finite group G. For each element $a \in G$, the set $aS = \{as \mid s \in S\}$ has the same number of elements as S. Moreover, if $a, b \in G$, then aS and bS are either disjoint or equal. To see this, suppose that aS and bS have an element c in common. Then $c = as = bt$ for some $s, t \in S$. From $a = bts^{-1}$ we get $aS \subseteq bS$, and from $b = ast^{-1}$ we get $bS \subseteq aS$, so $aS = bS$.

Thus either $aS \cap bS = \varnothing$ or $aS = bS$. Let $a_1 S, a_2 S, \ldots, a_m S$ be a complete list of the distinct sets aS. Then

$$G = a_1 S \cup a_2 S \cup \cdots \cup a_m S$$

because every element a of G is in the set aS, which must be equal to one of the sets in the list. Thus $o(G) = o(S) \times m$, so $o(S)$ divides $o(G)$. We have just proved

Theorem 9.33 (Lagrange's Theorem) *If S is a subgroup of the finite group G, then the order of S divides the order of G.*

In particular, if $S = \langle u \rangle$, then we get

Theorem 9.34 *If a is an element of a finite group G, the order of a divides the order of G.*

Let F be a field of order $m = p^n$. Then the multiplicative group F^* of F has $m - 1$ elements, so the order of each element of F^* divides $m - 1$. Thus each of the $m - 1$ elements in F^* is a root of the polynomial $x^m - 1$ from which we can conclude that

$$x^{m-1} - 1 = \prod_{i-1}^{m-1} (x - r_i)$$

where $r_1, r_2, \ldots, r_{m-1}$ are all the elements of F^*. More generally, if n divides $m - 1$, then $x^n - 1$ has exactly n roots in F. Why is that? That's because if n divides $m - 1$, then the polynomial $x^n - 1$ divides the polynomial $x^{m-1} - 1$ (see Problem 7) so $x^n - 1$ is a product of factors $x - r_i$ where i ranges over some subset of $\{1, 2, \ldots, m - 1\}$.

So F^* is a finite commutative group, and for each divisor n of the order of F^*, there are exactly n solutions in F^* to the equation $x^n = 1$. We want to show that such a group is cyclic.

Theorem 9.35 *If G is a finite commutative group, and for each divisor n of the order of G, there are exactly n solutions in G to the equation $x^n = 1$, then G is cyclic.*

Proof. We first show that the theorem is true if the order of G is a prime power, say p^k. It suffices to show there is an element in G of order p^k. As the order of any element of G divides p^k, if an element a in G does not have order p^k, then $a^{p^{k-1}} = 1$. But there are at most p^{k-1} solutions to the equation $x^{p^{k-1}} = 1$, and there are p^k elements of G, so some element of G must have order p^k.

Now suppose that p^k divides the order of G. The equation $x^{p^k} - 1$ has exactly p^k solutions in G and these p^k solutions clearly form a group H because if we multiply two of them, we get another one. Moreover, for each divisor n of p^k there are exactly n solutions in G to the equation $x^n = 1$, and these solutions lie in H. But we proved the theorem for groups of order a power of a prime, so H is cyclic.

Finally, write the order of G as a product of prime powers $p_i^{k_i}$ for distinct primes p_i. We have just seen that there is an element a_i in G of order $p_i^{k_i}$ for each i. The product of these elements will have order equal to the product of the numbers $p_i^{k_i}$ which is the order of G. Why is that?

It suffices to show that if α is an element of order c and β is an element of order d, and c and d are relatively prime, then $o(\alpha\beta) = cd$. For that, see Problem 5. ∎

In particular, the multiplicative group of a finite field is cyclic: there is an element ω such that every nonzero element in the field can be written in the form ω^k.

Definition 9.36 *A generator of the multiplicative group of GF_{p^n} is called a **primitive** element. A polynomial of degree n over $GF(p)$ is a **primitive polynomial** if it is the minimum polynomial of a primitive element in GF_{p^n}.*

You have seen many examples of the following lemma.

Lemma 9.37 *The minimal polynomial of an element of GF_{p^n} is irreducible.*

Proof. Let $\theta \in GF_{p^n}$, and let $m_\theta(x)$ be its minimal polynomial—that is, $m_\theta(x)$ is the monic polynomial of smallest degree with coefficients in GF_p such that $m_\theta(\theta) = 0$. Suppose $m_\theta(x) = a(x)b(x)$ for some $a(x), b(x) \in F_p[x]$. Then $m_\theta(\theta) = a(\theta)b(\theta) = 0$ implies that either $a(\theta) = 0$ or $b(\theta) = 0$. If $a(\theta) = 0$, $\theta(x)$ must have degree equal to the degree of $m_\theta(x)$, since $m_\theta(x)$ is the minimal polynomial of θ. Thus $b(x)$ has degree 0, that is, it is a constant. Similarly, if $b(\theta) = 0$ then $a(x)$ is a constant. Thus $m_\theta(x)$ is irreducible. ∎

We now show that the minimal polynomial over GF_p of any primitive element of GF_{p^n} is an irreducible polynomial of degree n.

Theorem 9.38 *Let p be a prime and n a positive integer. Then there is a monic irreducible polynomial in $GF_p[x]$ of degree n.*

Proof. Let ω be a primitive element of GF_{p^n}, and let $f(x)$ be its minimal polynomial over GF_p. From a previous lemma, we know that $f(x)$ is irreducible. We want to show that $f(x)$ has degree n.

Let $m = \deg f(x)$. The set of elements $g(\omega)$, where $g(x)$ ranges over polynomials with coefficients in GF_p, forms a field F containing ω. As ω is a primitive element of GF_{p^n}, the field F must be all of GF_{p^n}. But the order of F is p^m, so $m = n$. ∎

Thus Kronecker's construction will construct GF_{p^n} for every prime p and positive integer n.

We want to show that any two finite fields of order p^n are essentially the same. We do this by showing

Theorem 9.39 *If $f(x)$ is an irreducible polynomial of degree n over GF_p, then $f(x)$ divides $x^{p^n} - x$.*

Proof. By Kronecker's construction, there is a field of order p^n that contains an element θ such that $f(\theta) = 0$. As $f(x)$ is irreducible, it is the minimal polynomial of θ. As $\theta^{p^n} - \theta = 0$, the polynomial $f(x)$ must divide $x^{p^n} - x$. ∎

Now suppose K is any finite field of order p^n, and let $f(x)$ be a fixed irreducible polynomial of degree n over GF_p. We know that K consists exactly of the roots of $x^{p^n} - x$ and that $f(x)$ divides $x^{p^n} - x$. So there is an element θ of K whose minimal polynomial is $f(x)$. Therefore $GF_p[\theta]$ has order p^n, hence is equal to K. But $GF_p[\theta]$ can be thought of as Kronecker's construction from $f(x)$ and GF_p.

Example 9.40 To construct GF_{49}, we need an irreducible polynomial of degree 2 in $GF_7[x]$. We will show that $x^2 + 2$ has no roots in GF_7 and is therefore irreducible. We will also find a primitive root in GF_{49}. Define $g(x) = x^2 + 2$. Then

$$g(0) = 2 \quad g(1) = 3 \quad g(2) = 6 \quad g(3) = 4$$
$$g(4) = 4 \quad g(5) = 6 \quad g(6) = 3$$

As none of these values is zero, $x^2 + 2$ is irreducible over GF_7. To find a primitive root, recall that the order of an element of GF_{49} must be a divisor of 48. The divisors of 48 are

$$1, 2, 3, 4, 6, 8, 12, 16, 24, 48$$

First try the element $\theta = [x]$ which has the property that $\theta^2 = -2$. Then $\theta^4 = 4$, and $\theta^{12} = 4^3 = 64 = 1$, so θ has order at most 12. Undaunted, we go on and try the element $\beta = \theta + 1$.

$$
\begin{aligned}
\beta^2 &= (\theta + 1)^2 = \theta^2 + 2\theta + 1 = 2\theta - 1 = 2\beta - 3 \\
\beta^4 &= \left(\beta^2\right)^2 = (2\beta - 3)^2 = 4\beta^2 - 5\beta + 2 = 4(2\beta - 3) - 5\beta + 2 = 3\beta - 3 \\
\beta^8 &= \left(\beta^4\right)^2 = (3\beta - 3)^2 = 2\beta^2 - 4\beta + 2 = 2(2\beta - 3) - 4\beta + 2 = -4 \\
\beta^{16} &= \left(\beta^8\right)^2 = (-4)^2 = 2 \\
\beta^{24} &= \beta^8 \beta^{16} = (-4)\,2 = -1
\end{aligned}
$$

So β has order 48 in GF_{49}. That's because neither β^{16} nor β^{24} is equal to 1, but every proper divisor of 48 divides either 16 or 24. So β is a primitive element.

The polynomial $x^{49} - x$ factors over GF_7 into a 28-term product of irreducible polynomials:

$$
\begin{aligned}
x^{49} - x = {} & x\,(x + 1)\,(x + 2)\,(x + 3)\,(x + 4)\,(x + 5)\,(x + 6) \\
& \cdot (x^2 + 1)\,(x^2 + 2)\,(x^2 + 4)\,(x^2 + 5x + 3)\,(x^2 + 5x + 2) \\
& \cdot (x^2 + 2x + 3)\,(x^2 + 6x + 3)\,(x^2 + 6x + 6)\,(x^2 + 4x + 1) \\
& \cdot (x^2 + 3x + 5)\,(x^2 + 2x + 2)\,(x^2 + 3x + 1)\,(x^2 + x + 6) \\
& \cdot (x^2 + x + 4)\,(x^2 + 4x + 5)\,(x^2 + x + 3)\,(x^2 + 3x + 6) \\
& \cdot (x^2 + 2x + 5)\,(x^2 + 5x + 5)\,(x^2 + 6x + 4)\,(x^2 + 4x + 6)
\end{aligned}
$$

The order of every nonzero element of GF_{49} divides 48 and so is a root of $x^{48} - 1$.

So every element of GF_{49} is a root of one of the irreducible polynomials listed in the above factorization of $x^{49} - x$. The irreducible polynomial satisfied by an element γ in GF_{49} is called the minimal polynomial of γ.

Different irreducible polynomials of degree n result in addition and multiplication tables that look slightly different, but you can always relabel the elements to make the tables look exactly the same.

Problems 9.6

1. Verify that in the field GF_{13}, there are exactly $\varphi(d)$ elements of order d for each divisor d of 12.

2. Show that the set of permutations of the set $\mathbb{Z}_n$ together with the product $\sigma\tau$ obtained by first applying σ, then applying τ, is a (noncommutative) group. (See page 76.)

3. Let a be an element of a Group G. Show that the set

$$(a) = \{a^n \mid n \text{ is an integer}\}$$

is a subgroup of G.

4. Let a be an element of a finite group G. Show that the set

$$\langle a \rangle = \left\{ 1, a, a^2, a^3, ..., a^{o(a)-1} \right\}$$

is a subgroup of G.

5. Let a, b be elements of a finite group G.

   (a) If $a^n = 1$, show that $o(a) \mid n$.
   (b) If $o(a)$ and $o(b)$ are relatively prime and $ab = ba$, show that $o(ab) = o(a) o(b)$.

6. Verify that in the field GF_{17}, there are exactly $\varphi(d)$ elements of order d for each divisor d of 16.

7. Show that if n divides $m - 1$, then the polynomial $x^n - 1$ divides $x^{m-1} - 1$.

8. Show that $x^{20} - 1$ divides $x^{100} - 1$.

9. Find a primitive element of GF_{31}.

10. Find a primitive element of GF_{27} in terms of a root θ of the irreducible polynomial $x^3 - x - 1$.

9.7 Random Number Generators

In this section we will see how the field GF_p gives a way to generate random number sequences, or at least sequences that look like random number sequences.

To get a feel for a random number sequence, consider the following experiment. An *icosahedron* (see Figure 9.1) is a Platonic solid with 20 faces, each an equilateral triangle.

Figure 9.1   Icosahedron

Label the faces with the digits $0, 1, 2, \ldots, 9$ in such a way that opposite faces share the same label. Our intuition tells us that if we roll such an icosahedron 100 times and record the sequence of digits on the top face, then the resulting sequence should be random. The digits

$$
\begin{bmatrix}
2 & 6 & 5 & 9 & 0 & 1 & 5 & 3 & 3 & 8 \\
4 & 6 & 2 & 6 & 1 & 7 & 9 & 4 & 6 & 7 \\
5 & 8 & 9 & 7 & 4 & 5 & 0 & 3 & 6 & 1 \\
3 & 7 & 2 & 3 & 4 & 0 & 9 & 3 & 4 & 4 \\
8 & 9 & 0 & 0 & 5 & 8 & 8 & 5 & 1 & 5 \\
1 & 7 & 8 & 7 & 5 & 0 & 2 & 5 & 6 & 0 \\
5 & 6 & 8 & 1 & 8 & 5 & 7 & 0 & 1 & 6 \\
9 & 1 & 5 & 8 & 4 & 6 & 8 & 2 & 8 & 1 \\
9 & 2 & 1 & 3 & 9 & 4 & 4 & 0 & 6 & 0 \\
8 & 8 & 4 & 6 & 6 & 4 & 4 & 2 & 5 & 7
\end{bmatrix}
$$

were generated in this manner, with the first digit in the upper left corner and going from left to right, one row at a time.

The histogram

Digit	Frequency	
0	████████████	(10)
1	████████████	(10)
2	███████	(7)
3	███████	(7)
4	█████████████	(12)
5	██████████████	(13)
6	█████████████	(12)
7	████████	(8)
8	██████████████	(13)
9	████████	(8)

shows how often each digit appears.

We would expect longer sequences to be more evenly balanced. The *law of large numbers* says that the frequency of each digit should approach 10% (under ideal conditions such as that the icosahedron is perfectly cut and balanced). Furthermore, each two-digit pair (such as a 5 followed by a 7) should appear roughly 1% of the time. We could make similar statements for strings of length three, four, five, and so forth.

Here is another method for generating a sequence of 100 digits. Consider the prime $p = 101$. The matrix

$$
\begin{bmatrix}
26 & 70 & 2 & 52 & 39 & 4 & 3 & 78 & 8 & 6 \\
55 & 16 & 12 & 9 & 32 & 24 & 18 & 64 & 48 & 36 \\
27 & 96 & 72 & 54 & 91 & 43 & 7 & 81 & 86 & 14 \\
61 & 71 & 28 & 21 & 41 & 56 & 42 & 82 & 11 & 84 \\
63 & 22 & 67 & 25 & 44 & 33 & 50 & 88 & 66 & 100 \\
75 & 31 & 99 & 49 & 62 & 97 & 98 & 23 & 93 & 95 \\
46 & 85 & 89 & 92 & 69 & 77 & 83 & 37 & 53 & 65 \\
74 & 5 & 29 & 47 & 10 & 58 & 94 & 20 & 15 & 87 \\
40 & 30 & 73 & 80 & 60 & 45 & 59 & 19 & 90 & 17 \\
38 & 79 & 34 & 76 & 57 & 68 & 51 & 13 & 35 & 1
\end{bmatrix}
$$

is composed of the powers of 26 modulo 101. Since all of the entries are distinct, we know that 26 is a primitive element of GF_{101}. Each new entry is constructed by multiplying the previous entry by 26 and reducing the result modulo 101.

Reduce these numbers modulo 10 to get the digits

$$
\begin{bmatrix}
6 & 0 & 2 & 2 & 9 & 4 & 3 & 8 & 8 & 6 \\
5 & 6 & 2 & 9 & 2 & 4 & 8 & 4 & 8 & 6 \\
7 & 6 & 2 & 4 & 1 & 3 & 7 & 1 & 6 & 4 \\
1 & 1 & 8 & 1 & 1 & 6 & 2 & 2 & 1 & 4 \\
3 & 2 & 7 & 5 & 4 & 3 & 0 & 8 & 6 & 0 \\
5 & 1 & 9 & 9 & 2 & 7 & 8 & 3 & 3 & 5 \\
6 & 5 & 9 & 2 & 9 & 7 & 3 & 7 & 3 & 5 \\
4 & 5 & 9 & 7 & 0 & 8 & 4 & 0 & 5 & 7 \\
0 & 0 & 3 & 0 & 0 & 5 & 9 & 9 & 0 & 7 \\
8 & 9 & 4 & 6 & 7 & 8 & 1 & 3 & 5 & 1
\end{bmatrix}
$$

It turns out that "random" is a difficult term to define precisely. However, here are two criteria that might be used to describe a random sequence of digits:

1. There is no obvious pattern.

2. Each digit should appear about one-tenth of the time.

The sequence $\{6, 0, 2, 2, 9, 4, 3, 8, 8, 6, \ldots\}$ of 100 digits from the previous example satisfies the first criterion fairly well and satisfies the second exactly.

We can also use a seed to change the starting point. Using the seed 53, we multiply 26 by 53 and reduce modulo 101 to get the starting entry 65. Each new entry can then be generated by multiplying the previous entry by 26 and reducing the result modulo 101. This results in the numbers

$$
\begin{bmatrix}
65 & 74 & 5 & 29 & 47 & 10 & 58 & 94 & 20 & 15 \\
87 & 40 & 30 & 73 & 80 & 60 & 45 & 59 & 19 & 90 \\
17 & 38 & 79 & 34 & 76 & 57 & 68 & 51 & 13 & 35 \\
1 & 26 & 70 & 2 & 52 & 39 & 4 & 3 & 78 & 8 \\
6 & 55 & 16 & 12 & 9 & 32 & 24 & 18 & 64 & 48 \\
36 & 27 & 96 & 72 & 54 & 91 & 43 & 7 & 81 & 86 \\
14 & 61 & 71 & 28 & 21 & 41 & 56 & 42 & 82 & 11 \\
84 & 63 & 22 & 67 & 25 & 44 & 33 & 50 & 88 & 66 \\
100 & 75 & 31 & 99 & 49 & 62 & 97 & 98 & 23 & 93 \\
95 & 46 & 85 & 89 & 92 & 69 & 77 & 83 & 37 & 53
\end{bmatrix}
$$

If we try to extend this sequence, the next number is $26 \cdot 53 \bmod 101 = 65$, which is a repeat of the first number in the sequence. The **cycle length** of this sequence is 100: the sequence repeats after every 100 iterations.

This is an example of a **pseudorandom number generator**. Good pseudorandom number generators have long cycles.

In order to get long cycle lengths, start with the largest n-digit prime p and find a primitive element in GF_p. Convenient primes include $9\,999\,999\,967$, $99\,999\,999\,977$, and $999\,999\,999\,989$.

To show that $84\,906\,326$ is a primitive element of $GF_{999\,999\,999\,989}$, we could decide to examine the first $999\,999\,999\,988$ powers to see if they are all distinct (as we did previously to show that 26 was a primitive element of GF_{101}). However, there is an easier way. Factorization yields

$$999\,999\,999\,988 = 2^2 11 \times 124\,847 \times 182\,041$$

To show that $84\,906\,326$ is primitive, it is enough to show that

$$84\,906\,326^{999\,999\,999\,988/p} \bmod 999\,999\,999\,989 \neq 1$$

for each prime p that divides $999\,999\,999\,988 = 2^2 11 \times 124\,847 \times 182\,041$. Indeed, we see that

$$84906326^{999\,999\,999\,988/2} \bmod 999\,999\,999\,989 = 999\,999\,999\,988$$
$$84906326^{999\,999\,999\,988/11} \bmod 999\,999\,999\,989 = 744\,525\,462\,020$$
$$84906326^{999\,999\,999\,988/124\,847} \bmod 999\,999\,999\,989 = 498\,590\,406\,481$$
$$84906326^{999\,999\,999\,988/182\,041} \bmod 999\,999\,999\,989 = 958\,213\,467\,498$$

Thus $84\,906\,326$ is a primitive element of $GF_{999\,999\,999\,989}$.

A seed can be generated manually or using an internal function such as a clock. Starting with the seed $1\,973\,647$ and primitive element $84\,906\,326$, we get the sequence

$$1973\,647 \cdot 84\,906\,326 \bmod 999\,999\,999\,989 = 575\,115\,592\,759$$
$$575\,115\,592\,759 \cdot 84\,906\,326 \bmod 999\,999\,999\,989 = 7016\,033\,906$$
$$7016\,033\,906 \cdot 84\,906\,326 \bmod 999\,999\,999\,989 = 662\,056\,442\,111$$

The cycle length is $999\,999\,999\,988$. Every integer n in the range $1 \leq n \leq 999\,999\,999\,988$ will appear exactly once before the entire sequence repeats. (The only 12-digit integers that do not appear are the eleven integers $999\,999\,999\,989$, $999\,999\,999\,990, \ldots, 999\,999\,999\,999$.)

Several computer algebra systems use primitive elements modulo a large prime to generate pseudorandom number sequences.

Problems 9.7

1. Let p be the largest three-digit prime. Show that 525 is a primitive element of GF_p. Use the seed 789 to generate the first 10 terms of a pseudorandom number sequence.

2. Show that if a is a primitive element in GF_p, then so is a^k for any integer k such that $k \perp p - 1$.

3. In this section, we showed that $84\,906\,326$ is a primitive element of GF_p, where $p = 999\,999\,999\,989$. What is the average gap between primitive elements? (How hard is it actually to find a primitive element?)

4. A total of n darts are thrown at a rectangular grid with k small rectangles of equal size. (Each dart hits the grid somewhere; each rectangle has an equal chance of being hit.) Calculate the expected number of untouched small rectangles.

5. An unfinished window composed of a 4×4 grid of small square pieces of glass was left lying horizontally in the path of a hail storm, where 16 hail stones land inside the grid and break some of the glass squares. What is the expected number of unbroken squares? Use a pseudorandom number generator to simulate this problem with 1000 repetitions. What is the average number of unbroken squares? What was the smallest number of unbroken squares? The largest number of unbroken squares?

6. Six distinct baseball cards are distributed randomly, one per box of cereal. You must have all six. Use a random number generator to write a computer simulation to determine how many boxes you should expect to purchase before your collection is complete. What is the average if you do 10 000 simulations? What is the largest number of boxes required? What is the smallest number of boxes required?

7. Write a computer simulation to estimate the volume of a sphere of radius 1. The volume of a cube of edge 2 is $2^3 = 8$. By symmetry, generate points (x, y, z) where each coordinate is between 0 and 1 and test to see if $x^2 + y^2 + z^2 < 1$. Multiply the ratio of points inside to the total number of points by 8 to approximate the volume of a sphere. Compare your approximation with the exact value $4\pi/3 \approx 4.188\,790\,205$.

8. Write a computer simulation to estimate the volume of a four-dimensional sphere of radius 1. The volume of a four-dimensional cube of edge 2 is $2^4 = 16$. By symmetry, generate points (w, x, y, z) where each coordinate is between 0 and 1 and test to see if $w^2 + x^2 + y^2 + z^2 \leq 1$. Multiply the ratio of points inside to the total number of points by 16 to approximate the volume of a four-dimensional sphere.

9. Write a computer simulation to estimate the expected length of an interval assuming the endpoints are chosen uniformly in the unit interval.

10. Write a computer simulation to estimate the expected area of a rectangle with sides parallel to the coordinate axes inside the unit square.

11. Write a computer simulation to estimate the expected area of a square with sides parallel to the coordinate axes inside the unit square.

12. Write a computer simulation to estimate the expected area of a square (possibly rotated) chosen at random inside the unit square.

Chapter 10

Error-Correcting Codes

In this chapter we will develop codes that will correct multiple errors. Codes based on polynomial multiplication were developed independently by R. C. Bose and Ray-Chaudhuri (1960) and Hocquenghem (1959). They are known as **BCH codes**. A **Reed-Solomon code** is a particular type of BCH code that has many applications.

The one-dimensional bar codes listed in Chapter 5 use a lot of space for the amount of information they contain. A number of two-dimensional bar codes have more recently been developed to pack more information into small spaces. Typically, these codes use aggressive error-correction algorithms such as Hamming (page 120), BCH (page 242), and Reed-Solomon (page 258).[1]

Datastrip code appeared as **Softstrip** in early personal computer magazines as a method for downloading simple computer programs.

Figure 10.1   Datastrip code

Data density can range between 20 and 150 bytes per square centimeter, depending on the printing technology used to produce the strips. Parity bits on each line are used for error correction.

MaxiCode was developed by United Parcel Service in 1992 in order to provide more information on its packages.

---

[1]See http://www.adams1.com/pub/russadam/stack.html for a description of additional two-dimensional bar codes.

241

Figure 10.2   MaxiCode

A one-inch square symbol can contain approximately 100 ASCII characters. Error-correction symbologies are sufficient to recover information on a symbol with damage on 25% of its surface area.

PDF 417 (Portable Data File) is a stacked symbology that is in the public domain (see Figure 10.3).[2]

Figure 10.3   PDF 417

Error correction and detection can be set at various levels. The Reed-Solomon error-correction algorithm uses the fact that 3 is a primitive root of 1 modulo the prime 929.

An important application of PDF 417 is its use as a substitute for stamps on letters and packages (see Figure 10.4).

Figure 10.4   PC postage

10.1   BCH Codes

In a **polynomial code**, we identify a word with a polynomial $a(x)$, then multiply by a fixed polynomial $g(x)$ to get the codeword $g(x)a(x)$. The polynomial $g(x)$ is called the generator polynomial.

For example, if the word is a binary word $(a_r, a_{r-1}, \ldots, a_2, a_1, a_0)$, we form the polynomial

$$a(x) = a_r x^r + a_{r-1} x^{r-1} + \cdots + a_2 x^2 + a_1 x + a_0$$

---

[2]See http://www.linux.org/apps/AppId_2658.html for a free PDF 417 generator.

in $GF_2[x]$. The corresponding codeword polynomial is

$$a(x)g(x) \bmod 2 = \sum c_i x^i$$

If $\deg y(x) - s$, then the codeword is the word $(c_n, c_{n-1}, \ldots, c_2, c_1, c_0)$, where $n = r + s$.

A particularly efficient class of polynomial codes are the BCH codes. Here is how to construct a generator polynomial for a BCH code over GF_p that corrects at least t errors. See the reference [23, Theorem 75] for a proof.

Theorem 10.1 (Bose Chaudhuri Hocquenghem) *Let θ be a primitive element in GF_{p^n}, let $m_i(x)$ be the minimal polynomial of θ^i over GF_n for $i = 1, 2, \ldots, 2t$, and set*

$$g(x) = \text{lcm}\,[m_1(x), m_2(x), \ldots, m_{2t}(x)].$$

If $\deg g(x) = k$, then for any polynomial $a(x)$ of degree at most $p^n - k - 2$, the weight of the codeword $a(x)g(x)$ is at least $d = 2l + 1$, so at least t errors can be corrected.

Example 10.2 We will use GF_{16} to construct a code that will correct double errors. Let $q(x) = x^4 + x + 1$, and let α be a root of $q(x)$. Thus $\alpha^4 + \alpha + 1 = 0$. Since

$$\alpha^3 \neq 1$$
$$\alpha^5 - \alpha + \alpha^2 \neq 1$$
$$\alpha^{15} = 1$$

it follows that $o(\alpha) = 15$ and hence α is a primitive element of GF_{16}. We have that $m_1(x) = q(x) = x^4 + x + 1$. Notice that in GF_2 we have $(a + b)^2 = a^2 + 2ab + b^2 = a^2 + b^2$. From this it follows that

$$(a_1 + a_2 + \cdots + a_n)^2 = a_1^2 + a_2^2 + \cdots + a_n^2$$

In particular,

$$0 = 0^2 = (\alpha^4 + \alpha + 1)^2 = (\alpha^4)^2 + \alpha^2 + 1 = (\alpha^2)^4 + \alpha^2 + 1$$

and hence $m_1(\alpha^2) = 0$, which means that $m_2(x) = m_1(x)$. Similarly, $m_1(\alpha^4) = 0$ and hence $m_4(x) = m_1(x)$. In fact, $m_1(x) = m_2(x) = m_4(x) = m_8(x)$.

To calculate $m_3(x)$, we use again the observation that $g(\beta) = 0$ implies $g(\beta^2) = 0$. It follows that the zeros of $m_3(x)$ are α^3, α^6, α^{12}, and $\alpha^{24} = \alpha^9$ (because $\alpha^{15} = 1$). Thus,

$$m_3(x) = (x - \alpha^3)(x - \alpha^6)(x - \alpha^9)(x - \alpha^{12})$$
$$= x^4 + (\alpha^3 + \alpha^6 + \alpha^9 + \alpha^{12})x^3 + (\alpha^9 + \alpha^{12} + 1 + 1 + \alpha^3 + \alpha^6)x^2$$
$$+ (\alpha^3 + \alpha^6 + \alpha^9 + \alpha^{12})x + 1$$
$$= x^4 + (\alpha^3 + \alpha^6 + \alpha^9 + \alpha^{12})x^3 + (\alpha^3 + \alpha^6 + \alpha^9 + \alpha^{12})x^2$$
$$+ (\alpha^3 + \alpha^6 + \alpha^9 + \alpha^{12})x + 1$$

Calculating, we have

$$\alpha^3 + \alpha^6 + \alpha^9 + \alpha^{12} = \alpha^3 + (\alpha^3 + \alpha^2) + (\alpha^3 + \alpha) + (\alpha^3 + \alpha^2 + \alpha + 1)$$
$$= 1$$

and hence

$$m_3(x) = x^4 + x^3 + x^2 + x + 1$$

It follows that

$$g(x) = \text{lcm}\,(m_1(x), m_2(x), m_3(x), m_4(x))$$
$$= (x^4 + x + 1)(x^4 + x^3 + x^2 + x + 1) \bmod 2$$
$$= x^8 + x^7 + x^6 + x^4 + 1$$

Since $g(x)$ has degree 8, it follows that plaintext polynomials can have degree at most 6. In particular, if

$$(a_6, a_5, a_4, a_3, a_2, a_1, a_0)_2 = (1, 0, 1, 1, 0, 0, 1)_2$$

then $a(x) = x^6 + x^4 + x^3 + 1$ and

$$a(x)g(x) = x^{14} + x^{13} + x^{10} + x^9 + x^3 + 1$$

and hence the codeword is transmitted as

$$(1, 1, 0, 0, 1, 1, 0, 0, 0, 0, 0, 1, 0, 0, 1)_2$$

Example 10.3 With a bit more effort we can construct a three-error-correcting code. We proceed as in the previous example, but add in the polynomial $m_5(x)$. To compute $m_5(x)$, we note that if α^5 is a root of $m_5(x)$, then so is α^{10}. Since $\alpha^{20} = \alpha^5$ (why?) we can stop there and write

$$m_5(x) = (x - \alpha^5)(x - \alpha^{10})$$
$$= x^2 + (\alpha^5 + \alpha^{10})x + 1$$
$$= x^2 + x + 1$$

The last equality holds because

$$\alpha^5 + \alpha^{10} \bmod 2 = (\alpha^2 + \alpha) + (\alpha^2 + \alpha + 1) \bmod 2 = 1$$

Thus,

$$g(x) = \text{lcm}\,(m_1(x), m_2(x), m_3(x), m_4(x), m_5(x), m_6(x))$$
$$= (x^4 + x + 1)(x^4 + x^3 + x^2 + x + 1)(x^2 + x + 1)$$
$$= x^{10} + x^8 + x^5 + x^4 + x^2 + x + 1$$

The downside of this example is that the plaintext polynomial must be of degree at most 4, so although triple error correction is possible, the code carries with it a lot of overhead. We are always on the lookout for codes that require only a few extra check bits but that correct a high rate of errors.

In order to get an improvement of the information rate, we need to work with larger fields.

Example 10.4 We will construct a four-error-correcting code, using the field GF_{64}. To construct GF_{64}, we need an irreducible polynomial of degree 6 in $GF_2[x]$. The naive guess is $q(x) = x^6 + x + 1$. If we take α to be a root of $q(x)$, then α has order 63 since

$$\alpha^7 \bmod \alpha^6 + \alpha + 1 = \alpha^2 + \alpha$$
$$\alpha^{21} \bmod \alpha^6 + \alpha + 1 = \alpha^5 + \alpha^4 + \alpha^3 + \alpha + 1$$
$$\alpha^{63} \bmod \alpha^6 + \alpha + 1 = 1$$

and hence $q(x) = x^6 + x + 1$ is indeed irreducible since α is primitive. We have $m_1(x) = m_2(x) = m_4(x) = m_8(x) = m_{16}(x) = m_{32}(x)$. The minimal polynomial of α^3 is given by

$$m_3(x) = (x - \alpha^3)(x - \alpha^6)(x - \alpha^{12})(x - \alpha^{24})(x - \alpha^{48})(x - \alpha^{33})$$

since $\alpha^{63} = 1$. Note that

$(m_3(x) \bmod q(\alpha)) \bmod 2$

$= \left(9206x - 123\,182x^2 + 584\,160 - 37\,032x^3 + 3790x^4 - 46x^5\right) \alpha^5$

$+ \left(-309\,296x + 395\,780 - 2628x^3 - 200\,706x^2 + 3716x^4 - 66x^5\right) \alpha^4$

$+ \left(-60x^5 + 2220x^4 + 41\,944x^3 - 223\,310x^2 - 596\,584x + 7800\right) \alpha^3$

$+ \left(-487\,004 - 735\,524x - 167\,640x^2 + 82\,738x^3 - 518x^4 - 32x^5\right) \bmod 2$

$= x^6 + x^4 + x^2 + x + 1$

Similarly, we compute $m_5(x)$, and $m_7(x)$ as

$$m_5(x) = (x - \alpha^5)(x - \alpha^{10})(x - \alpha^{20})(x - \alpha^{40})(x - \alpha^{17})(x - \alpha^{34})$$
$$= x^6 + x^5 + x^2 + x + 1$$
$$m_7(x) = (x - \alpha^7)(x - \alpha^{14})(x - \alpha^{28})(x - \alpha^{56})(x - \alpha^{49})(x - \alpha^{35})$$
$$= x^6 + x^3 + 1$$

Then

$$g(x) = \text{lcm}\,[m_1(x), m_2(x), m_3(x), m_4(x), m_5(x), m_6(x), m_7(x), m_8(x)]$$
$$= m_1(x)m_3(x)m_5(x)m_7(x)$$
$$= (x^6 + x + 1)(x^6 + x^4 + x^2 + x + 1)$$
$$\cdot (x^6 + x^5 + x^2 + x + 1)(x^6 + x^3 + 1)$$
$$= x^{24} + x^{23} + x^{22} + x^{20} + x^{19} + x^{17} + x^{16} + x^{13}$$
$$+ x^{10} + x^9 + x^8 + x^6 + x^5 + x^4 + x^2 + x + 1$$

This is a large polynomial, but the plaintext polynomials can also be large. In this case, allowable plaintext can be of degree $62 - 24 = 38$.

In calculating minimal polynomials, we used the fact that

$$(a + b)^2 = a^2 + b^2$$

for a and b in GF_{2^n}. A similar equation holds for GF_{p^n}.

Theorem 10.5 *If a and b are elements of GF_{p^n} then*

$$(a + b)^p = a^p + b^p$$

Proof. Expanding $(a + b)^p$ gives

$$(a + b)^p = \sum_{k=0}^{p} \binom{p}{k} a^k b^{p-k}$$

For $1 \leq k \leq p - 1$, the binomial coefficient is

$$\binom{p}{k} = \frac{p!}{k!\,(p - k)!}$$

which is an integer multiple of p, so

$$\binom{p}{k} \bmod p = 0$$

Thus

$$(a + b)^p = \sum_{k=0}^{p} \binom{p}{k} a^k b^{p-k} = a^p + b^p$$

in GF_{p^n}.   ∎

The theorem $(a + b)^p = a^p + b^p$ can be used to prove, by induction on k, that

$$(a_1 + a_2 + \cdots + a_k)^p = a_1^p + a_2^p + \cdots + a_k^p$$

An interesting consequence of this theorem is that every element of GF_{p^n} is a pth power. To see this, note that if $a^p = b^p$, then $(a - b)^p = a^p - b^p = 0$, so $a - b = 0$ as GF_{p^n} is a field. That is, raising to the pth power is a one-to-one map from GF_{p^n} to itself. As GF_{p^n} is finite, the map must also be onto, that is, every element of GF_{p^n} has a unique pth root.

Finally we observe that if a is in GF_{p^n}, then a is in GF_p if and only if $a^p = a$. That is, the elements of GF_p are exactly the fixed points of the map taking a to a^p. That $a^p = a$ for a in GF_p is Fermat's little theorem. Conversely, if $a^p = a$, then $a^p - a = 0$. But the equation $x^p - x = 0$ has at most p distinct solutions, and we already know that the p elements of GF_p satisfy it.

Example 10.6 The polynomial $q(x) = x^3 + x + 1$ is irreducible over $GF_5 = \{0, 1, 2, 3, 4\}$ because

$$q(0) \bmod 5 = 1 \qquad q(1) \bmod 5 = 3 \qquad q(2) \bmod 5 = 1$$
$$q(3) \bmod 5 = 1 \qquad q(4) \bmod 5 = 4$$

so $q(x)$ has no root in GF_5. Let θ be a root of $q(x)$ in GF_{5^3}. The minimal polynomial of θ^2 is

$$\left((x - \theta^2)(x - \theta^{10})(x - \theta^{50}) \bmod \theta^3 + \theta + 1\right) \bmod 5 = x^3 + 2x^2 + x + 4$$

Note that $\theta^{250} = \theta^2$ since $\theta^{5^3 - 1} = \theta^{124} = 1$.

If β in GF_{p^n} is a root of an irreducible polynomial $m(x) = x^k + c_{k-1}x^{k-1} + \cdots + c_1 x + c_0$ in $GF_p[x]$, then

$$0 = m(\theta)^p$$
$$= \left(\beta^k + c_{k-1}\beta^{k-1} + \cdots + c_1\beta + c_0\right)^p$$
$$= \left(\beta^k\right)^p + c_{k-1}^p \left(\beta^{k-1}\right)^p + \cdots + c_1^p (\beta)^p + (c_0)^p$$
$$= \beta^{pk} + c_{k-1}\beta^{p(k-1)} + \cdots + c_1\beta^p + c_0$$
$$= \left(\beta^p\right)^k + c_{k-1}\left(\beta^p\right)^{k-1} + \cdots + c_1\beta^p + c_0$$

since $a^p = a$ for a in GF_p by Fermat's little theorem. Thus β^p is also a root of $m(x)$. Repeating this argument we see that $(\beta^p)^p = \beta^{p^2}$ is also a root of $m(x)$, and similarly β^{p^3}, β^{p^4}, and so on are roots of $m(x)$. As GF_{p^n} is finite, there must be a positive integer h so that $\beta^{p^h} = \beta^{p^i}$ for some $i < h$. Let h be the first time that happens. Then $\beta, \beta^2, \ldots, \beta^{p^{h-1}}$ are all distinct, so, in particular, $h \le k$. As raising to the pth power is a one-to-one map, we must have $\beta^{p^h} = \beta$. Now consider the polynomial

$$g(x) = (x - \beta)(x - \beta^p)\left(x - \beta^{p^2}\right) \cdots \left(x - \beta^{p^{h-1}}\right)$$

which has degree k. Raising to the pth power permutes the elements $\beta, \beta^2, \ldots$, $\beta^{p^{h-1}}$, so

$$g(x)^p = (x^p - \beta)(x^p - \beta^p)\left(x^p - \beta^{p^2}\right) \cdots \left(x^p - \beta^{p^{h-1}}\right) = g(x^p)$$

That means that every coefficient of $g(x)$ is its own pth power, hence lies in GF_p. So $g(x)$ is in $GF_p[x]$ and $g(\beta) = 0$. Therefore $g(x)$ is divisible by $m(x)$. But $h \le k$, that is, $\deg g(x) \le \deg m(x)$. Therefore $g(x) = m(x)$ so $h = k$ and we can write $m(x)$ as

$$m(x) = (x - \beta)(x - \beta^p)\left(x - \beta^{p^2}\right) \cdots \left(x - \beta^{p^{k-1}}\right)$$

Problems 10.1

1. Show that $x^3 + x + 1$ is irreducible over GF_2.

2. Let α be a root of $x^3 + x + 1$ in GF_8. Construct the minimal polynomial of α^3.

3. Use the polynomial $x^3 + x + 1$ to construct a BCH code that corrects a single error. List all of the plaintext and codeword polynomials and observe that the minimum distance between codewords is three.

4. Show that $x^2 + 1$ is irreducible over GF_3.

5. Let α be a root of $x^2 + 1$ in GF_9. Construct the minimal polynomial of $\alpha + 1$.

6. Show that $x^4 + x^3 + 1$ is irreducible over GF_2 by showing that if α is a root of $x^4 + x^3 + 1$, then the order of α is 15.

7. Let θ be a root of $x^4 + x^3 + 1$ in GF_{16}. Construct the minimal polynomial of $\theta^3 + \theta^2$.

8. Show that $x^5 + x^3 + x + 1$ is irreducible over GF_3 by showing that if θ is a root of $x^5 + x^3 + x + 1$, then the smallest field containing θ has at least $3^5 = 243$ elements.

9. Let θ be a root of $x^5 + x^3 + x + 1$ in the field GF_{3^5}. Construct the minimal polynomial of $\theta^4 + 2\theta$.

10. Show that $x^2 - 2$ is irreducible over GF_5.

11. Let θ be a root of $x^2 - 2$ in the field $GF_{5^2} = GF_{25}$. Construct the minimal polynomial of $3\theta + 2$.

12. From the equation $(a + b)^P = a^P + b^P$ in GF_{p^n} (see Theorem 10.5), derive the equation $(a_1 + a_2 + \cdots + a_k)^P = a_1^P + a_2^P + \cdots + a_k^P$ by induction on k.

13. From the equation $(a + b)^P = a^P + b^P$ in GF_{p^n}, derive the equation $(a - b)^P = a^P - b^P$.

14. The code that encodes a word of length 5 by repeating it three times is a $(15, 5)$ polynomial code. What is its generator polynomial?

15. The triple-repetition code (see page 102) is not a polynomial code but can be described as encoding the polynomial $p(x)$ by the polynomial $(1 + x + x^2) p(x^3)$. Verify both parts of this statement.

10.2   A BCH Decoder

Consider a t-error correcting BCH code. A cumbersome way to decode is to calculate all the codewords $c(x) = a(x)g(x)$ and for each codeword $c(x)$, calculate $c(x) + e(x)$ for all possible error polynomials, where each $e(x)$ has at most t nonzero terms. This produces a large table, and we search until we locate $c(x) + e(x)$ in the table, and convert this to $c(x)$. Polynomial division by $g(x)$ then yields $a(x)$, or we may simply include $a(x)$ in the table.

The following example gives an alternative method for correcting a single error.

Example 10.7 Consider the double-error correcting BCH code described in Example 10.2. The irreducible polynomial $q(x) = x^4 + x + 1$ with root θ yields the generator polynomial $g(x) = x^8 + x^7 + x^6 + x^4 + 1$. Suppose that a single error occurs so that the error polynomial is $e(x) = x^k$, and consider

$$c(x) + x^k = x^{14} + x^{13} + x^{12} + x^{10} + x^6 + x^4 + x^3 + x^2 + 1$$

Since $c(\theta) = 0$, we can evaluate the polynomial $c(x) + x^k$ at θ to get

$$\theta^k = \left(\theta^{14} + \theta^{13} + \theta^{12} + \theta^{10} + \theta^6 + \theta^4 + \theta^3 + \theta^2 + 1 \bmod q(\theta)\right) \bmod 2$$
$$= \theta^3 + \theta^2 + \theta$$

The challenge is to figure out k. But knowing θ^k is enough to determine k. In this case,

$$\theta^3 + \theta^2 + \theta = \theta^{11}$$

so $k = 11$. We correct the polynomial $c(x)$ by adding x^{11} to get

$$c(x) = x^{14} + x^{13} + x^{12} + x^{11} + x^{10} + x^6 + x^4 + x^3 + x^2 + 1$$

Dividing by $g(x) = x^8 + x^7 + x^6 + x^4 + 1$, we get

$$\frac{c(x)}{g(x)} = x^6 + x^3 + x^2 + 1$$

Example 10.8 The case where $e(x) = x^j + x^k$ is more challenging. Suppose that

$$r(x) = c(x) + e(x)$$
$$= x^{14} + x^{11} + x^{10} + x^5 + x + 1$$

where

$$e(x) = x^j + x^k$$

for some (unknown) integers j and k. We figure out j and k by computing the polynomial

$$\left(x\theta^j + 1\right)\left(x\theta^k + 1\right) = x^2\theta^{j+k} + x\left(\theta^j + \theta^k\right) + 1$$

where
$$r(\theta) = \theta^j + \theta^k$$

Evaluating $r(x) = c(x) + e(x)$ at powers of θ yields

$$\theta^j + \theta^k = r(\theta)$$
$$= \left(\theta^{14} + \theta^{11} + \theta^{10} + \theta^5 + \theta + 1 \bmod \theta^4 + \theta + 1\right) \bmod 2$$
$$= 1 + \theta^2$$

$$\left(\theta^2\right)^j + \left(\theta^2\right)^k = r\left(\theta^2\right)$$
$$= \left(\theta^{28} + \theta^{22} + \theta^{20} + \theta^{10} + \theta^2 + 1 \bmod \theta^4 + \theta + 1\right) \bmod 2$$
$$= \theta$$

$$\left(\theta^3\right)^j + \left(\theta^3\right)^k = r\left(\theta^3\right)$$
$$= \left(\theta^{42} + \theta^{33} + \theta^{30} + \theta^{15} + \theta^3 + 1 \bmod \theta^4 + \theta + 1\right) \bmod 2$$
$$= \theta^3 + \theta^2 + \theta$$

$$\left(\theta^4\right)^j + \left(\theta^4\right)^k = r\left(\theta^4\right)$$
$$= \left(\theta^{56} + \theta^{44} + \theta^{40} + \theta^{20} + \theta^4 + 1 \bmod \theta^4 + \theta + 1\right) \bmod 2$$
$$= \theta^2$$

We rewrite these equations as

$$\theta^j + \theta^k = 1 + \theta^2$$
$$\theta^{2j} + \theta^{2k} = \theta$$
$$\theta^{3j} + \theta^{3k} = \theta^3 + \theta^2 + \theta$$
$$\theta^{4j} + \theta^{4k} = \theta^2$$

Notice that

$$\left(\theta^j + \theta^k\right)\theta^{j+k} + \left(\theta^{2j} + \theta^{2k}\right)\left(\theta^j + \theta^k\right) = \theta^{3j} + \theta^{3k}$$
$$\left(\theta^{2j} + \theta^{2k}\right)\theta^{j+k} + \left(\theta^{3j} + \theta^{3k}\right)\left(\theta^j + \theta^k\right) = \theta^{4j} + \theta^{4k}$$

This system can be rewritten as a matrix equation

$$\begin{pmatrix} \theta^j + \theta^k & \theta^{2j} + \theta^{2k} \\ \theta^{2j} + \theta^{2k} & \theta^{3j} + \theta^{3k} \end{pmatrix} \begin{pmatrix} \theta^{j+k} \\ \theta^j + \theta^k \end{pmatrix} = \begin{pmatrix} \theta^{3j} + \theta^{3k} \\ \theta^{4j} + \theta^{4k} \end{pmatrix}$$

and substitution yields

$$\begin{pmatrix} 1 + \theta^2 & \theta \\ \theta & \theta^3 + \theta^2 + \theta \end{pmatrix} \begin{pmatrix} \theta^{j+k} \\ \theta^j + \theta^k \end{pmatrix} = \begin{pmatrix} \theta^3 + \theta^2 + \theta \\ \theta^2 \end{pmatrix}$$

Since $\theta^4 + \theta + 1 = 0$ and $\theta^{15} = 1$, the matrix equation can be rewritten in the form

$$\begin{pmatrix} \theta^8 & \theta \\ \theta & \theta^{11} \end{pmatrix} \begin{pmatrix} \theta^{j+k} \\ \theta^j + \theta^k \end{pmatrix} = \begin{pmatrix} \theta^{11} \\ \theta^2 \end{pmatrix}$$

The augmented matrix

$$\begin{pmatrix} \theta^8 & \theta & \theta^{11} \\ \theta & \theta^{11} & \theta^2 \end{pmatrix}$$

can be reduced using elementary row operations:

$$\begin{pmatrix} \theta^8 & \theta & \theta^{11} \\ \theta & \theta^{11} & \theta^2 \end{pmatrix} \xrightarrow[R_2 \cdot \theta^{-1}]{R_1 \cdot \theta^7} \begin{pmatrix} 1 & \theta^8 & \theta^3 \\ 1 & \theta^{10} & \theta \end{pmatrix}$$

$$\xrightarrow{R_2 - R_1} \begin{pmatrix} 1 & \theta^8 & \theta^3 \\ 0 & \theta & \theta^9 \end{pmatrix}$$

$$\xrightarrow{R_2 \cdot \theta^{-1}} \begin{pmatrix} 1 & \theta^8 & \theta^3 \\ 0 & 1 & \theta^8 \end{pmatrix}$$

$$\xrightarrow{R_1 - \theta^8 R_2} \begin{pmatrix} 1 & 0 & \theta^9 \\ 0 & 1 & \theta^8 \end{pmatrix}$$

Thus

$$\theta^{j+k} = \theta^9$$

$$\theta^j + \theta^k = \theta^8$$

Setting $a = \theta^j$ and $b = \theta^k$ we search the table

a	b	ab	$a+b$
1	θ^9	θ^9	θ^7
θ	θ^8	θ^9	θ^{10}
θ^2	θ^7	θ^9	θ^{12}
θ^3	θ^6	θ^9	θ^2
θ^4	θ^5	θ^9	θ^8

and observe that $e(x) = x^4 + x^5$. Thus

$$c(x) = x^{14} + x^{11} + x^{10} + x^4 + x + 1$$

and long division yields

$$\frac{x^{14} + x^{11} + x^{10} + x + 1 + x^4}{x^8 + x^7 + x^6 + x^4 + 1} \bmod 2 = x^6 - x^5 + 2x^3 - 2x^2 + x + 1 \bmod 2$$

$$= x^6 + x^5 + x + 1$$

The single-error-correcting BCH code described in Problem 10.3 uses the codeword $x^3 + x + 1$ and message words $a(x)$ of degree at most 3 over GF_2. In Table 10.1 a polynomial of the form $b_6 x^6 + b_5 x^5 + b_4 x^4 + b_3 x^3 + b_2 x^2 + b_1 x + b_0$ has been converted to the vector $b_6 b_5 b_4 b_3 b_2 b_1 b_0$. A message with a single error

$e(x) = b_i x^i$ can be corrected by locating the received message in Table 10.1a or 10.1b and recovering $a(x)$ in the first column of that row.

$a(x)$	$a(x)q(x)$	b_0	b_1	b_2	b_3
0000	0000000	0000001	0000010	0000100	0001000
0001	0001011	0001010	0001001	0001111	0000011
0010	0010101	0010100	0010111	0010001	0011101
0011	0011101	0011100	0011111	0011001	0010101
0100	0101100	0101101	0101110	0101000	0100100
0101	0100111	0100110	0100101	0100011	0101111
0110	0111010	0111011	0111000	0111110	0110010
0111	0110001	0110000	0110011	0110101	0111001
1000	1011000	1011001	1011010	1011100	1010000
1001	1010011	1010010	1010001	1010111	1011011
1010	1001110	1001111	1001100	1001010	1000110
1011	1000101	1000100	1000111	1000001	1001101
1100	1110100	1110101	1110110	1110000	1111100
1101	1111111	1111110	1111101	1111011	1110111
1110	1100010	1100011	1100000	1100110	1101010
1111	1101001	1101000	1101011	1101101	1100001

Table 10.1a   Using a table for single-error correction

$a(x)$	$a(x)g(r)$	b_4	b_5	b_6
0000	0000000	0010000	0100000	1000000
0001	0001011	0011011	0101011	0101011
0010	0010101	0000101	0110101	1010101
0011	0011101	0001101	0111101	1011101
0100	0101100	0111100	0001100	1101100
0101	0100111	0110111	0000111	1100111
0110	0111010	0101010	0011010	1111010
0111	0110001	0100001	0010001	1110001
1000	1011000	1001000	1111000	0011000
1001	1010011	1000011	1110011	0010011
1010	1001110	1011110	1101110	0001110
1011	1000101	1010101	1100101	0000101
1100	1110100	1100100	1010100	0110100
1101	1111111	1101111	1011111	0111111
1110	1100010	1110010	1000010	0100010
1111	1101001	1111001	1001001	0101001

Table 10.1b   Using a table for single-error correction

Finding efficient methods for BCH error correction has been an active area of research. The widespread use of BCH codes attests to the success of these efforts.

The following is a general decoder for a binary BCH code. If no error occurs,

the plaintext can be recovered by the polynomial division

$$a(x) = \frac{c(x)}{g(x)}$$

If $c(x) \bmod g(x) \neq 0$ then it is necessary to locate the errors. Error correction algorithms are more complicated than the encoding algorithm which is just polynomial multiplication. The following scheme is typical. Let

$$c(x) = c_0 + c_1 x + c_2 x^2 + \cdots + c_{n-1} x^{n-1}$$

be the encoded text, and suppose

$$r(x) = r_0 + r_1 x + r_2 x^2 + \cdots + r_{n-1} x^{n-1}$$

is received, so the error pattern is

$$e(x) = r(x) - c(x) = x^{e_1} + x^{e_2} + \cdots + x^{e_u}$$

and u is the number of errors.

Since $c\left(\theta^k\right) = 0$ for $k = 1, 2, 3, \ldots, 2t$, it follows that $r\left(\theta^k\right) = e\left(\theta^k\right)$ for $k = 1, 2, 3, \ldots, 2t$. Define

$$s_1 = r(\theta) = e(\theta) = \theta^{e_1} + \theta^{e_2} + \cdots + \theta^{e_u}$$
$$s_2 = r\left(\theta^2\right) = e\left(\theta^2\right) = \left(\theta^{e_1}\right)^2 + \left(\theta^{e_2}\right)^2 + \cdots + \left(\theta^{e_u}\right)^2$$
$$\vdots$$
$$s_{2t} = r\left(\theta^{2t}\right) = e\left(\theta^{2t}\right) = \left(\theta^{e_1}\right)^{2t} + \left(\theta^{e_2}\right)^{2t} + \cdots + \left(\theta^{e_u}\right)^{2t}$$

The challenge is to locate $e_1, e_2, \ldots, e_u$. Start with the assumption $u = t$, the maximum number of errors that can be corrected. If the matrix

$$S = \begin{pmatrix} s_1 & s_2 & \cdots & s_u \\ s_2 & s_3 & \cdots & s_{u+1} \\ \vdots & \vdots & \ddots & \vdots \\ s_u & s_{u+1} & \cdots & s_{2u-1} \end{pmatrix}$$

is singular, then subtract 1 from u and repeat with smaller sized S until a nonsingular matrix is found. This can be accomplished just as well while solving the system

$$\begin{pmatrix} s_1 & s_2 & \cdots & s_u \\ s_2 & s_3 & \cdots & s_{u+1} \\ \vdots & \vdots & \ddots & \vdots \\ s_u & s_{u+1} & \cdots & s_{2u-1} \end{pmatrix} \begin{pmatrix} y_u \\ y_{u-1} \\ \vdots \\ y_1 \end{pmatrix} = \begin{pmatrix} s_{u+1} \\ s_{u+2} \\ \vdots \\ s_{2u} \end{pmatrix}$$

Then factor the polynomial

$$y_u x^u + y_{u-1} x^{u-1} + \cdots + y_1 x + 1 - \prod_{i=1}^{u} \left(x \theta^{e_i} + 1\right)$$

The errors occur at $x^{e_1}, x^{e_2}, \ldots, x^{e_u}$. To recover $a(x)$, first set $c(x) = r(x) + e(x)$, then divide by $g(x)$.

Example 10.9 We give an example of a triple-error-correcting code that uses GF_{32} and how triple errors can be corrected. We first need an irreducible polynomial of degree 5 in $GF_2[x]$. In Table 10.2, we conclude that $q(x) = x^5 + x^2 + 1$ is irreducible by showing that if θ is a root of $q(x)$, then the order of θ modulo $\theta^5 + \theta^2 + 1$ is 31.

θ	$\theta^{11} = \theta^2 + \theta + 1$	$\theta^{21} = \theta^4 + \theta^3$
θ^2	$\theta^{12} = \theta^3 + \theta^2 + \theta$	$\theta^{22} = \theta^4 + \theta^2 + 1$
θ^3	$\theta^{13} = \theta^4 + \theta^3 + \theta^2$	$\theta^{23} = \theta^3 + \theta^2 + \theta + 1$
θ^4	$\theta^{14} = \theta^4 + \theta^3 + \theta^2 + 1$	$\theta^{24} = \theta^4 + \theta^3 + \theta^2 + \theta$
$\theta^5 = \theta^2 + 1$	$\theta^{15} = \theta^4 + \theta^3 + \theta^2 + \theta + 1$	$\theta^{25} = \theta^4 + \theta^3 + 1$
$\theta^6 = \theta^3 + \theta$	$\theta^{16} = \theta^4 + \theta^3 + \theta + 1$	$\theta^{26} = \theta^4 + \theta^2 + \theta + 1$
$\theta^7 = \theta^4 + \theta^2$	$\theta^{17} = \theta^4 + \theta + 1$	$\theta^{27} = \theta^3 + \theta + 1$
$\theta^8 = \theta^3 + \theta^2 + 1$	$\theta^{18} = \theta + 1$	$\theta^{28} = \theta^4 + \theta^2 + \theta$
$\theta^9 = \theta^4 + \theta^3 + \theta$	$\theta^{19} = \theta^2 + \theta$	$\theta^{29} = \theta^3 + 1$
$\theta^{10} = \theta^4 + 1$	$\theta^{20} = \theta^3 + \theta^2$	$\theta^{30} = \theta^4 + \theta$
		$\theta^{31} = 1$

Table 10.2   Powers of θ

We next locate the minimal polynomial for θ^3 by defining

$$m_3(x) = \left(x - \theta^3\right)\left(x - \theta^6\right)\left(x - \theta^{12}\right)\left(x - \theta^{24}\right)\left(x - \theta^{17}\right)$$

and reducing $m_3(x)$ modulo $\theta^5 + \theta^2 + 1$ to get

$$m_3(x) = x^5 + x^4 + x^3 + x^2 + 1$$

The minimal polynomial of θ^5 is given by

$$m_5(x) = \left(x - \theta^5\right)\left(x - \theta^{10}\right)\left(x - \theta^{20}\right)\left(x - \theta^9\right)\left(x - \theta^{18}\right)$$

Reduce the coefficients modulo $\theta^5 + \theta^2 + 1$ to get

$$m_5(x) = x^5 + x^4 + x^2 + x + 1$$

Thus the generating polynomial is given by

$$
\begin{aligned}
g(x) &= \mathrm{lcm}\left[m_1(x), m_2(x), m_3(x), m_4(x), m_5(x)\, m_6(x)\right] \\
&= \left(x^5 + x^2 + 1\right)\left(x^5 + x^4 + x^3 + x^2 + 1\right)\left(x^5 + x^4 + x^2 + x + 1\right) \\
&= x^{15} + x^{11} + x^{10} + x^9 + x^8 + x^7 + x^5 + x^3 + x^2 + x + 1
\end{aligned}
$$

Plaintext can be a polynomial of degree at most

$$2^5 - \deg(g(x)) - 2 = 32 - 15 - 2 = 15$$

Suppose the plaintext is given by

$$a(x) = x^{15} + x^{11} + x^{10} + x^3 + x + 1$$

so that

$$
\begin{aligned}
c(x) &= a(x)\,g(x) \\
&= x^{30} + x^{24} + x^{23} + x^{20} + x^{16} + x^{15} \\
&\quad + x^{13} + x^{11} + x^8 + x^7 + x^3 + 1
\end{aligned}
$$

Introduce the triple error

$$e(x) = x^{11} + x^9 + x^5$$

so that

$$
\begin{aligned}
r(x) &= c(x) + e(x) \\
r(x) &= x^{30} + x^{24} + x^{23} + x^{20} + x^{16} + x^{15} \\
&\quad + x^{13} + x^9 + x^8 + x^7 + x^5 + x^3 + 1
\end{aligned}
$$

To compute the matrix

$$
S = \begin{pmatrix}
s_{11} & s_{12} & s_{13} & s_{14} \\
s_{21} & s_{22} & s_{23} & s_{24} \\
s_{31} & s_{32} & s_{33} & s_{34}
\end{pmatrix}
$$

we set

$$s_{ij} = r\left(\theta^{i+j-1}\right)$$

and reduce the coefficients modulo $\theta^5 + \theta^2 + 1$ to get

$$
S = \begin{pmatrix}
\theta^4 + \theta^3 & \theta^2 + \theta + 1 & \theta^4 & \theta^4 + \theta^2 + 1 \\
\theta^2 + \theta + 1 & \theta^4 & \theta^4 + \theta^2 + 1 & \theta^4 + \theta^3 + \theta \\
\theta^4 & \theta^4 + \theta^2 + 1 & \theta^4 + \theta^3 + \theta & \theta^3 + \theta^2 + 1
\end{pmatrix}
$$

Since $\theta^4 + \theta^3 = \theta^{21}$ it follows that $\left(\theta^4 + \theta^3\right)^{-1} = \theta^{10}$. Similarly, $\theta^2 + \theta + 1 = \theta^{11}$ implies $\left(\theta^2 + \theta + 1\right)^{-1} = \theta^{20}$ and $\left(\theta^4\right)^{-1} = \theta^{27}$. Multiplication of the first row by θ^{10}, the second row by θ^{20}, and the third row by θ^{27} yields the matrix

$$
\begin{pmatrix}
1 & \theta^4 + \theta^3 & \theta^4 + \theta^3 + \theta^2 + 1 & \theta \\
1 & \theta^4 + \theta^3 + \theta^2 + \theta & \theta^2 + \theta + 1 & \theta^3 + 1 \\
1 & \theta + 1 & \theta^2 + 1 & \theta^4
\end{pmatrix}
$$

Add the first row to each of the second and third rows to get

$$\begin{pmatrix} 1 & \theta^4 + \theta^3 & \theta^4 + \theta^3 + \theta^2 + 1 & \theta \\ 0 & \theta^2 + \theta & \theta^4 + \theta^3 + \theta & \theta^3 + \theta + 1 \\ 0 & \theta^4 + \theta^3 + \theta + 1 & \theta^4 + \theta^3 & \theta^4 + \theta \end{pmatrix}$$

Multiply the second row by $\left(\theta^2 + \theta\right)^{-1} = \theta^{12}$ to get

$$\begin{pmatrix} 1 & \theta^4 + \theta^3 & \theta^4 + \theta^3 + 1 + \theta^2 & \theta \\ 0 & 1 & \theta^4 + \theta^3 & \theta^3 + \theta^2 + 1 \\ 0 & \theta^4 + \theta^3 + \theta + 1 & \theta^4 + \theta^3 & \theta^4 + \theta \end{pmatrix}$$

Now add appropriate multiples of the second row to the first and third rows to get

$$\begin{pmatrix} 1 & 0 & \theta^4 + \theta^3 + \theta & \theta^3 + \theta + 1 \\ 0 & 1 & \theta^4 + \theta^3 & \theta^3 + \theta^2 + 1 \\ 0 & 0 & \theta^4 + \theta & \theta^3 + \theta^2 \end{pmatrix}$$

Multiply the third row by $\left(\theta^4 + \theta\right)^{-1} = \theta$ to get

$$\begin{pmatrix} 1 & 0 & \theta^3 + \theta^2 + 1 & \theta^2 + \theta + 1 \\ 0 & 1 & \theta^4 + \theta^3 & \theta^3 + \theta^2 + 1 \\ 0 & 0 & 1 & \theta^4 + \theta^3 \end{pmatrix}$$

Add appropriate multiples of the third row to the first and second rows to get

$$\begin{pmatrix} 1 & 0 & 0 & \theta^4 + \theta^3 + 1 \\ 0 & 1 & 0 & \theta^3 + \theta \\ 0 & 0 & 1 & \theta^4 + \theta^3 \end{pmatrix} = \begin{pmatrix} 1 & 0 & 0 & \theta^{25} \\ 0 & 1 & 0 & \theta^6 \\ 0 & 0 & 1 & \theta^{21} \end{pmatrix}$$

Now look for a factorization

$$\left(\alpha^i x + 1\right)\left(\alpha^j x + 1\right)\left(\alpha^k x + 1\right) = \alpha^{i+j+k} x^3 + \left(\alpha^{i+j} + \alpha^{i+k} + \alpha^{j+k}\right) x^2$$
$$+ \left(\alpha^i + \alpha^j + \alpha^k\right) x + 1$$
$$= \alpha^{25} x^3 + \alpha^6 x^2 + \alpha^{21} x + 1$$

This implies

$$\alpha^{i+j+k} = \alpha^{25}$$
$$\alpha^{i+j} + \alpha^{i+k} + \alpha^{j+k} = \alpha^6$$
$$\alpha^i + \alpha^j + \alpha^k = \alpha^{21}$$

A search locates the solution

$$i = 5, \ j = 9, \ k = 11$$

and hence $e(x) = x^5 + x^9 + x^{11}$. This yields

$$c(x) = r(x) + e(x)$$
$$= x^{30} + x^{24} + x^{23} + x^{20} + x^{16} + x^{15} + x^{13} + x^{11} + x^8 + x^7 + x^3 + 1$$

so the plaintext is

$$a(x) = \frac{c(x)}{g(x)} = x^{15} + x^{11} + x^{10} + x^3 + x + 1$$

Problems 10.2

1. The polynomial $q(x) = x^3 + x + 1$ is used to construct a BCH code that corrects a single error with plaintext polynomials of the form $a(x) = a_3 x^3 + a_2 x^2 + a_1 x + a_0 \in GF_2[x]$. If the message $x^5 + x^4 + x^3 + 1$ is received, what was the plaintext?

2. The polynomial $q(x) = x^4 + x + 1$ is used to construct a BCH code that corrects a single error with plaintext polynomials $a(x) \in GF_2[x]$. What is the largest possible degree for $a(x)$? If the polynomial $x^{12} + x^{10} + x^8 + x^6 + x^2 + x$ is received, what was the plaintext?

3. Let θ be a root of $x^2 + x + 2$ in GF_{32} and calculate $g(x) = m_1(x) m_2(x)$.

4. If the generator polynomial defined in problem 3 produced a codeword that was received as $2x^7 + 2x^4 + 2x^3 + x^2 + 2x + 2$, what was the plaintext?

   In problems 5-7, let θ be a root of $q(x) = x^5 + x^2 + 1$ in GF_{32}.

5. Show that $q(x)$ is a primitive polynomial.

6. Construct a double-error-correcting BCH code that uses the generator polynomial $g(x) = \operatorname{lcm}(m_1(x), m_2(x), m_3(x), m_4(x))$, where $m_i(x)$ is the minimal polynomial of θ^i. What is the largest possible degree for a plaintext polynomial $a(x) \in GF_2[x]$?

7. The polynomial $r(x) = x^{29} + x^{28} + x^{27} + x^{25} + x^{24} + x^{22} + x^{20} + x^{18} + x^{16} + x^{15} + x^{13} + x^{12} + x^{11} + x^{10} + x^8 + x^7 + x^5 + x^4 + x^3 + 2x + 1$ is received using the BCH code from the previous problem. What is the plaintext polynomial?

   In problems 8-11, let θ be a root of $q(x) = x^3 + 2x + 1$ in GF_{27}.

8. Show that $q(x)$ is a primitive polynomial.

9. Construct a single-error-correcting BCH code that uses the generator polynomial $p(x) = \operatorname{lcm}(m_1(x), m_2(x))$, where $m_i(x)$ is the minimal polynomial of θ^i.

10. The polynomial $r(x) = 2x^{25} + 2x^{24} + 2x^{22} + 2x^{21} + x^{19} + 2x^{18} + x^{17} + x^{16} + 2x^{14} + 2x^{12} + 2x^{11} + x^{10} + 2x^9 + x^8 + 2x^7 + 2x^5 + 2x^3 + x^2 + 2x + 2$ is received after being encoded by the BCH code from the previous problem. What was the plaintext?

11. Construct a double-error-correcting BCH code that uses the generator polynomial $g(x) = \text{lcm}(m_1(x), m_2(x), m_3(x), m_4(x))$, where $m_i(x)$ is the minimal polynomial of θ^i. What is the largest possible degree for a plaintext polynomial $a(x) \in GF_3[x]$?

10.3   Reed-Solomon Codes

A Reed-Solomon code is a BCH code where the degree of the irreducible polynomial $f(x)$ is 1. Let θ be a primitive element in GF_p. The minimal polynomial of θ^i is $m_i(x) = x - \theta^i$, and hence the encoding polynomial is given by

$$g(x) = (x - \theta)(x - \theta^2)(x - \theta^3) \cdots (x - \theta^{2t})$$

Let $a(x)$ denote a plaintext polynomial of degree at most $p - 2t - 2$. Then the minimum weight of a nonzero codeword $a(x)g(x)$ is at least $d = 2t + 1$, and hence at least t errors can be corrected.

Example 10.10 Consider the prime $p = 929$. Then $\theta = 3$ is a primitive element. To see this, note that $p - 1 = 928 = 2^5 29$, and

$$3^{928} \mod 929 = 1$$

$$3^{928/2} \mod 929 = 928$$
$$3^{928/29} \mod 929 = 347$$

Consider the polynomial

$$g(x) = (x - 3)(x - 3^2)(x - 3^3)(x - 3^4)(x - 3^5)(x - 3^6)$$
$$= x^6 + 766x^5 + 17x^4 + 803x^3 + 19x^2 + 285x + 861$$

over GF_{929}. The plaintext polynomial can have degree up to

$$p - 2t - 2 = 929 - 6 - 2 = 921$$

with coefficients in GF_{929}, and up to three errors can be corrected. The message

$$(523, 13, 794, 24, 3, 563)$$

is represented in $GF_{929}[x]$ as the polynomial

$$m(x) = 523x^5 + 13x^4 + 794x^3 + 24x^2 + 3x + 563$$

The corresponding codeword is

$$m(x) g(x) \bmod 929 = 523x^{11} + 232x^{10} + 134x^9 + 15x^8$$
$$+ 237x^7 + 503x^6 + 894x^5 + 18x^4$$
$$+ 879x^3 + 630x^2 + 463x + 734$$

Example 10.11 Consider the field $GF_{2^6} = GF_{64}$. To show that the polynomial $q(x) = x^6 + x + 1$ is irreducible, it is sufficient to show that a root θ of $q(x)$ has order 63. That a root θ has order $63 = 3^2 7$ follows from the evaluations

$$\left(\theta^{63} \bmod p(x)\right) \bmod 2 = 1$$
$$\left(\theta^{63/3} \bmod p(x)\right) \bmod 2 = \theta^6 + \theta^5 + \theta^4 + \theta^3 \neq 1$$
$$\left(\theta^{63/7} \bmod p(x)\right) \bmod 2 = \theta^4 + \theta^3 \neq 1$$

Consider the generating polynomial

$$g(x) = (x - \theta)\left(x - \theta^2\right)\left(x - \theta^3\right)\left(x - \theta^4\right)$$
$$= x^4 + \left(\theta^4 + \theta^3 + \theta^2 + \theta\right) x^3$$
$$+ \left(\theta^4 + \theta^3 + \theta^2 + 1\right) x^2 + \left(\theta^4 + 1\right) x + \theta^5 + \theta^4$$

Plaintext can be a polynomial in $GF_2[x]$ of degree

$$p - 2t - 2 = 64 - 4 - 2 = 58$$

The Reed-Solomon code (58, 62) corrects any two errors.

Problems 10.3

1. Find a primitive element in GF_{16}.

2. Construct a generating polynomial for a two-error-correcting Reed-Solomon code over GF_{16}.

3. Find a primitive element in GF_{17}.

4. Construct a generating polynomial for a two-error-correcting Reed-Solomon code over GF_{17}.

5. Find a primitive element in GF_{257}.

6. Construct a generating polynomial for a five-error-correcting Reed-Solomon code over GF_{257}.

7. Find a primitive element in GF_{256}.

8. Construct a generating polynomial for a five-error-correcting Reed-Solomon code over GF_{256}.

Chapter 11

Advanced Encryption Standard

In 1997 the National Institute of Standards and Technology held a three-year competition to develop an advanced encryption standard (AES) to protect sensitive information in federal computer systems. Many businesses were expected to use the AES as well.

> *Once final, this standard will serve as a critical computer security tool supporting the rapid growth of electronic commerce. This is a very significant step toward creating a more secure digital economy. It will allow e-commerce and e-government to flourish safely, creating new opportunities for all Americans.*
>
> Norman Y. Mineta, Secretary of Commerce

The effort to develop the AES reflected the dramatic transformation that cryptography had undergone. Before 1950, cryptography was used primarily by governments to protect state and military secrets. Today, millions of Americans use cryptography without knowing it. Automated tellers use it to encrypt personal identification numbers. Internet transactions use it to identify users and to protect personal information like credit card numbers.

Many encryption devices used Data Encryption Standard (DES) or Triple DES. Such systems had become universal in the financial services industry. Consequently, the selection of the AES affected millions of consumers and businesses.

The National Institute of Standards and Technology (NIST) requested proposals for the AES in 1997. The criteria for selection included

1. **Good security**. This was the primary quality required of the winning formula.

2. **Key sizes**. The algorithm must support key sizes of 128, 192, and 256 bits. There are

$$2^{128} = 340\,282\,366\,920\,938\,463\,463\,374\,607\,431\,768\,211\,456$$

possible keys for the 128-bit key size.

3. **Speed**. Other algorithms such as RSA have good security but run slowly.

4. **Versatility**. The algorithm must run efficiently on large computers, desktop computers, and even small devices such as smart cards.

All five finalist algorithms were found to have a very high degree of security. Rijndael was selected for its combination of security, performance, efficiency, ease of implementation, and flexibility.

Rijndael was approved as the AES standard on December 6, 2001 and became effective on May 26, 2002.[1]

11.1   Data Encryption Standard

In 1973, the National Bureau of Standards solicited proposals for cryptographic algorithms to be used by federal agencies for sensitive, unclassified information. IBM submitted an algorithm with the code name LUCIFER. In 1977, a modified version of LUCIFER was adopted as the Data Encryption Standard (DES). In addition to use by federal agencies, DES and a variant called Triple DES are used in the financial services industry.

The DES algorithm is a block cipher that uses a 56-bit key to encode a 64-bit data block. The 64 data bits, labeled 1 through 64, are put into an 8 by 8 matrix

$$P = \begin{pmatrix} 58 & 50 & 42 & 34 & 26 & 18 & 10 & 2 \\ 60 & 52 & 44 & 36 & 28 & 20 & 12 & 4 \\ 62 & 54 & 46 & 38 & 30 & 22 & 14 & 6 \\ 64 & 56 & 48 & 40 & 32 & 24 & 16 & 8 \\ 57 & 49 & 41 & 33 & 25 & 17 & 9 & 1 \\ 59 & 51 & 43 & 35 & 27 & 19 & 11 & 3 \\ 61 & 53 & 45 & 37 & 29 & 21 & 13 & 5 \\ 63 & 55 & 47 & 39 & 31 & 23 & 15 & 7 \end{pmatrix}$$

That is, data bit 58 appears in row 1 column 1 and data bit 7 appears in row 8 column 8. Then they are read off by rows, bit 2 becomes bit 8, bit 60 becomes bit 9, and so on. In the notation introduced on page 76, the incoming data bits are subject to the permutation

$$P = \begin{pmatrix} 1 & 2 & 3 & \cdots & 64 \\ 40 & 8 & 48 & \cdots & 25 \end{pmatrix}$$

---

[1]See http://csrc.nist.gov/publications/fips/fips197/fips-197.pdf for the announcement in the Federal Register.

Written as a product of disjoint cycles, P is given by

$$P = (1, 40, 28, 13, 55, 58) \, (2, 8, 32, 29, 53, 50) \, (3, 48, 27, 45, 51, 42)$$
$$\cdot \, (4, 16, 31, 61, 49, 34) \, (5, 56, 26) \, (6, 24, 30, 21, 54, 18) \, (7, 64, 25, 37, 52, 10)$$
$$\cdot \, (9, 39, 60) \, (11, 47, 59, 41, 35, 44) \, (12, 15, 63, 57, 33, 36) \, (14, 23, 62, 17, 38, 20)$$
$$\cdot \, (19, 46) \, (22) \, (43)$$

The inverse permutation has a somewhat more obvious structure

$$P^{-1} = \begin{pmatrix}
40 & 8 & 48 & 16 & 56 & 24 & 64 & 32 \\
39 & 7 & 47 & 15 & 55 & 23 & 63 & 31 \\
38 & 6 & 46 & 14 & 54 & 22 & 62 & 30 \\
37 & 5 & 45 & 13 & 53 & 21 & 61 & 29 \\
36 & 4 & 44 & 12 & 52 & 20 & 60 & 28 \\
35 & 3 & 43 & 11 & 51 & 19 & 59 & 27 \\
34 & 2 & 42 & 10 & 50 & 18 & 58 & 26 \\
33 & 1 & 41 & 9 & 49 & 17 & 57 & 25
\end{pmatrix}$$

Written in the usual notation, it is

$$P^{-1} = \begin{pmatrix} 1 & 2 & 3 & 4 & 5 & \ldots & 64 \\ 58 & 50 & 42 & 34 & 26 & \ldots & 7 \end{pmatrix}$$

This inverse permutation is applied to the output after the real cryptographic work is done.

The key consists of 56 bits. These are held in eight 8-bit words with the last bit in each word serving as a parity check bit. Referring to these key bits with the numbers $1, 2, \ldots, 64$, the bits numbered 8, 16, 24, 32, 40, 48, 56, and 64 are parity check bits. The key is permuted by putting it into a matrix

$$K = \begin{pmatrix}
57 & 49 & 41 & 33 & 25 & 17 & 9 \\
1 & 58 & 50 & 42 & 34 & 26 & 18 \\
10 & 2 & 59 & 51 & 43 & 35 & 27 \\
19 & 11 & 3 & 60 & 52 & 44 & 36 \\
63 & 55 & 47 & 39 & 31 & 23 & 15 \\
7 & 62 & 54 & 46 & 38 & 30 & 22 \\
14 & 6 & 61 & 53 & 45 & 37 & 29 \\
21 & 13 & 5 & 28 & 20 & 12 & 4
\end{pmatrix}$$

with bit 57 in the first column and first row, and bit 4 in the seventh column and eighth row. Notice that the parity check bits do not appear. This matrix is then read off in rows to get the permuted 56-bit key.

Given a $2n$-bit word, let L denote the left n bits and R the right n bits. For f any function that takes n bits as input and produces n bits of output, define

$$F(L, R) = (R, L + f(R))$$

where addition is bitwise addition modulo 2. The function F takes a $2n$-bit word LR to the $2n$-bit word $R(L + f(R))$. It is invertible with inverse

$$G(L, R) = (R + f(L), L)$$

because

$$\begin{aligned}
G \circ F(L, R) &= G(R, L + f(R)) \\
&= (L + f(R) + f(R), R) \\
&= (L, R)
\end{aligned}$$

The DES algorithm uses the key matrix K to produce a sequence of functions $f_1, f_2, \ldots, f_{16}$. The sixteen corresponding functions

$$F_i(L, R) = (R, L + f_i(R))$$

are successively applied to the 64-bit data block. This type of algorithm is called a **Feistel network**. Most of the security of DES resides in the complexity of how the functions f_i are derived from the original 56-bit key.

Special-purpose computers built specifically to break DES have made DES nearly obsolete, although triple DES is still in common use.

Problems 11.1

1. Let $u_i \in \{0, 1\}$ and define

   $$f(u_0, u_1, u_2, u_3) = (u_0 + u_1, u_0, u_1 + u_2, u_2 + u_3) \bmod 2$$

   Given an 8-bit word, let L denote the left 4 bits and R the right 4 bits. Define
   $$F(L, R) = (R, L + f(R))$$
   Compute the first four iterations of F applied to the bytes (11111111) and (11101111).

2. One measure of how well an algorithm scrambles data is how far apart two codewords are that begin at a Hamming distance of one apart. How far apart are the codewords generated in problem 1?

3. What is the inverse of the function F defined in problem 1?

4. Let $u_i \in \{0, 1\}$ and define

   $$f(u_0, u_1, u_2, u_3) = (u_0, u_0 u_1 + u_2, u_0 u_1 u_2 + u_3, u_0 u_1 u_2 u_3) \bmod 2$$

   Given an 8-bit word, let L denote the left 4 bits and R the right 4 bits. Define
   $$F(L, R) = (R, L + f(R))$$
   Compute the first four iterations of F applied to the bytes (11111111) and (11101111).

5. One measure of how well an algorithm scrambles data is how far apart two codewords are that begin at a Hamming distance of one apart. How far apart are the codewords generated in problem 4?

6. Let

$$f_1(u_0, u_1, u_2, u_3) = (u_0 + u_1, u_0, u_1 + u_2, u_2 + u_3) \bmod 2$$
$$f_2(u_0, u_1, u_2, u_3) = (u_0, u_0 u_1 + u_2, u_0 u_1 u_2 + u_3, u_0 u_1 u_2 u_3) \bmod 2$$

and set

$$F_i(L, R) = (R, L + f_i(R)) \qquad (i = 1, 2)$$

Start with the byte (11111111) and first apply F_1, then apply F_2. Repeat with the byte (11101111).

11.2   The Galois Field GF_{256}

The polynomial

$$q(x) = x^8 + x^4 + x^3 + x + 1$$

is irreducible over GF_2. To verify this, consider the ring R of equivalence classes of polynomials in $GF_2[x]$ modulo $q(x)$, and let θ be the equivalence class of x in R. We will show that $\theta + 1$ has order $255 = 3 \times 5 \times 17$ in R. Indeed

$$(\theta + 1)^{255} \bmod \theta^8 + \theta^4 + \theta^3 + \theta + 1 = 1$$

so $\theta + 1$ and all of its powers are invertible. Moreover,

$$(\theta + 1)^{255/3} \bmod \theta^8 + \theta^4 + \theta^3 + \theta + 1 = \theta^7 + \theta^5 + \theta^4 + \theta^3 + \theta^2 + 1$$
$$(\theta + 1)^{255/5} \bmod \theta^8 + \theta^4 + \theta^3 + \theta + 1 = \theta^3 + \theta^2$$
$$(\theta + 1)^{255/17} \bmod \theta^8 + \theta^4 + \theta^3 + \theta + 1 = 1 + \theta^5 + \theta^4 + \theta^2$$

so $\theta + 1$ has order exactly 255. There are 255 polynomials of degree less than 8 in θ that have at least one nonzero coefficient. So each one of them must be a power of $\theta + 1$, hence invertible. If $q(x)$ were reducible, then we could write $q(x) = a(x) b(x)$ with the degrees of $a(x)$ and $b(x)$ less than 8. Then $0 = a(\theta) b(\theta)$ but $a(\theta)$ and $b(\theta)$ are invertible. So $q(x)$ is irreducible.

It follows that R is GF_{256}, and each of its elements written uniquely as a polynomial

$$b_7 \theta^7 + b_6 \theta^6 + b_5 \theta^5 + b_4 \theta^4 + b_3 \theta^3 + b_2 \theta^2 + b_1 \theta + b_0$$

in θ with coefficients in GF_2. Addition is straightforward and multiplication is followed by reduction modulo $q(\theta)$.

It is convenient to represent the 256 elements of GF_{256} by hexadecimal numbers. Thus, the element

$$\theta^7 + \theta^6 + \theta^4 + \theta^2 + \theta = 1\theta^7 + 1\theta^6 + 0\theta^5 + 1\theta^4 + 0\theta^3 + 1\theta^2 + 1\theta + 0$$

can be represented by the binary string 11010110 which can be thought of as the binary number $(11010110)_2$ or the hexadecimal number $D6$. Similarly, the element

$$\theta^5 + \theta^3 + \theta^2 = 0\theta^7 + 0\theta^6 + 1\theta^5 + 0\theta^4 + 1\theta^3 + 1\theta^2 + 0\theta + 0$$

can be represented by the binary string 00101100 which can be thought of as the binary number $(00101100)_2$ or the hexadecimal number $2C$.

Multiplication is given by

$$
\begin{aligned}
D6 \cdot 2C &= \left((\theta^7 + \theta^6 + \theta^4 + \theta^2 + \theta)\,(\theta^5 + \theta^3 + \theta^2)\, \mathrm{mod}\, q\,(\theta) \right) \mathrm{mod}\, 2 \\
&= \theta^4 + \theta^3 + \theta \\
&= 0\theta^7 + 0\theta^6 + 0\theta^5 + 1\theta^4 + 1\theta^3 + 0\theta^2 + 1\theta + 0 \\
&= 1A
\end{aligned}
$$

whereas addition is ordinary polynomial addition with coefficients reduced modulo 2, so that

$$
\begin{aligned}
D6 + 2C &= \left(\theta^7 + \theta^6 + \theta^4 + \theta^2 + \theta\right) + \left(\theta^5 + \theta^3 + \theta^2\right) \mathrm{mod}\, 2 \\
&= \theta^7 + \theta^6 + \theta^5 + \theta^4 + \theta^3 + \theta \\
&= 1\theta^7 + 1\theta^6 + 1\theta^5 + 1\theta^4 + 1\theta^3 + 0\theta^2 + 1\theta + 0 \\
&= FA
\end{aligned}
$$

The element $a\,(\theta)\,b\,(\theta)\,\mathrm{mod}\,q\,(\theta)$ can be computed using repeated multiplication by θ. If the degree of $b\,(\theta)$ is less than 7, then multiplication by θ is a simple shift. For example,

$$\theta\left(\theta^5 + \theta^4 + \theta + 1\right) = \theta^6 + \theta^5 + \theta^2 + \theta$$

For polynomials of degree 7, multiplication by θ is a shift of the lower-order terms followed by addition by $\theta^4 + \theta^3 + \theta + 1$. In particular,

$$
\begin{aligned}
\theta\left(\theta^7 + \theta^5 + \theta^3 + \theta\right) &= \theta^8 + \theta^6 + \theta^4 + \theta^2 \\
&= \left(\theta^4 + \theta^3 + \theta + 1\right) + \left(\theta^6 + \theta^4 + \theta^2\right) \\
&= \theta^6 + \theta^3 + \theta^2 + \theta + 1
\end{aligned}
$$

since

$$\left(\theta^8 \,\mathrm{mod}\, \theta^8 + \theta^4 + \theta^3 + \theta + 1\right) \mathrm{mod}\, 2 = \theta^4 + \theta^3 + \theta + 1$$

Multiplication by higher powers of θ is repeated multiplication by θ. Multiplication of one polynomial by another can be broken down into multiplication by θ followed by addition of polynomials modulo 2. For example,

$$
\begin{aligned}
\left(\theta^5 + \theta^3 + \theta^2\right)\left(\theta^7 + \theta^6 + \theta^4 + \theta^2 + \theta\right) &= \theta^5 \left(\theta^7 + \theta^6 + \theta^4 + \theta^2 + \theta\right) \\
&+ \theta^3 \left(\theta^7 + \theta^6 + \theta^4 + \theta^2 + \theta\right) \\
&+ \theta^2 \left(\theta^7 + \theta^6 + \theta^4 + \theta^2 + \theta\right)
\end{aligned}
$$

where

$$\theta^2 \left(\theta^7 + \theta^6 + \theta^4 + \theta^2 + \theta\right) = \theta \cdot \theta \left(\theta^7 + \theta^6 + \theta^4 + \theta^2 + \theta\right)$$
$$= \theta \left(\left(\theta^4 + \theta^3 \mid \theta + 1\right) + \left(\theta^7 + \theta^5 + \theta^3 + \theta^2\right)\right)$$
$$= \theta \left(\theta^7 + \theta^5 + \theta^4 + \theta^2 + \theta + 1\right)$$
$$= \left(\theta^4 + \theta^3 + \theta + 1\right) + \theta^6 + \theta^5 + \theta^3 + \theta^2 + \theta$$
$$= \theta^6 \mid \theta^5 \mid \theta^4 + \theta^2 + 1$$

$$\theta^3 \left(\theta^7 + \theta^6 + \theta^4 + \theta^2 + \theta\right) = \theta \left(\theta^6 + \theta^5 + \theta^4 + \theta^2 + 1\right)$$
$$= \theta^7 + \theta^6 + \theta^5 + \theta^3 + \theta$$

$$\theta^5 \left(\theta^7 + \theta^6 + \theta^4 + \theta^2 + \theta\right) = \theta \cdot \theta \left(\theta^7 + \theta^6 + \theta^5 + \theta^3 + \theta\right)$$
$$= \theta \left(\left(\theta^4 + \theta^3 + \theta \mid 1\right) + \theta^7 + \theta^6 + \theta^4 + \theta^2\right)$$
$$= \theta \left(\theta^7 + \theta^6 + \theta^3 + \theta^2 + \theta + 1\right)$$
$$= \left(\theta^4 + \theta^3 + \theta + 1\right) + \left(\theta^7 + \theta^4 + \theta^3 + \theta^2 + \theta\right)$$
$$= \theta^7 + \theta^2 + 1$$

so

$$\left(\theta^5 + \theta^3 + \theta^2\right)\left(\theta^7 + \theta^6 + \theta^4 + \theta^2 + \theta\right) = \left(\theta^6 + \theta^5 + \theta^4 + \theta^2 + 1\right)$$
$$+ \left(\theta^7 + \theta^6 + \theta^5 + \theta^3 + \theta\right)$$
$$+ \left(\theta^7 + \theta^2 + 1\right)$$
$$= \theta^4 + \theta^3 + \theta$$

Multiplication by θ can be easily translated into computer code acting on 8-bit bytes. Thus

$$\theta = (00000010)$$
$$\theta^2 = (00000100)$$
$$\theta^3 = (00001000)$$
$$\theta^4 = (00010000)$$
$$\theta^5 = (00100000)$$
$$\theta^6 = (01000000)$$
$$\theta^7 = (10000000)$$

so multiplication by θ is a shift or a shift plus addition modulo 2 by the vector

(00011011). Repeated multiplication by θ produces

$$\theta\left(\theta^7 + \theta^6 + \theta^4 + \theta^2 + \theta\right) = (00000010) \cdot (11010110)$$
$$= (10101100) + (00011011)$$
$$= (10110111)$$

$$\theta^2\left(\theta^7 + \theta^6 + \theta^4 + \theta^2 + \theta\right) = (00000100) \cdot (11010110)$$
$$= (00000010) \cdot (10110111)$$
$$= (01101110) + (00011011)$$
$$= (01110101)$$

$$\theta^3\left(\theta^7 + \theta^6 + \theta^4 + \theta^2 + \theta\right) = (00001000) \cdot (11010110)$$
$$= (00000010) \cdot (01110101)$$
$$= (11101010)$$

$$\theta^4\left(\theta^7 + \theta^6 + \theta^4 + \theta^2 + \theta\right) = (00010000) \cdot (11010110)$$
$$= (00000010) \cdot (11101010)$$
$$= (11010100) + (00011011)$$
$$= (11001111)$$

$$\theta^5\left(\theta^7 + \theta^6 + \theta^4 + \theta^2 + \theta\right) = (00100000) \cdot (11010110)$$
$$= (00000010) \cdot (11001111)$$
$$= (10011110) + (00011011)$$
$$= (10000101)$$

and hence

$$\left(\theta^5 + \theta^3 + \theta^2\right)\left(\theta^7 + \theta^6 + \theta^4 + \theta^2 + \theta\right)$$
$$= (10000101) + (11101010) + (01110101)$$
$$= (00011010)$$

Since $u^{255} = 1$ for every nonzero u in GF_{256}, the inverse of u is u^{254}. In particular,

$$\left(\theta^4 + \theta\right)^{-1} = \left(\left(\theta^4 + \theta\right)^{254} \bmod \theta^8 + \theta^4 + \theta^3 + \theta + 1\right) \bmod 2 = \theta^7 + \theta^5 + \theta^3 + \theta$$

As a check, note that

$$\left(\left(\theta^4 + \theta\right)\left(\theta^7 + \theta^5 + \theta^3 + \theta\right) \bmod \theta^8 + \theta^4 + \theta^3 + \theta + 1\right) \bmod 2 = 1$$

Polynomials Over GF_{256}

Let θ be a root of the polynomial $x^8 + x^4 + x^3 + x + 1$. Elements of GF_{256} can be represented as polynomials in θ of degree at most 7. Polynomials in $GF_{256}[x]$ can be written in the form

$$b_n x^n + b_{n-1} x^{n-1} + \cdots + b_1 x + b_0$$

where $b_i \in GF_{256}$.

Example 11.1 To compute the product $a(x) \cdot b(x)$ where

$$a(x) = \left(\theta^5 + \theta^3 + 1\right)x^4 + \left(\theta^7 + \theta\right)x^2 + \left(\theta^4 + 1\right)x$$
$$b(x) = \left(\theta^6 + \theta^5\right)x^3 + \left(\theta^4 + \theta + 1\right)x + \theta^3$$

do polynomial multiplication and reduce the coefficients modulo $\theta^8 + \theta^4 + \theta^3 + \theta + 1$. Thus,

$$\begin{aligned}
a(x) \cdot b(x) = & \left(\theta^7 + \theta^6 + \theta^5 + \theta^4 + \theta^3 + 1\right)x^7 \\
& + \left(\theta^7 + \theta^6 + \theta^5 + \theta^4 + \theta^3 + \theta + 1\right)x^5 \\
& + \left(\theta^6 + \theta^5 + \theta^3 + 1\right)x^4 \\
& + \left(\theta^6 + \theta^5 + \theta^2 + 1\right)x^3 \\
& + \left(\theta^6 + \theta^2\right)x^2 \\
& + \left(\theta^7 + \theta^3\right)x
\end{aligned}$$

Problems 11.2

In problems 1-6, let θ be a root of $x^8 + x^4 + x^3 + x + 1$ in GF_{256}.

1. Write $\left(\theta^6 + \theta^3 + \theta\right)\left(\theta^7 + \theta^2 + 1\right)$ as a polynomial of degree at most 7 in θ.

2. Verify that the element $\theta + 1$ has order 255 in GF_{256}.

3. Find an irreducible polynomial with root $\theta + 1$.

   [*Hint*: Simplify the polynomial

$$\prod_{k=0}^{7}\left(x - (\theta + 1)^{2^k}\right)$$

   modulo $\theta^8 + \theta^4 + \theta^3 + \theta + 1$.]

4. Find the inverse of $\theta^7 + \theta^3 + 1$ in GF_{256}.

5. Solve the system

$$\begin{aligned}
\left(\theta^5 + \theta\right)x + \left(\theta^4 + \theta + 1\right)y &= \theta^2 + 1 \\
\left(\theta^7 + \theta^6\right)x + \left(\theta^3 + \theta^2\right)y &= \theta^6 + \theta^3
\end{aligned}$$

   of two equations and two unknowns for x, $y \in GF_{256}$.

6. Calculate the product in $GF_{256}[x]$ of the two polynomials

$$p(x) = \left(\theta^5 + \theta\right)x^3 + \left(\theta^4 + 1\right)x + \left(\theta^7 + \theta + 1\right)$$
$$q(x) = \left(\theta^7 + \theta^3 + \theta^2\right)x^2 + \left(\theta^5 + \theta^2\right)x + \left(\theta^3 + \theta^2\right)$$

7. Find the inverse of the matrix

$$A = \begin{pmatrix} \theta^5 + \theta & \theta^2 + \theta \\ \theta & 1 \end{pmatrix}$$

with entries in the field GF_{256}.

8. Solve the system

$$
\begin{aligned}
(\theta + 1)\, x + \left(\theta^2 + \theta\right) y + \left(\theta^5 + \theta^4\right) z &= \theta^7 \\
\left(\theta^2 + \theta\right) x + \left(\theta^4 + \theta^3\right) y + \theta^5 z &= \theta^4 + \theta \\
\left(\theta^6 + \theta^5\right) x + \left(\theta^7 + \theta + 1\right) y + \left(\theta^3 + 1\right) z &= \theta^3 + \theta
\end{aligned}
$$

of three equations and three unknowns for $x,\, y,\, z \in GF_{256}$.

11.3   The Rijndael Block Cipher

This page is dedicated to the fans of the Rijndael Block Cipher, whose selection, in an upset of Karelinean proportions, as the National Institute of Standards and Technology's proposed Advanced Encryption Standard has brought down the Feistel cipher dynasty. It's a classic story of two guys in a garage taking on the establishment and winning.

From the home page of the Rijndael Fan Club

Belgian mathematicians Joan Daemen and Vincent Rijmen submitted the winning cipher in the Advanced Encryption Standard open competition, held in 2000, to establish a new technique for protecting computerized information.

The Rijndael Block Cipher uses keys of 128 bits, 192 bits, or 256 bits. The 128-bit version will be described here. The algorithm includes a series of operations on 4×4 matrices whose entries are bytes (8 bits). Each such matrix contains $4 \cdot 4 \cdot 8 = 128$ bits. The 192- and 256-bit versions use 4×6 and 4×8 matrices.

The 128-bit cipher is based upon the use of 10 rounds: each round consists of a Galois field operation on the bytes (ByteSub), an operation on the rows (ShiftRows), an operation on the columns (MixColumns), and a vector sum with a scrambled key (AddRoundKey).

Throughout the remainder of this section, let θ be a fixed root of the polynomial $x^8 + x^4 + x^3 + x + 1$ in the Galois field GF_{256}.

ByteSub

Each element of the 4×4 matrix is a byte, which can be considered as an element of the field GF_{256}. The **ByteSub** transformation consists of two steps:

1. Replace each nonzero field element u with its inverse u^{-1} in GF_{256}. (The zero element is fixed.) This operation is certainly reversible. Call the new field element

$$b_7\theta^7 + b_6\theta^6 + b_5\theta^5 + b_4\theta^4 + b_3\theta^3 + b_2\theta^2 + b_1\theta + b_0$$
$$= (b_7, b_6, b_5, b_4, b_3, b_2, b_1, b_0)$$

2. Apply the affine transformation

$$\begin{pmatrix} 1 & 0 & 0 & 0 & 1 & 1 & 1 & 1 \\ 1 & 1 & 0 & 0 & 0 & 1 & 1 & 1 \\ 1 & 1 & 1 & 0 & 0 & 0 & 1 & 1 \\ 1 & 1 & 1 & 1 & 0 & 0 & 0 & 1 \\ 1 & 1 & 1 & 1 & 1 & 0 & 0 & 0 \\ 0 & 1 & 1 & 1 & 1 & 1 & 0 & 0 \\ 0 & 0 & 1 & 1 & 1 & 1 & 1 & 0 \\ 0 & 0 & 0 & 1 & 1 & 1 & 1 & 1 \end{pmatrix} \begin{pmatrix} b_0 \\ b_1 \\ b_2 \\ b_3 \\ b_4 \\ b_5 \\ b_6 \\ b_7 \end{pmatrix} + \begin{pmatrix} 1 \\ 1 \\ 0 \\ 0 \\ 0 \\ 1 \\ 1 \\ 0 \end{pmatrix} \bmod 2 = \begin{pmatrix} b_0' \\ b_1' \\ b_2' \\ b_3' \\ b_4' \\ b_5' \\ b_6' \\ b_7' \end{pmatrix}$$

Notice that

$$\begin{pmatrix} 1 & 0 & 0 & 0 & 1 & 1 & 1 & 1 \\ 1 & 1 & 0 & 0 & 0 & 1 & 1 & 1 \\ 1 & 1 & 1 & 0 & 0 & 0 & 1 & 1 \\ 1 & 1 & 1 & 1 & 0 & 0 & 0 & 1 \\ 1 & 1 & 1 & 1 & 1 & 0 & 0 & 0 \\ 0 & 1 & 1 & 1 & 1 & 1 & 0 & 0 \\ 0 & 0 & 1 & 1 & 1 & 1 & 1 & 0 \\ 0 & 0 & 0 & 1 & 1 & 1 & 1 & 1 \end{pmatrix}^{-1} \bmod 2 = \begin{pmatrix} 0 & 0 & 1 & 0 & 0 & 1 & 0 & 1 \\ 1 & 0 & 0 & 1 & 0 & 0 & 1 & 0 \\ 0 & 1 & 0 & 0 & 1 & 0 & 0 & 1 \\ 1 & 0 & 1 & 0 & 0 & 1 & 0 & 0 \\ 0 & 1 & 0 & 1 & 0 & 0 & 1 & 0 \\ 0 & 0 & 1 & 0 & 1 & 0 & 0 & 1 \\ 1 & 0 & 0 & 1 & 0 & 1 & 0 & 0 \\ 0 & 1 & 0 & 0 & 1 & 0 & 1 & 0 \end{pmatrix}$$

so the inverse affine transformation is given by

$$\begin{pmatrix} 0 & 0 & 1 & 0 & 0 & 1 & 0 & 1 \\ 1 & 0 & 0 & 1 & 0 & 0 & 1 & 0 \\ 0 & 1 & 0 & 0 & 1 & 0 & 0 & 1 \\ 1 & 0 & 1 & 0 & 0 & 1 & 0 & 0 \\ 0 & 1 & 0 & 1 & 0 & 0 & 1 & 0 \\ 0 & 0 & 1 & 0 & 1 & 0 & 0 & 1 \\ 1 & 0 & 0 & 1 & 0 & 1 & 0 & 0 \\ 0 & 1 & 0 & 0 & 1 & 0 & 1 & 0 \end{pmatrix} \left(\begin{pmatrix} b_0' \\ b_1' \\ b_2' \\ b_3' \\ b_4' \\ b_5' \\ b_6' \\ b_7' \end{pmatrix} + \begin{pmatrix} 1 \\ 1 \\ 0 \\ 0 \\ 0 \\ 1 \\ 1 \\ 0 \end{pmatrix} \right) = \begin{pmatrix} b_0 \\ b_1 \\ b_2 \\ b_3 \\ b_4 \\ b_5 \\ b_6 \\ b_7 \end{pmatrix}$$

ByteSub replaces each byte of the 4×4 matrix with a transformed byte.

ShiftRows

The **ShiftRows** transformation is given by

$$
\text{ShiftRows}
\begin{pmatrix}
s_{00} & s_{01} & s_{02} & s_{03} \\
s_{10} & s_{11} & s_{12} & s_{13} \\
s_{20} & s_{21} & s_{22} & s_{23} \\
s_{30} & s_{31} & s_{32} & s_{33}
\end{pmatrix}
=
\begin{pmatrix}
s_{00} & s_{01} & s_{02} & s_{03} \\
s_{11} & s_{12} & s_{13} & s_{10} \\
s_{22} & s_{23} & s_{20} & s_{21} \\
s_{33} & s_{30} & s_{31} & s_{32}
\end{pmatrix}
$$

The first row is fixed, the second row is shifted one cell to the left with wrap around, the third row is shifted two cells to the left with wrap around, and the fourth row is shifted three cells to the left with wrap around. The inverse transformation shifts the rows to the right.

MixColumns

MixColumns is the linear transformation

$$
\text{MixColumns}
\begin{pmatrix}
s_{11} & s_{12} & s_{13} & s_{14} \\
s_{21} & s_{22} & s_{23} & s_{24} \\
s_{31} & s_{32} & s_{33} & s_{34} \\
s_{41} & s_{42} & s_{43} & s_{44}
\end{pmatrix}
$$

$$
=
\begin{pmatrix}
\theta & \theta+1 & 1 & 1 \\
1 & \theta & \theta+1 & 1 \\
1 & 1 & \theta & \theta+1 \\
\theta+1 & 1 & 1 & \theta
\end{pmatrix}
\begin{pmatrix}
s_{11} & s_{12} & s_{13} & s_{14} \\
s_{21} & s_{22} & s_{23} & s_{24} \\
s_{31} & s_{32} & s_{33} & s_{34} \\
s_{41} & s_{42} & s_{43} & s_{44}
\end{pmatrix}
$$

This is an invertible linear transformation since

$$
\begin{pmatrix}
\theta & \theta+1 & 1 & 1 \\
1 & \theta & \theta+1 & 1 \\
1 & 1 & \theta & \theta+1 \\
\theta+1 & 1 & 1 & \theta
\end{pmatrix}^{-1}
$$

$$
=
\begin{pmatrix}
\theta^3+\theta^2+\theta & \theta^3+\theta+1 & \theta^3+\theta^2+1 & \theta^3+1 \\
\theta^3+1 & \theta^3+\theta^2+\theta & \theta^3+\theta+1 & \theta^3+\theta^2+1 \\
\theta^3+\theta^2+1 & \theta^3+1 & \theta^3+\theta^2+\theta & \theta^3+\theta+1 \\
\theta^3+\theta+1 & \theta^3+\theta^2+1 & \theta^3+1 & \theta^3+\theta^2+\theta
\end{pmatrix}
$$

over GF_{256}.

AddRoundKey

AddRoundKey, the final transformation in each of the ten rounds, is a matrix sum

$$\text{AddRoundKey} \begin{pmatrix} s_{00} & s_{01} & s_{02} & s_{03} \\ s_{10} & s_{11} & s_{12} & s_{13} \\ s_{20} & s_{21} & s_{22} & s_{23} \\ s_{30} & s_{31} & s_{32} & s_{33} \end{pmatrix}$$

$$= \begin{pmatrix} s_{00} & s_{01} & s_{02} & s_{03} \\ s_{10} & s_{11} & s_{12} & s_{13} \\ s_{20} & s_{21} & s_{22} & s_{23} \\ s_{30} & s_{31} & s_{32} & s_{33} \end{pmatrix} + \begin{pmatrix} k_{00} & k_{01} & k_{02} & k_{03} \\ k_{10} & k_{11} & k_{12} & k_{13} \\ k_{20} & k_{21} & k_{22} & k_{23} \\ k_{30} & k_{31} & k_{32} & k_{33} \end{pmatrix}$$

where a modified key matrix (k_{ij}) is generated by the key expansion algorithm for each round. Since $u + u = 0$ in GF_{256}, it follows that the inverse transformation adds the same matrix.

Key Expansions

The first key matrix is constructed directly from the 128-bit key

$$(k_0, k_1, k_2, k_3, k_4, k_5, k_6, k_7, k_8, k_9, k_A, k_B, k_C, k_D, k_E, k_F)$$

where each k_i is one byte, as follows

$$\begin{pmatrix} k_{00} & k_{01} & k_{02} & k_{03} \\ k_{10} & k_{11} & k_{12} & k_{13} \\ k_{20} & k_{21} & k_{22} & k_{23} \\ k_{30} & k_{31} & k_{32} & k_{33} \end{pmatrix} = \begin{pmatrix} k_0 & k_4 & k_8 & k_C \\ k_1 & k_5 & k_9 & k_D \\ k_2 & k_6 & k_A & k_E \\ k_3 & k_7 & k_B & k_F \end{pmatrix}$$

Ten additional key matrices are constructed by a series of transformations on the columns. The leftmost column of the each additional key matrix is constructed by the following four steps:

1. Apply a rotation

$$\begin{pmatrix} k_{03} \\ k_{13} \\ k_{23} \\ k_{33} \end{pmatrix} \rightarrow \begin{pmatrix} k_{13} \\ k_{23} \\ k_{33} \\ k_{03} \end{pmatrix}$$

to the rightmost column of the previous key matrix. This can also be calculated as the matrix product

$$\begin{pmatrix} 0 & 1 & 0 & 0 \\ 0 & 0 & 1 & 0 \\ 0 & 0 & 0 & 1 \\ 1 & 0 & 0 & 0 \end{pmatrix} \begin{pmatrix} k_{03} \\ k_{13} \\ k_{23} \\ k_{33} \end{pmatrix} = \begin{pmatrix} k_{13} \\ k_{23} \\ k_{33} \\ k_{03} \end{pmatrix}$$

2. Apply the ByteSub transformation to each byte.

3. Add

$$\begin{pmatrix} \alpha^{i-1} \\ 0 \\ 0 \\ 0 \end{pmatrix}$$

where i is the number of the round.

4. Add the leftmost column of the previous key matrix. This gives the left-most column of the current key matrix.

The remaining columns of the key matrix are constructed, from left to right, by adding the column to the left to the corresponding column from the previous key matrix. If we let K_i^j denote column j of the ith key matrix, then for $j = 2, 3, 4$ we have

$$K_i^j = K_i^{j-1} + K_{i-1}^j$$

Problems 11.3

Let θ be a (fixed) root of $x^8 + x^4 + x^3 + x + 1$ in GF_{256}.

1. Verify that

$$\begin{pmatrix} 1 & 0 & 0 & 0 & 1 & 1 & 1 & 1 \\ 1 & 1 & 0 & 0 & 0 & 1 & 1 & 1 \\ 1 & 1 & 1 & 0 & 0 & 0 & 1 & 1 \\ 1 & 1 & 1 & 1 & 0 & 0 & 0 & 1 \\ 1 & 1 & 1 & 1 & 1 & 0 & 0 & 0 \\ 0 & 1 & 1 & 1 & 1 & 1 & 0 & 0 \\ 0 & 0 & 1 & 1 & 1 & 1 & 1 & 0 \\ 0 & 0 & 0 & 1 & 1 & 1 & 1 & 1 \end{pmatrix}^{-1} \mod 2 = \begin{pmatrix} 0 & 0 & 1 & 0 & 0 & 1 & 0 & 1 \\ 1 & 0 & 0 & 1 & 0 & 0 & 1 & 0 \\ 0 & 1 & 0 & 0 & 1 & 0 & 0 & 1 \\ 1 & 0 & 1 & 0 & 0 & 1 & 0 & 0 \\ 0 & 1 & 0 & 1 & 0 & 0 & 1 & 0 \\ 0 & 0 & 1 & 0 & 1 & 0 & 0 & 1 \\ 1 & 0 & 0 & 1 & 0 & 1 & 0 & 0 \\ 0 & 1 & 0 & 0 & 1 & 0 & 1 & 0 \end{pmatrix}$$

2. Verify that $\left(\theta^4 + \theta\right)^{-1} = \theta^7 + \theta^5 + \theta^3 + \theta$ in GF_{256}.

3. Use ByteSub to transform the byte $\theta^4 + \theta$.

4. Use ShiftRows to transform the matrix

$$\begin{pmatrix} 1 & \theta & \theta^2 & \theta^3 \\ 1 & \theta^2 & \theta^4 & \theta^6 \\ 1 & \theta^3 & \theta^6 & \theta^9 \\ 1 & \theta^4 & \theta^8 & \theta^{12} \end{pmatrix}$$

5. Use MixColumns to transform the matrix

$$\begin{pmatrix} 1 & \theta & \theta^2 & \theta^3 \\ 1 & \theta^2 & \theta^4 & \theta^6 \\ 1 & \theta^3 & \theta^6 & \theta^9 \\ 1 & \theta^4 & \theta^8 & \theta^{12} \end{pmatrix}$$

6. Assume that the ith key is given by

$$\begin{pmatrix} 1 & \theta & \theta^2 & \theta^3 \\ \theta & \theta^2 & \theta^3 & \theta^4 \\ \theta^3 & \theta^4 & \theta^5 & \theta^6 \\ \theta^5 & \theta^6 & \theta^7 & 0 \end{pmatrix}$$

Use the key expansion algorithm to generate key $i + 1$.

7. For any nonzero element $u \in GF_{p^n}$, show that $u^{p^n - 1} = 1$.

8. For any nonzero element $u \in GF_{p^n}$, show that $u^{-1} = u^{p^n - 2}$.

Chapter 12

Polynomial Algorithms and Fast Fourier Transforms

In this chapter, we look at different representations of polynomials and at methods for converting from one representation to another such as the fast Fourier transform (FFT).

12.1 Lagrange Interpolation Formula

A polynomial with integer coefficients is **primitive** if the greatest common divisor of its coefficients is 1. So $9x^2 - 4x + 6$ is primitive while $10x^2 - 4x + 6$ is not.

Theorem 12.1 *Any nonzero polynomial in* $\mathbb{Q}[x]$ *can be written uniquely as a positive rational number times a primitive polynomial in* $\mathbb{Z}[x]$.

Proof. Let $f(x)$ be a nonzero polynomial in $\mathbb{Q}[x]$. Choose nonzero $d \in \mathbb{Z}$ so that $g(x) = df(x)$ has integer coefficients. Let c be the greatest common divisor of the coefficients of $g(x)$. Clearly $g(x)/c$ is primitive and

$$f(x) = \frac{1}{d}g(x) = \frac{c}{d}\left(\frac{g(x)}{c}\right)$$

So we can write any nonzero polynomial that way.

Why is there only one way? Suppose we could write

$$
\begin{aligned}
f(x) &= rg(x) \\
f(x) &= sh(x)
\end{aligned}
$$

where r and s are positive rational numbers and $g(x)$ and $h(x)$ are primitive polynomials in $\mathbb{Z}[x]$. Then

$$h(x) = \frac{r}{s}g(x) = \frac{m}{n}g(x)$$

277

where m and n are relatively prime positive integers. So

$$nh(x) = mg(x)$$

Because m and n are relatively prime, n must divide each coefficient of $g(x)$ and m must divide every coefficient of $h(x)$. But $g(x)$ and $h(x)$ are primitive, so $m = n = 1$. Thus $r/s = 1$ so $r = s$. It follows that $g(x) = h(x)$. ∎

For example, the polynomial $\frac{2}{3}x^4 - \frac{4}{5}x - 6$ can be written as

$$\frac{2}{3}x^4 - \frac{4}{5}x - 6 = \frac{2}{15}\left(5x^4 - 6x - 45\right)$$

as the product of the positive rational $2/15$ and the primitive polynomial $5x^4 - 6x - 45$. The positive rational number is called the **content** of the polynomial. So the content of

$$\frac{2}{3}x^4 - \frac{4}{5}x - 6$$

is $2/15$. The content of a polynomial is an integer exactly when the polynomial has integer coefficients. Why is that? Also, the content of a polynomial is 1 exactly when the polynomial is a primitive polynomial in $\mathbb{Z}[x]$.

We want to show that the content of a product of polynomials is the product of their contents. Since a polynomial is primitive exactly when its content is 1, this says, in particular, that the product of two primitive polynomials is primitive. That result is known as Gauss's lemma.

Lemma 12.2 (Gauss's lemma) *The product of two primitive polynomials in* $\mathbb{Z}[x]$ *is primitive.*

Proof. Let $f(x)$ and $g(x)$ be primitive polynomials in $\mathbb{Z}[x]$. To show that $f(x)g(x)$ is primitive, it suffices to show that if p is any prime, then p does not divide some coefficient of the product $f(x)g(x)$. Note that because $f(x)$ and $g(x)$ are primitive, then p does not divide some coefficient of $f(x)$, and p does not divide some coefficient of $g(x)$.

This means that, considering $f(x)$ and $g(x)$ as polynomials over the field of integers modulo p, each of them is nonzero. But the product of two nonzero polynomials over a field is not zero, so $f(x)g(x)$ is not zero considered as a polynomial over the field of integers modulo p. That means that p does not divide some coefficient of $f(x)g(x)$. ∎

With Gauss's lemma and the preceding theorem in hand, it is easy to see that the content of the product of two polynomials is the product of their contents. Let $f_1(x)$ and $f_2(x)$ be two nonzero polynomials in $\mathbb{Q}[x]$. Write $f_1(x) = r_1 p_1(x)$ and $f_2(x) = r_2 p_2(x)$ where r_1 and r_2 are positive rational numbers and $p_1(x)$ and $p_2(x)$ are primitive polynomials in $\mathbb{Z}[x]$. Then

$$f_1(x) f_2(x) = r_1 r_2 p_1(x) p_2(x)$$

and $r_1 r_2$ is a positive rational number and $p_1(x) p_2(x)$ is a primitive polynomial in $\mathbb{Z}[x]$. So $r_1 r_2$ is the content of $f_1(x) f_2(x)$. But r_1 is the content of $f_1(x)$ and r_2 is the content of $f_2(x)$.

There are only a few possibilities for a rational root of a given polynomial with integer coefficients.

Theorem 12.3 (Rational root theorem) *If a polynomial with integer coefficients has a rational root p/q, written in lowest terms, then p divides the constant term and q divides the leading coefficient.*

Proof. If the polynomial $f(x)$ has the root p/q, then by the division algorithm we can write

$$f(x) = (qx - p) g(x) + r$$

where r is a constant. Plugging in p/q gives $0 = 0 \cdot g(p/q) + r$, so $r = 0$. The content of $qx - p$ is 1 so the content of $f(x)$ is the same as the content of $g(x)$. Thus the content of $g(x)$ is an integer, so $g(x)$ has integer coefficients. As

$$f(x) = (qx - p) g(x) = qxg(x) - pg(x)$$

it is clear that q divides the leading coefficient of $f(x)$ and p divides the constant term of $f(x)$. (See also problem 8.) ∎

The formula

$$p(x) = \sum_{i=0}^{k} d_i \prod_{j \neq i} \frac{x - r_j}{r_i - r_j}$$

is called the **Lagrange interpolation formula** after Joseph-Louis Lagrange (1736–1813). It gives the unique polynomial $p(x)$ of degree at most k whose graph passes through the $k+1$ points (r_i, d_i) in the plane. Of course we must have $r_i \neq r_j$ for $i \neq j$.

You can easily see that the polynomial

$$L_i(x) = \prod_{j \neq i} \frac{x - r_j}{r_i - r_j}$$

has degree k with $L_i(r_i) = 1$ and $L_i(r_j) = 0$ for each $j \neq i$. So

$$p(x) = \sum_{i=0}^{k} d_i L_i(x)$$

is a polynomial of degree at most k such that $p(r_i) = d_i$ for each i.

Example 12.4 To use the formula

$$p(x) = \sum_{i=0}^{k} d_i \prod_{j \neq i} \frac{x - r_j}{r_i - r_j}$$

to construct a polynomial of degree at most 5 that interpolates the data

n	x_n	y_n
0	0	2
1	1	15
2	2	146
3	3	743
4	4	2514
5	5	6657

it is sufficient to define the polynomials

$$p_i(x) = y_i \prod_{j \neq i} \frac{x - x_j}{x_i - x_j}$$

for $i = 0, 1, 2, 3, 4, 5$. These polynomials are given by

$$p_0(x) = 2 \cdot \frac{(x-1)(x-2)(x-3)(x-4)(x-5)}{(0-1)(0-2)(0-3)(0-4)(0-5)}$$

$$= -\frac{1}{60}x^5 + \frac{1}{4}x^4 - \frac{17}{12}x^3 + \frac{15}{4}x^2 - \frac{137}{30}x + 2$$

$$p_1(x) = 15 \cdot \frac{(x-0)(x-2)(x-3)(x-4)(x-5)}{(1-0)(1-2)(1-3)(1-4)(1-5)}$$

$$= \frac{5}{8}x^5 - \frac{35}{4}x^4 + \frac{355}{8}x^3 - \frac{385}{4}x^2 + 75x$$

$$p_2(x) = 146 \cdot \frac{(x-0)(x-1)(x-3)(x-4)(x-5)}{(2-0)(2-1)(2-3)(2-4)(2-5)}$$

$$= -\frac{73}{6}x^5 + \frac{949}{6}x^4 - \frac{4307}{6}x^3 + \frac{7811}{6}x^2 - 730x$$

$$p_3(x) = 743 \cdot \frac{(x-0)(x-1)(x-2)(x-4)(x-5)}{(3-0)(3-1)(3-2)(3-4)(3-5)}$$

$$= \frac{743}{12}x^5 - 743x^4 + \frac{36\,407}{12}x^3 - \frac{9659}{2}x^2 + \frac{7430}{3}x$$

$$p_4(x) = 2514 \cdot \frac{(x-0)(x-1)(x-2)(x-3)(x-5)}{(4-0)(4-1)(4-2)(4-3)(4-5)}$$

$$= -\frac{419}{4}x^5 + \frac{4609}{4}x^4 - \frac{17\,179}{4}x^3 + \frac{25\,559}{4}x^2 - \frac{6285}{2}x$$

$$p_5(x) = 6657 \cdot \frac{(x-0)(x-1)(x-2)(x-3)(x-4)}{(5-0)(5-1)(5-2)(5-3)(5-4)}$$

$$= \frac{2219}{40}x^5 - \frac{2219}{4}x^4 + \frac{15\,533}{8}x^3 - \frac{11\,095}{4}x^2 + \frac{6657}{5}x$$

Now form the sum

$$p_0(x) + p_1(x) + p_2(x) + p_3(x) + p_4(x) + p_5(x)$$

$$= 2 + 6x - \frac{25}{6}x^2 + \frac{71}{12}x^3 + \frac{25}{6}x^4 + \frac{13}{12}x^5$$

Note that the polynomial $p_0(x)$ has value y_0 at x_0 and is zero at x_1, x_2, x_3, x_4, and x_5. In fact, each p_i has the property that $p_i(x_i) = y_i$, and for $i \neq j$, $p_i(x_j) = 0$.

Problems 12.1

1. Find the content, and the corresponding primitive polynomial, of the following polynomials.

   a. $2x^2 - \dfrac{1}{3}x$ b. $\dfrac{4}{15}x^3 - \dfrac{6}{7}x^2 + 10$

   c. $\dfrac{1}{3}x + \dfrac{1}{2}$ d. $14x^2 - 21x + 35$

2. Use the Lagrange interpolation formula to find an equation of the line through the points $(1, 2)$ and $(-3, 5)$.

3. Use the Lagrange interpolation formula to find an equation of the line through the points (x_0, y_0) and (x_1, y_1). Show that the equation reduces to the two-point form

$$y = y_0 + \frac{y_1 - y_0}{x_1 - x_0}(x - x_0)$$

   for the line through the points (x_0, y_0) and (x_1, y_1).

4. Use the Lagrange interpolation formula to find an equation of the parabola that goes through the three points $(0, 1)$, $(1, -2)$, $(2, 2)$.

5. Find the polynomial of lowest degree that goes through the points $(2, 7)$, $(4, 10)$, $(7, 23)$, $(11, 5)$ by defining a generic polynomial

$$p(x) = a + bx + cx^2 + dx^3$$

and solving the system

$$p(2) = 7,\ p(4) = 10,\ p(7) = 23,\ p(11) = 5$$

for the unknown coefficients a, b, c, d.

6. Find a polynomial over the integers modulo 101 of lowest degree that goes through the points $(1, 3)$, $(2, 51)$, $(4, 78)$, $(7, 23)$ by defining a generic polynomial

$$p(x) = a + bx + cx^2 + dx^3$$

solving the system

$$p(1) = 3,\ p(2) = 51,\ p(4) = 78,\ p(7) = 23$$

for the unknown coefficients $\{a, b, c, d\}$ and reducing the coefficients modulo 101.

7. Find the polynomial of lowest degree that goes through the points $(-2, 5)$, $(-1, -10)$, $(0, 8)$, $(1, -5)$, $(2, 10)$.

8. Prove the rational root theorem directly by plugging in a root p/q in lowest terms and clearing fractions.

9. Verify that the polynomial

$$p(x) = \sum_{i=0}^{k} d_i \prod_{j \neq i} \frac{x - r_j}{r_i - r_j}$$

satisfies $p(r_i) = d_i$ for $i = 0, 1, \ldots, k$.

12.2   Kronecker's Algorithm

Using Theorem 12.3, we can find all the rational roots of a polynomial with integer coefficients by constructing a finite set of candidates and testing each one. The same idea can be used to find all the factors of such a polynomial.

Suppose $f(x) = g(x)h(x)$, with $g(x)$ and $h(x)$ polynomials with integer coefficients. If m is any integer, then $f(m)$ and $g(m)$ are integers and $g(m) \mid f(m)$ because $f(m) = g(m)h(m)$.

To find a factor of degree at most k of a given polynomial $f(x)$, if there is one, let $r_0, r_1, \ldots, r_k$ be any $k + 1$ distinct integers, for example, $r_i = i$. If $f(r_i) = 0$, then $x - r_i$ is a factor of $f(x)$. Otherwise, let D_i be the finite set of divisors of the integer $f(r_i)$. If $g(x)$ is a factor of $f(x)$, then $g(r_i) \in D_i$ for $i = 0, 1, 2, \ldots, k$.

For each i select an integer $d_i \in D_i$ and consider the polynomial

$$g(x) = \sum_{i=0}^{k} d_i \prod_{j \neq i} \frac{x - r_j}{r_i - r_j}$$

which is the unique polynomial of degree at most k that satisfies $g(r_i) = d_i$ for $i = 0, 1, \ldots, k$. If $g(x)$ has integer coefficients, we test it to see if it divides $f(x)$. Any factor of $f(x)$ of degree at most k will be constructed as such a polynomial $g(x)$ from some sequence $d_0, d_1, \ldots, d_k$ with $d_i \in D_i$. This idea was named after Leopold Kronecker (1823–1891). (See Algorithm 12.1.) The algorithm is usually very inefficient because there are so many cases to check.

Algorithm 12.1   Kronecker's algorithm

Input: A positive integer k and a polynomial $f(x) \in \mathbb{Z}[x]$
Output: A divisor $p(x)$ of $f(x)$ of degree at most k or the message "fail"

For i **from** 0 **to** k **do**
    $y_i = f(i)$
    If $y_i = 0$ **then**
        Return $x - i$
    End If
End For
For d_i **dividing** y_i **do**
$$p(x) = \sum_{i=0}^{k} d_i \prod_{j \neq i} \frac{x - j}{i - j}$$
    If $p(x) \in \mathbb{Z}[x]$ **and** $p(x)$ **divides** $f(x)$ **then**
        Return $p(x)$
    End If
End For
Return "fail"

Here is an example of Kronecker's algorithm in action.

Example 12.5 Let $f(x) = x^5 + x^4 + 1$. There are no linear factors because the only possible rational roots are 1 and -1, and $f(1) = 3$, $f(-1) = 1$. If $f(x)$ factors, then it must be the product of a polynomial of degree 2 with a polynomial of degree 3. So we need only look for a factor of degree 2.

Suppose $g(x)$ is a factor of degree 2. Using the integers $r_0 = 0$, $r_1 = 1$, and $r_2 = -1$, we see that $f(-1) = 1$, $f(0) = 1$, and $f(1) = 3$. It follows that

$$D_0 = \{-1, 1\}$$
$$D_1 = \{-1, 1\}$$
$$D_2 = \{-1, 1, -3, 3\}$$

Thus there are $2 \times 2 \times 4 = 16$ choices for d_0, d_1, and d_2, so 16 possible polynomial divisors of degree 2. Each such polynomial divisor must be of the form

$$g(x) = d_0 \frac{(x - 0)(x - 1)}{(-1 - 0)(-1 - 1)} + d_1 \frac{(x + 1)(x - 1)}{(0 + 1)(0 - 1)} + d_2 \frac{(x - 0)(x + 1)}{(1 - 0)(1 + 1)}$$

Selecting $d_0 = 1$, $d_1 = 1$, and $d_2 = 1$, we get the possible factor

$$\frac{(x - 0)(x - 1)}{(-1 - 0)(-1 - 1)} + \frac{(x + 1)(x - 1)}{(0 + 1)(0 - 1)} + \frac{(x - 0)(x + 1)}{(1 - 0)(1 + 1)} = 1$$

which is not even a polynomial of degree 2 (but it is a factor!).

The choices $d_0 = 1$, $d_1 = 1$, and $d_2 = 3$ yield

$$\frac{(x-0)(x-1)}{(-1-0)(-1-1)} + \frac{(x+1)(x-1)}{(0+1)(0-1)} + 3\frac{(x-0)(x+1)}{(1-0)(1+1)} = x^2 + x + 1$$

and long division shows that indeed

$$x^5 + x^4 + 1 = \left(x^2 + x + 1\right)\left(x^3 - x + 1\right)$$

This method works very well for polynomials of small degree. However, the amount of time it takes is an exponential function of the degree, so it is not a practical method for polynomials of large degree.

We look at one more example to show the added complexity for higher-degree polynomials.

Example 12.6 Consider the polynomial

$$f(x) = 3x^9 + 12x^7 + 3x^6 + 25x^5 + 24x^4 + 26x^3 + 48x^2 + 22x + 6$$

If $f(x)$ factors then at least one of the factors is of degree at most 4, so we need only look for a factor of degree at most 4. Choose $r_0 = 0$, $r_1 = 1$, $r_2 = 2$, $r_3 = -1$, and $r_4 = -2$. Then

$$f(0) = 6$$
$$f(1) = 169 = 13^2$$
$$f(2) = 4898 = 2 \times 31 \times 79$$
$$f(-1) = -7$$
$$f(-2) = -3350 = -2 \times 5^2 \times 67$$

so

$$D_0 = \{\pm 1, \pm 2, \pm 3, \pm 6\}$$
$$D_1 = \{\pm 1, \pm 13, \pm 169\}$$
$$D_2 = \{\pm 1, \pm 2, \pm 31, \pm 62, \pm 79, \pm 158, \pm 2449, \pm 4898\}$$
$$D_3 = \{\pm 1, \pm 7\}$$
$$D_4 = \{\pm 1, \pm 2, \pm 5, \pm 10, \pm 25, \pm 50, \pm 67, \pm 134, \pm 335, \pm 670, \pm 1675, \pm 3350\}$$

The number of different choices for d_0, d_1, d_2, d_3, d_4 is $8 \cdot 6 \cdot 16 \cdot 4 \cdot 24 = 73728$. After many failures, we will try

$$d_0 = 1, \, d_1 = 13, \, d_2 = 79, \, d_3 = 7, \, d_4 = 67$$

and get the polynomial

$$g(x) = 1\frac{(x-1)(x-2)(x+1)(x+2)}{(-1)(-2)(1)(2)} + 13\frac{x(x-2)(x+1)(x+2)}{1(1-2)(1+1)(1+2)}$$

$$+ 79\frac{x(x-1)(x+1)(x+2)}{2(2-1)(2+1)(2+2)} + 7\frac{x(x-1)(x-2)(x+2)}{(-1)(-1-1)(-1-2)(-1+2)}$$

$$+ 67\frac{x(x-1)(x-2)(x+1)}{(-2)(-2-1)(-2-2)(-2+1)}$$

$$= 3x^4 + 6x^2 + 3x + 1$$

Long division then shows that indeed

$$f(x) = \left(3x^4 + 6x^2 + 3x + 1\right)\left(x^5 + 2x^3 + 4x + 6\right)$$

Problems 12.2

1. Factor the polynomial $f(x) = x^4 + x^2 + 1$ in $Z[x]$ by using Kronecker's algorithm.

2. Factor the polynomial $f(x) = 2x^5 + 4x^4 + 11x^3 + 30x^2 + 25x + 5$.

3. Factor the polynomial $f(x) = 2x^6 + 2x^5 + 3x^4 + 3x^3 + 3x^2 + x + 1$.

4. Verify that the polynomial $f(x) - x^4 + x^3 + x^2 + x + 1$ is irreducible in $Z[x]$ by showing that $f(x)$ generates GF_{16} over GF_2.

5. Verify that $f(x) = x^4 + 2x + 2$ is irreducible.

6. Factor the polynomial $f(x) = 6x^4 + 7x^3 + 9x^2 + 4x + 2$.

7. Factor the polynomial $f(x) = x^5 + 3x^4 + 3x^3 + 3x^2 + 3x + 2$.

8. Factor the polynomial $f(x) = x^6 + 2x^5 + 3x^4 + 3x^3 + 3x^2 + 2x + 1$.

9. Verify that $f(x) = x^5 - x + 1$ is irreducible in $Z[x]$ by using Kronecker's algorithm.

10. Verify that $f(x) = x^5 - x + 1$ is irreducible in $Z[x]$ by showing that $f(x)$ generates GF_{243} over GF_3.

12.3 Neville's Iterated Interpolation Algorithm

Kronecker's algorithm uses the Lagrange interpolation formula for constructing the interpolating polynomials. In this section we look at another algorithm for constructing an interpolating polynomial. This method first constructs interpolating lines through pairs of points, then uses these lines to construct interpolating quadratic polynomials through triples of points, and so forth until eventually the polynomial of degree n is constructed that interpolates all $n + 1$ points.

Theorem 12.7 *Let* $\{x_0, x_1, \ldots, x_{n-1}, x_n\}$ *be any* $n + 1$ *distinct numbers. Assume* $p(x)$ *and* $q(x)$ *are functions such that*

$$p(x_i) = f(x_i) \qquad (i = 0, 1, \ldots, n - 1)$$
$$q(x_i) = f(x_i) \qquad (i = 1, \ldots, n - 1, n)$$

and define g *by*

$$g(x) = \frac{(x - x_0) q(x) - (x - x_n) p(x)}{x_n - x_0}$$

Then

$$g(x_i) = f(x_i) \qquad (i = 0, 1, \ldots, n)$$

Proof. Note that

$$g(x_0) = \frac{(x_0 - x_0) q(x_0) - (x_0 - x_n) p(x_0)}{x_n - x_0}$$
$$= -\frac{x_0 - x_n}{x_n - x_0} p(x_0) = f(x_0)$$
$$g(x_n) = \frac{(x_n - x_0) q(x_n) - (x_n - x_n) p(x_n)}{x_n - x_0}$$
$$= \frac{x_n - x_0}{x_n - x_0} q(x_n) = f(x_n)$$

and for $1 \le i \le n - 1$ we have

$$g(x_i) = \frac{(x_i - x_0) q(x_i) - (x_i - x_n) p(x_i)}{x_n - x_0}$$
$$= \frac{(x_i - x_0) f(x_i) - (x_i - x_n) f(x_i)}{x_n - x_0}$$
$$= \frac{x_n - x_0}{x_n - x_0} f(x_i) = f(x_i)$$

as claimed. ∎

Neville's iterated interpolation algorithm is based on the preceding theorem.

Algorithm 12.2   Neville's iterated interpolation algorithm

Input: Integers n and $\{x_i\}_{i=0}^{n}$ and function values $q(i, 0) = f(x_i)$
Output: Interpolating Polynomial $q(n, n)$

For i **from** 1 **to** n **do**
    For j **from** 1 **to** i **do**
        Set $q(i, j) = \dfrac{(x - x_{i-j}) q(i, j - 1) - (x - x_i) q(i - 1, j - 1)}{x_i - x_{i-j}}$
    Next j
Next i

Return $q(n, n)$

Example 12.8 Set $x_0 = -1$, $x_1 = 1$, $x_2 = 3$, $x_3 = 5$, and $x_4 = 10$, and then initialize $q(0,0) = 76$, $q(1,0) = 36$, $q(2,0) = 116$, $q(3,0) = -20$, and $q(4,0) = -45$. The interpolating polynomial can then be constructed by typing

$$q(1,1) = \frac{(x - x_0)\, q(1,0) - (x - x_1)\, q(0,0)}{x_1 - (x_0)}$$

$$= \frac{(x - (-1))\, 36 - (x - 1)\, 76}{1 - (-1)} = -20x + 56$$

$$q(2,1) = \frac{(x - x_1)\, q(2,0) - (x - x_2)\, q(1,0)}{x_2 - x_1}$$

$$= \frac{(x - 1)\, 116 - (x - 3)\, 36}{3 - 1} = 40x - 4$$

$$q(3,1) = \frac{(x - x_2)\, q(3,0) - (x - x_3)\, q(2,0)}{x_3 - x_2}$$

$$= \frac{(x - 3)\,(-20) - (x - 5)\, 116}{5 - 3} = -68x + 320$$

$$q(4,1) = \frac{(x - x_3)\, q(4,0) - (x - x_4)\, q(3,0)}{x_4 - x_3}$$

$$= \frac{(x - 5)\,(-45) - (x - 10)\,(-20)}{10 - 5} = -5x + 5$$

to generate the intermediate polynomials of degree 1, and typing

$$q(2,2) = \frac{(x - x_0)\, q(2,1) - (x - x_2)\, q(1,1)}{x_2 - x_0}$$

$$= \frac{(x + 1)\,(40x - 4) - (x - 3)\,(-20x + 56)}{3 + 1}$$

$$= 15x^2 - 20x + 41$$

$$q(3,2) = \frac{(x - x_1)\, q(3,1) - (x - x_3)\, q(2,1)}{x_3 - x_1}$$

$$= \frac{(x - 1)\,(-68x + 320) - (x - 5)\,(40x - 4)}{5 - 1}$$

$$= -27x^2 + 148x - 85$$

$$q(4,2) = \frac{(x - x_2)\, q(4,1) - (x - x_4)\, q(3,1)}{x_4 - x_2}$$

$$= \frac{(x - 3)\,(-5x + 5) - (x - 10)\,(-68x + 320)}{10 - 3}$$

$$= 9x^2 - 140x + 455$$

to generate the polynomials of degree 2, and typing

$$q(3,3) = \frac{(x - x_0)\, q(3,2) - (x - x_3)\, q(2,2)}{x_3 - x_0}$$

$$= \frac{(x + 1)\left(-27x^2 + 148x - 85\right) - (x - 5)\left(15x^2 - 20x + 41\right)}{5 + 1}$$

$$= -7x^3 + 36x^2 - 13x + 20$$

$$q(4,3) = \frac{(x - x_1)\, q(4,2) - (x - x_4)\, q(3,2)}{x_4 - x_1}$$

$$= \frac{(x - 1)\left(9x^2 - 140x + 455\right) - (x - 10)\left(-27x^2 + 148x - 85\right)}{10 - 1}$$

$$= 4x^3 - 63x^2 + 240x - 145$$

for the polynomials of degree 3, and finally, for the interpolating polynomial of degree 4,

$$q(4,4) = \frac{(x - x_0)\, q(4,3) - (x - x_4)\, q(3,3)}{x_4 - x_0}$$

$$= \frac{(x + 1)\left(4x^3 - 63x^2 + 240x - 145\right)}{10 + 1}$$

$$- \frac{(x - 10)\left(-7x^3 + 36x^2 - 13x + 20\right)}{10 + 1}$$

$$= x^4 - 15x^3 + 50x^2 - 5x + 5$$

The graph of the interpolating polynomial is shown in Figure 12.1. The linear polynomials $q(i,1)$ interpolate each pair of consecutive points as illustrated in Figure 12.2. The quadratic polynomials $q(i,2)$ interpolate each of three consecutive points (see Figure 12.3) and the cubic polynomials $q(i,3)$ interpolate each of four consecutive points (see Figure 12.4).

Example 12.9

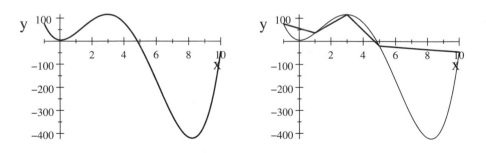

Figure 12.1 Interpolating polynomial   **Figure 12.2** Linear polynomials

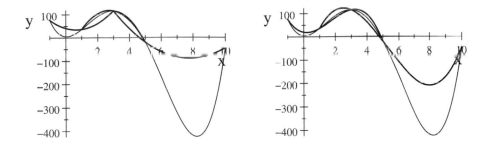

Figure 12.3 Quadratic polynomials   **Figure 12.4** Cubic polynomials

Problems 12.3

1. Given two points (x_0, y_0) and (x_1, y_1) with $x_0 \neq x_1$, show that $q(1, 1)$ given in Neville's iterated interpolation algorithm reduces to the usual two-point form

$$y = y_0 + \frac{y_1 - y_0}{x_1 - x_0} (x - x_0)$$

for a line through the two points (x_0, y_0) and (x_1, y_1).

2. Find an equation of the polynomial of degree 2 that goes through the three points $(-3, -2)$, $(1, 5)$, and $(2, 1)$, using Neville's iterated interpolation algorithm. Plot the parabola $p(x)$, together with the two lines $q(1, 1)$ and $q(2, 1)$.

3. Find the polynomial $p(x)$ of smallest degree that satisfies $p(-2) = 3$, $p(-1) = -5$, $p(1) = 1$, $p(2) = 1$, using Neville's iterated interpolation algorithm. Plot the graph of $p(x)$ and the polynomials $q(i, j)$ on one graph and label each of these polynomials.

4. Recall that $\sin 0 = 0$, $\sin\left(\frac{\pi}{2}\right) = \cos\left(\frac{\pi}{2}\right) = \frac{1}{\sqrt{2}}$, and $\cos\left(\frac{\pi}{2}\right) = 0$. Show that

$$f(x) = \frac{(x - 0)\cos x - \left(x - \frac{\pi}{2}\right)\sin x}{\frac{\pi}{2} - 0}$$

satisfies $f(0) = 0$ and $f\left(\frac{\pi}{4}\right) = \frac{1}{\sqrt{2}}$ and $f\left(\frac{\pi}{2}\right) = 0$.

5. Use Neville's iterated interpolation algorithm to construct a polynomial $p(x)$ of degree 2 in $GF_7[x]$ that interpolates the points $(0, 2)$, $(1, 3)$, $(2, 5)$.

6. Let θ be a root of $x^4 + x + 1$ in GF_{16}. Use Neville's iterated interpolation algorithm to find a polynomial $p(x) \in GF_{16}[x]$ that interpolates the three points $(0, \theta + 1)$, $(1, \theta^3 + \theta)$, $(0, \theta^2 + \theta + 1)$.

12.4   Secure Multiparty Protocols

First Trust Bank trusts its employees but not their judgment. The bank requires that n tellers must sign off on any large transaction. Employee schedules change constantly, so it is difficult to predict which n tellers might be working at any one time. Thus the bank would like any group of n tellers to have the authority to approve a large transaction, but no smaller group should have that authority. Each teller is assigned an integer-valued password, and when each of n tellers submits his or her password, a central computer does a calculation based on these passwords.

The following scheme allows any group of n tellers to generate a hidden key, but fewer than n tellers, even if they share their passwords, have absolutely no clue about what the hidden key might be.

Let

$$p(x) = a_0 + a_1 x + a_2 x^2 + \cdots + a_{n-1} x^{n-1}$$

be a polynomial of degree $n - 1$ with integer coefficients. The hidden key is the number $p(0)$. The ith teller is given the secret number $p(i)$. Since any collection of n distinct points determines a polynomial of degree $n - 1$, any group of n tellers can collectively compute $p(0)$. However, fewer than n tellers will be unable to determine the value of $p(0)$.

Example 12.10 The coefficients of the polynomial

$$p(x) = 965\,785\,106\,674 + 314\,427\,379\,054x$$
$$+ 353\,796\,761\,912x^2 + 940\,773\,378\,958x^3$$

were generated by a pseudorandom number generator modulo $999\,999\,999\,989$. Ten tellers were given the passwords

$$p(1) \bmod 999\,999\,999\,989 = 574\,782\,626\,620$$
$$p(2) \bmod 999\,999\,999\,989 = 536\,013\,944\,204$$
$$p(3) \bmod 999\,999\,999\,989 = 494\,119\,333\,240$$
$$p(4) \bmod 999\,999\,999\,989 = 93\,739\,067\,542$$

$$p(5) \bmod 999\,999\,999\,989 = 979\,513\,420\,902$$
$$p(6) \bmod 999\,999\,999\,989 = 796\,082\,667\,156$$
$$p(7) \bmod 999\,999\,999\,989 = 188\,087\,080\,107$$
$$p(8) \bmod 999\,999\,999\,989 = 800\,166\,933\,547$$
$$p(9) \bmod 999\,999\,999\,989 = 276\,962\,501\,312$$
$$p(10) \bmod 999\,999\,999\,989 = 263\,114\,057\,194$$

and were told to guard these numbers with their lives. The four tellers 2, 5, 6, and 8 submitted their secret numbers and the networked computer produced the calculation

$$p(0) \bmod 999\,999\,999\,989 = 965\,785\,106\,674$$

using Neville's iterated interpolation algorithm. The next day, tellers 1, 2, 4, and 8 submitted their secret numbers and the networked computer produced the calculation

$$p(0) \bmod 999\,999\,999\,989 = 965\,785\,106\,674$$

Given n equations of the form $q(i,0) = p(x_i)$, it is not always necessary to determine all of the coefficients of the polynomial $p(x)$. The modified Neville's iterated interpolation algorithm (see Algorithm 12.3) produces the value of $p(0)$ directly.

Algorithm 12.3 Modified Neville's iterated interpolation algorithm

Input: Integers n and nonzero $\{x_i\}_{i=0}^{n}$ and function values $q(i,0) = p(x_i)$

Output: $q(n,n) = p(0)$

For i **from** 1 **to** n **do**
$\quad$ **For** j **from** 1 **to** i **do**
$$\text{Set } q(i,j) = \frac{x_i q(i-1,j-1) - x_{i-j} q(i,j-1)}{x_i - x_{i\,j}}$$
$\quad$ **Next** j
Next i

Return $q(n,n)$

Example 12.11 We use Algorithm 12.3 to find the y-intercept of the polynomial of degree 2 that goes through the three points $(1,5)$, $(3,8)$, and $(6,2)$. First set $x_0 = 1$, $x_1 = 3$, $x_2 = 6$ and $q(0,0) = 5$, $q(1,0) = 8$, $q(2,0) = 2$. Then

$$q(1,1) = \frac{x_1 q(0,0) - x_0 q(1,0)}{x_1 - x_0} = \frac{3 \cdot 5 - 1 \cdot 8}{3 - 1} = \frac{7}{2}$$

$$q(2,1) = \frac{x_2 q(1,0) - x_1 q(2,0)}{x_2 - x_1} = \frac{6 \cdot 8 - 3 \cdot 2}{6 - 3} = 14$$

$$q(2,2) = \frac{x_2 q(1,1) - x_0 q(2,1)}{x_2 - x_0} = \frac{6 \cdot \frac{7}{2} - 1 \cdot 14}{6 - 1} = \frac{7}{5}$$

and hence the parabola goes through the point $\left(0, \frac{7}{5}\right)$.

Problems 12.4

1. Assume $p(1) = 2$, $p(2) = 4$, and $p(3) = 1$. Construct two polynomials $f(x)$ and $g(x)$ of degree 3 such that $f(k) = g(k) = p(k)$ for $k = 1, 2, 3$, but $f(0) = 0$ and $g(0) = 5$. Plot the graphs of f and g.

2. Four soldiers are stationed in a missile silo. Any two soldiers can authorize a missile to be fired. Use a polynomial $p(x) = ax + b$ with coefficients chosen at random in the range $0 \leq a, b < 999\,999\,999\,989$ to assign the four hidden numbers

$$p(1) \bmod 999\,999\,999\,989, \; p(2) \bmod 999\,999\,999\,989,$$
$$p(3) \bmod 999\,999\,999\,989, \; p(4) \bmod 999\,999\,999\,989$$

Show that soldiers 1 and 2 working together generate the same value of

$$p(0) \bmod 999\,999\,999\,989$$

as do soldiers 3 and 4.

3. Show that a conspiracy by $n-1$ tellers can learn nothing about the value $p(0)$. That is, if $p(x)$ is a polynomial of degree $n-1$ and the $n-1$ values $p(1), p(2), \ldots, p(n-1)$ are known, then $p(0)$ could have any possible integer value in the range $0 \leq p(0) < 999\,999\,999\,989$.

4. Use the modified Neville's iterated interpolation algorithm to find the y-intercept of the polynomial $p(x)$ of degree 3 that satisfies $p(2) = 8$, $p(3) = 4$, $p(5) = 7$, $p(9) = 2$.

12.5   Discrete Fourier Transforms

The Lagrange interpolation formula and Neville's iterated interpolation algorithm are two ways of constructing a polynomial that takes on prescribed values at certain points. Prescribing $n+1$ values determines a unique polynomial of degree at most n.

We usually describe a polynomial by listing its coefficients. However, it can be useful to describe a polynomial in other ways. The polynomial $p(x) = (x+1)(x-1)(x-2)$ is determined by any of the following descriptions:

1. The polynomial whose coefficients are 1, -2, -1, and 2: the polynomial $p(x) = x^3 - 2x^2 - x + 2$

2. The monic polynomial of degree 3 whose roots are -1, 1, and 2

3. The polynomial $p(x)$ of degree 3 such that $p(0) = 2$, $p(1) = 0$, $p(2) = 0$, and $p(3) = 8$

4. The polynomial $p(x)$ of degree 3 such that $p(i) = -2i + 4$, $p(i^2) = 0$, $p(i^3) = 2i + 4$, and $p(i^4) = 0$, where $i^2 = -1$

It takes four parameters to determine a polynomial of degree 3. Note that, in the second description, the fourth parameter is the leading coefficient, which is taken to be 1.

Polynomial interpolation takes us from a representation of a polynomial in terms of its function values (item 3 in the preceding list) to a representation in terms of its coefficients (item 1). In this section we will present a very fast algorithm for converting between the coefficients (item 1) and the function values at powers of a primitive root of 1. What is a primitive root of 1?

In the complex numbers, $i^2 = -1$, so $i^4 = (-1)^2 = 1$. We say that i is a fourth root of 1. The complex number i is a **primitive** fourth root of 1 because $i^n \neq 1$ for $n = 1, 2, 3$.

Instead of the complex number i, we will use a primitive root of 1 in a Galois field GF_q.

The algorithm for converting from coefficients to function values is called the **fast Fourier transform** (FFT). The algorithm for converting from function values to coefficients is called **fast Fourier interpolation** (FFI).

For each sequence

$$x_0, \ x_1, \ldots, \ x_{n-1}$$

from GF_q, we can form the polynomial

$$f(t) = x_0 + x_1 t + x_2 t^2 + \cdots + x_{n-1} t^{n-1}$$

We want to compute its values at the nth roots of unity in GF_q. If ω is a primitive nth root of unity, then the nth roots of unity are $1, \omega, \omega^2, \ldots, \omega^{n-1}$. The values of $f(t)$ at the nth roots of unity are

$$f(1) = x_0 + x_1 + x_2 + \cdots + x_{n-1}$$
$$f(\omega) = x_0 + x_1\omega + x_2\omega^2 + \cdots + x_{n-1}\omega^{n-1}$$
$$f(\omega^2) = x_0 + x_1\omega^2 + x_2\omega^4 + \cdots + x_{n-1}\omega^{2(n-1)}$$

$$\vdots$$

$$f(\omega^{n-1}) = x_0 + x_1\omega^{n-1} + x_2\omega^{2(n-1)} + \cdots + x_{n-1}\omega^{(n-1)(n-1)}$$

If we let $y_k = f(\omega^k)$, then these equations can be written in the matrix form

$$
\begin{pmatrix}
1 & 1 & 1 & \cdots & 1 \\
1 & \omega & \omega^2 & \cdots & \omega^{n-1} \\
1 & \omega^2 & \omega^4 & \cdots & \omega^{2(n-1)} \\
\vdots & \vdots & \vdots & \ddots & \vdots \\
1 & \omega^{n-1} & \omega^{2(n-1)} & \cdots & \omega^{(n-1)(n-1)}
\end{pmatrix}
\begin{pmatrix}
x_0 \\ x_1 \\ x_2 \\ \vdots \\ x_n
\end{pmatrix}
-
\begin{pmatrix}
y_0 \\ y_1 \\ y_2 \\ \vdots \\ y_n
\end{pmatrix}
$$

The square matrix is a *Vandermonde matrix*. The general **Vandermonde matrix** is an n by n matrix of the form

$$
V(\alpha_0, \alpha_1, \alpha_2, \ldots, \alpha_{n-1}) =
\begin{pmatrix}
1 & \alpha_0 & \alpha_0^2 & \cdots & \alpha_0^{n-1} \\
1 & \alpha_1 & \alpha_1^2 & \cdots & \alpha_1^{n-1} \\
1 & \alpha_2 & \alpha_2^2 & \cdots & \alpha_2^{n-1} \\
\vdots & \vdots & \vdots & \ddots & \vdots \\
1 & \alpha_{n-1} & \alpha_{n-1}^2 & \cdots & \alpha_{n-1}^{n-1}
\end{pmatrix}
$$

In our case this matrix is

$$V = V(\omega)$$
$$= V\left(1, \omega, \omega^2, \ldots, \omega^{n-1}\right)$$
$$= \begin{pmatrix} 1 & 1 & 1 & \cdots & 1 \\ 1 & \omega & \omega^2 & \cdots & \omega^{n-1} \\ 1 & \omega^2 & \omega^4 & \cdots & \omega^{2(n-1)} \\ \vdots & \vdots & \vdots & \ddots & \vdots \\ 1 & \omega^{n-1} & \omega^{2(n-1)} & \cdots & \omega^{(n-1)(n-1)} \end{pmatrix}$$

The matrix equation $\mathbf{y} = V\mathbf{x}$ gives the values $y_0, y_1, \ldots, y_{n-1}$ of the polynomial f at the nth roots of 1 in terms of its coefficients $x_0, x_1, x_2, \ldots, x_{n-1}$. Polynomial interpolation (calculating the coefficients of a polynomial from its values at selected points) amounts to solving the matrix equation $\mathbf{y} = V\mathbf{x}$ for $\mathbf{x}$. If V^{-1} is the inverse matrix of V, then $\mathbf{x} = V^{-1}\mathbf{y}$. The Vandermonde matrix has a particularly simple inverse.

In fact, let $\zeta = \omega^{-1}$. Then $1, \zeta, \zeta^2, \ldots, \zeta^{n-1}$ are also nth roots of 1 and are equal to $1, \omega, \omega^2, \ldots, \omega^{n-1}$ in some order. In fact, $\zeta^i = \omega^{n-i}$. With a moderate amount of calculation (see problem 7) you can see that

$$\begin{pmatrix} 1 & 1 & \cdots & 1 \\ 1 & \omega & \cdots & \omega^{n-1} \\ \vdots & \vdots & \ddots & \vdots \\ 1 & \omega^{n-1} & \cdots & \omega^{(n-1)(n-1)} \end{pmatrix} \begin{pmatrix} 1 & 1 & \cdots & 1 \\ 1 & \zeta & \cdots & \zeta^{n-1} \\ \vdots & \vdots & \ddots & \vdots \\ 1 & \zeta^{n-1} & \cdots & \zeta^{(n-1)(n-1)} \end{pmatrix}$$
$$= \begin{pmatrix} n & 0 & \cdots & 0 \\ 0 & n & \cdots & 0 \\ \vdots & \vdots & \ddots & \vdots \\ 0 & 0 & \cdots & n \end{pmatrix}$$

As $\zeta = \omega^{-1}$, it is fairly easy to see that the diagonal entries of the product are all equal to n. That the off-diagonal entries are all 0 is a consequence of the fact that ω is a primitive nth root of 1. The off-diagonal entries have the form

$$1 + \theta + \theta^2 + \theta^3 + \cdots + \theta^{n-1}$$

where $\theta \neq 1$ is an nth root of 1. But

$$0 = 1 - \theta^n = (1 - \theta)\left(1 + \theta + \theta^2 + \theta^3 + \cdots + \theta^{n-1}\right)$$

so $0 = 1 + \theta + \theta^2 + \theta^3 + \cdots + \theta^{n-1}$.

It follows that $(V(\omega))^{-1} = n^{-1}V(\omega^{-1})$, so $\mathbf{x} = n^{-1}V(\omega^{-1})\mathbf{y}$. Thus interpolation is closely related to polynomial evaluation.

In practice, we choose n to be a power of 2, say $n = 2^m$. If ω is a primitive

nth root of unity, then

$$f(\omega) = \sum_{i=0}^{n-1} x_i \omega^i = \sum_{i=0}^{n/2-1} x_{2i} \omega^{2i} + \omega \sum_{i=0}^{n/2-1} x_{2i+1} \omega^{2i}$$

$$= \sum_{i=0}^{n/2-1} x_{2i} \left(\omega^2\right)^i + \omega \sum_{i=0}^{n/2-1} x_{2i+1} \left(\omega^2\right)^i$$

So $f(\omega)$ can be computed by calculating two sums, each of which contains only half the original number of terms. We break the original sum into its even and odd terms, then factor ω out of the odd terms. In the final algorithm, this process is repeated until each part contains only one term, which is easily evaluated. Notice that ω^2 is a primitive $(n/2)$th root of 1. The resulting algorithm is recursive.

Example 12.12 We will use the preceding equations to convert from coefficients to function values at the 8th roots of 1 in GF_{17}. Notice that $\omega = 2$ is a primitive 8th root of 1 because $\omega^0 = 1$, $\omega^1 = 2$, $\omega^2 = 4$, $\omega^3 = 8$, $\omega^4 = 16$, $\omega^5 = 15$, $\omega^6 = 13$, $\omega^7 = 9$, and $\omega^8 = 1$ in GF_{17}. We will evaluate the polynomial

$$f(x) = 1 + x + 2x^2 + x^3 + 3x^4 + 2x^5 + x^6 + 2x^7$$

at ω^i, for $i = 0, 1, 2, 3, 4, 5, 6, 7$. By substituting $y = x^2$ and $z = y^2$ into $f(x)$, we get

$$f(x) = 1 + x + 2x^2 + x^3 + 3x^4 + 2x^5 + x^6 + 2x^7$$
$$= 1 + 2y + 3y^2 + y^3 + x\left(1 + y + 2y^2 + 2y^3\right)$$
$$= 1 + 3y^2 + y\left(2 + y^2\right) + x\left(1 + 2y^2 + y\left(1 + 2y^2\right)\right)$$
$$= 1 + 3z + y\left(2 + z\right) + x\left(1 + 2z + y\left(1 + 2z\right)\right)$$

We evaluate each of the four polynomials $1 + 3z$, $2 + z$, $1 + 2z$, and $1 + 2z$ at $z = 1$ and $z = -1$ to get

	$1 + 3z$	$2 + z$	$1 + 2z$	$1 + 2z$
$\omega^0 = 1$	$1 + 3 = 4$	$2 + 1 = 3$	$1 + 2 = 3$	$1 + 2 = 3$
$\omega^4 = -1$	$1 - 3 = -2$	$2 - 1 = 1$	$1 - 2 = -1$	$1 - 2 = -1$

At the next layer, the two polynomials $1 + 2y + 3y^2 + y^3 = 1 + 3z + y\left(2 + z\right)$ and $1 + y + 2y^2 + 2y^3 = 1 + 2z + y\left(1 + 2z\right)$ are evaluated at $y = \omega^0 = 1$ (using $z = 1$), $y = \omega^2 = 4$ (using $z = -1$), $y - \omega^4 = -1$ (using $z = 1$), and $y = \omega^6 = -4$ (using $z = -1$) to get

		$1 + 3z + y\left(2 + z\right)$		$1 + 2z + y\left(1 + 2z\right)$	
ω^0	$= 1$	$4 + 1\,(3)$	$= 7$	$3 + 1\,(3)$	$= 6$
ω^2	$= 4$	$-2 + 4\,(1)$	$= 2$	$-1 + 4\,(-1)$	$= -5$
ω^4	$= -1$	$4 - 1\,(3)$	$= 1$	$3 - 1\,(3)$	$= 0$
ω^6	$= -4$	$-2 - 4\,(1)$	$= -6$	$-1 - 4\,(-1)$	$= 3$

Finally, we evaluate the original polynomial

$$f(x) = 1 + x + 2x^2 + x^3 + 3x^4 + 2x^5 + x^6 + 2x^7$$
$$= 1 + 3z + y(2 + z) + x(1 + 2z + y(1 + 2z))$$

at $x = \omega^i$, for $i = 0, 1, 2, 3, 4, 5, 6, 7$, to get

x			$f(x)$		
ω^0	$=$	1	$7 + 6$	$=$	13
ω^1	$=$	2	$2 + 2(12)$	$=$	9
ω^2	$=$	4	$1 + 4(0)$	$=$	1
ω^3	$=$	8	$11 + 8(3)$	$=$	1
ω^4	$=$	-1	$7 - 6$	$=$	1
ω^5	$=$	-2	$2 - 2(12)$	$=$	12
ω^6	$=$	-4	$1 - 4(0)$	$=$	1
ω^7	$=$	-8	$11 - 8(3)$	$=$	4

The following FFT algorithm uses the fact that $\omega^{n+k} = -\omega^k$ if ω is a primitive $2n$-th root of 1.

Algorithm 12.4   FFT algorithm

Input: N a power of two, ω a primitive Nth root of one,
$$\mathbf{x} = (x_0, x_1, x_2, \ldots, x_{N-1})$$
Output: $\mathbf{y} = (y_0, y_1, y_2, \ldots, y_{N-1})$ where
$$y_i = x_0 + x_1 \omega^i + x_2 (\omega^i)^2 + \cdots + x_{N-1} (\omega^i)^{N-1}$$
Function FFT$(N, \mathbf{x}, \omega)$
If $N = 1$
   then
      Set $y_0 = x_0$
   else
      Set $n = N/2$
      Set xeven $= (x_0, x_2, x_4, \ldots, x_{2n-2})$
      Set xodd $= (x_1, x_3, x_5, \ldots, x_{2n-1})$
      Set yeven $=$ FFT $(n, \text{xeven}, \omega^2)$
      Set yodd $=$ FFT$(n, \text{xodd}, \omega^2)$
      Do for $k = 0$ **to** $n - 1$
         $y_k = \text{yeven}_k + \omega^k \text{yodd}_k$
         $y_{n+k} = \text{yeven}_k - \omega^k \text{yodd}_k$
      Loop
End If
Return $\mathbf{y} = (y_0, y_1, y_2, \ldots, y_{N-1})$

The FFT algorithm we gave is recursive so a lot of bookkeeping was done for you. To implement this algorithm without recursion requires some care. We illustrate with $n = 8$. We start with the polynomial

$$f(t) = x_0 + x_1 t + x_2 t^2 + x_3 t^3 + x_4 t^4 + x_5 t^5 + x_6 t^6 + x_7 t^7$$

and want to compute the function values

$$f(1) = x_0 + x_1 + x_2 + x_3 + x_4 + x_5 + x_6 + x_7$$
$$f(\omega) = x_0 + x_1\omega + x_2\omega^2 + x_3\omega^3 + x_4\omega^4 + x_5\omega^5 + x_6\omega^6 + x_7\omega^7$$
$$f(\omega^2) = x_0 + x_1\omega^2 + x_2\omega^4 + x_3\omega^6 + x_4 + x_5\omega^2 + x_6\omega^4 + x_7\omega^6$$
$$f(\omega^3) = x_0 + x_1\omega^3 + x_2\omega^6 + x_3\omega^9 + x_4\omega^4 + x_5\omega^7 + x_6\omega^2 + x_7\omega^5$$
$$f(\omega^4) = x_0 + x_1\omega^4 + x_2 + x_3\omega^4 + x_4 + x_5\omega^4 + x_6 + x_7\omega^4$$
$$f(\omega^5) = x_0 + x_1\omega^5 + x_2\omega^2 + x_3\omega^7 + x_4\omega^4 + x_5\omega + x_6\omega^6 + x_6\omega^3$$
$$f(\omega^6) = x_0 + x_1\omega^6 + x_2\omega^4 + x_3\omega^2 + x_4 + x_5\omega^6 + x_6\omega^4 + x_6\omega^2$$
$$f(\omega^7) = x_0 + x_1\omega^7 + x_2\omega^6 + x_3\omega^5 + x_4\omega^4 + x_5\omega^3 + x_6\omega^2 + x_6\omega$$

The algorithm has two phases: a permuting phase and a combining phase. In the permuting phase, we rearrange (permute) the sequence $x_0, x_1, x_2, x_3, x_4, x_5, x_6, x_7$ so that it becomes $x_0, x_4, x_2, x_6, x_1, x_5, x_3, x_7$ as shown in Figure 12.5.

Permute

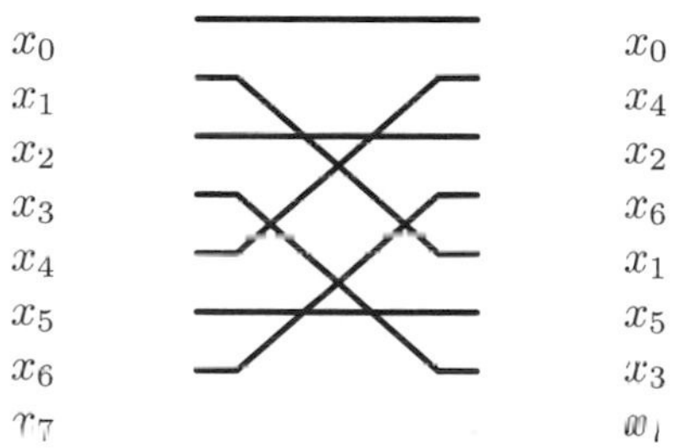

Figure 12.5   Permutation phase

The permuted indexes, $0, 4, 2, 6, 1, 5, 3, 7$ are obtained from the original indexes, $0, 1, 2, 3, 4, 5, 6, 7$, by reversing their binary representations: $1 = 001_2$ becomes $100_2 = 4$, while $2 = 010_2$ becomes $010_2 = 2$, and so on. For $n = 16$, we use four bits for the binary representation, so the sequence

$$0, 1, 2, 3, 4, 5, 6, 7, 8, 9, 10, 11, 12, 13, 14, 15$$

becomes

$$0, 8, 4, 12, 2, 10, 6, 14, 1, 9, 5, 13, 3, 11, 1, 15$$

In the combining phase, we work on a sequence $y_0, y_1, y_2, y_3, y_4, y_5, y_6, y_7$ which is initially equal to $x_0, x_4, x_2, x_0, x_1, x_5, x_3, x_7$ and is then repeatedly modified. There are $\log_2 n$ steps. For $n = 8$, there are three steps. In step k, the two entries

$$y_i \quad \text{and} \quad y_{i+2^{k-1}}$$

are replaced by

$$y_i + \omega^{in/2^k} y_{i+2^{k-1}} \quad \text{and} \quad y_i - \omega^{in/2^k} y_{i+2^{k-1}}$$

for each i such that $i \bmod 2^k < 2^{k-1}$. Note that $\omega^m = 1$ if n divides m. The first three steps are

Step 1 $(y_i, y_{i+1}) \longrightarrow (y_i + y_{i+1}, \; y_i - y_{i+1})$ for $i \bmod 2 = 0$
Step 2 $(y_i, y_{i+2}) \longrightarrow \left(y_i + \omega^{in/4} y_{i+2}, \; y_i - \omega^{in/4} y_{i+2}\right)$ for $i \bmod 4 = 0, 1$
Step 3 $(y_i, y_{i+4}) \longrightarrow \left(y_i + \omega^{in/8} y_{i+4}, \; y_i - \omega^{in/8} y_{i+4}\right)$ for $i \bmod 8 = 0, 1, 2, 3$

For $n = 8$, this is

Step 1 $(y_i, y_{i+1}) \longrightarrow (y_i + y_{i+1}, \; y_i - y_{i+1})$ for $i \bmod 2 = 0$
Step 2 $(y_i, y_{i+2}) \longrightarrow \left(y_i + \omega^{2i} y_{i+2}, \; y_i - \omega^{2i} y_{i+2}\right)$ for $i \bmod 4 = 0, 1$
Step 3 $(y_i, y_{i+4}) \longrightarrow \left(y_i + \omega^i y_{i+4}, \; y_i - \omega^i y_{i+4}\right)$ for $i \bmod 8 = 0, 1, 2, 3$

Here is a picture of the three combining steps for $n = 8$:

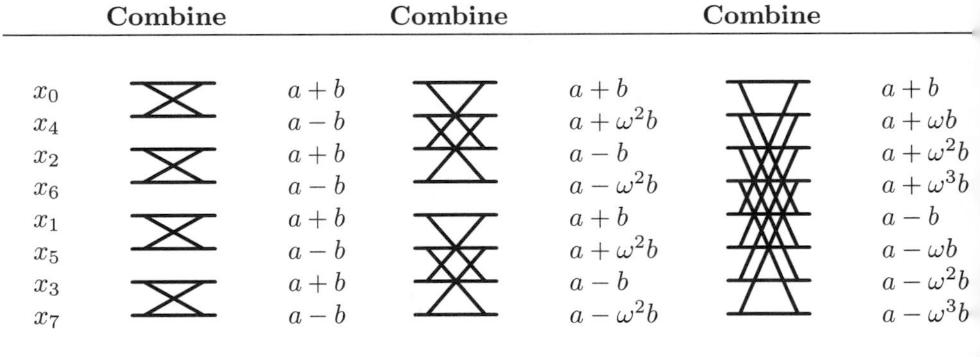

Figure 12.6   Combination phase

Here are the contents of $y_0, y_1, y_2, y_3, y_4, y_5, y_6, y_7$ after each step:

		Step 1	Step 2
y_0	x_0	$x_0 + x_4$	$x_0 + x_2 + x_4 + x_6$
y_1	x_4	$x_0 - x_4$	$x_0 + \omega^2 x_2 - x_4 - \omega^2 x_6$
y_2	x_2	$x_2 + x_6$	$x_0 - x_2 + x_4 - x_6$
y_3	x_6	$x_2 - x_6$	$x_0 - \omega^2 x_2 - x_4 + \omega^2 x_6$
y_4	x_1	$x_1 + x_5$	$x_1 + x_3 + x_5 + x_7$
y_5	x_5	$x_1 - x_5$	$x_1 + \omega^2 x_3 - x_5 - \omega^2 x_7$
y_6	x_3	$x_3 + x_7$	$x_1 - x_3 + x_5 - x_7$
y_7	x_7	$x_3 - x_7$	$x_1 - \omega^2 x_3 - x_5 + \omega^2 x_7$

	Step 3
y_0	$x_0 + x_1 + x_2 + x_3 + x_4 + x_5 + x_6 + x_7$
y_1	$x_0 + x_1\omega + x_2\omega^2 + x_3\omega^3 + x_4\omega^4 + x_5\omega^5 + x_6\omega^6 + x_7\omega^7$
y_2	$x_0 + x_1\omega^2 + x_2\omega^4 + x_3\omega^6 + x_4 + x_5\omega^2 + x_6\omega^4 + x_7\omega^6$
y_3	$x_0 + x_1\omega^3 + x_2\omega^6 + x_3\omega^9 + x_4\omega^4 + x_5\omega^7 + x_6\omega^2 + x_7\omega^5$
y_4	$x_0 + x_1\omega^4 + x_2 + x_3\omega^4 + x_4 + x_5\omega^4 + x_6 + x_7\omega^4$
y_5	$x_0 + x_1\omega^5 + x_2\omega^2 + x_3\omega^7 + x_4\omega^4 + x_5\omega + x_6\omega^6 + x_6\omega^3$
y_6	$x_0 + x_1\omega^6 + x_2\omega^4 + x_3\omega^2 + x_4 + x_5\omega^6 + x_6\omega^4 + x_6\omega^2$
y_7	$x_0 + x_1\omega^7 + x_2\omega^6 + x_3\omega^5 + x_4\omega^4 + x_5\omega^3 + x_6\omega^2 + x_6\omega$

We repeat the previous example.

Example 12.13 Let $p = 17$ and notice that $\omega = 2$ is a primitive 8th root of 1 modulo p. To evaluate the polynomial

$$x(t) = 1 + t + 2t^2 + t^3 + 3t^4 + 2t^5 + t^6 + 2t^7$$

at each of the powers $\omega^0 = 1$, $\omega^1 = 2$, $\omega^2 = 4$, $\omega^3 = 8$, $\omega^4 = 16$, $\omega^5 = 15$, $\omega^6 = 13$, $\omega^7 = 9$ we proceed as follows:

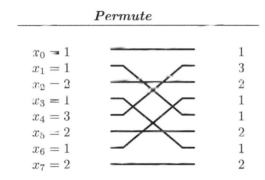

Figure 12.7 Permutation Phase

Figure 12.8 Combination phase

We conclude that

$$f\left(\omega^0\right) = 13 \quad f\left(\omega^1\right) = 9 \quad f\left(\omega^2\right) = 1 \quad f\left(\omega^3\right) = 1$$
$$f\left(\omega^4\right) = 1 \quad f\left(\omega^5\right) = 12 \quad f\left(\omega^6\right) = 1 \quad f\left(\omega^7\right) = 4$$

Problems 12.5

1. Describe the polynomial $f\left(x\right) = 3\left(x+1\right)\left(x-2\right)\left(x-5\right)$ in terms of

   (a) its coefficients,

   (b) its values at $x = 0, 1, 2, 3$,

   (c) its values at $x = i, i^2, i^3, i^4$, where $i^2 = -1$,

   (d) its zeros and its leading coefficient.

2. Repeat the preceding problem with $f\left(x\right) = x - x^3$.

3. Use the nonrecursive FFT algorithm to compute $f\left(\omega^i\right)$ for $i = 0, 1, 2, 3, 4,$
   $5, 6, 7$ where $f\left(t\right) = t$ and ω is a primitive 8-th root of 1. Verify your
   answer by direct evaluation.

4. Repeat the preceding problem with $f\left(t\right) = t^2$.

5. Use the nonrecursive FFT algorithm to compute $f\left(\omega^i\right)$ for $i = 0, 1, 2, 3, 4,$
   $5, 6, 7$ in GF_{17} for $\omega = 2$ and $f\left(t\right) = 3 + 6t + 4t^2 + 5t^4 + t^6 + 8t^7$. Verify
   your answer by direct evaluation modulo 17.

6. Repeat the preceding problem in GF_{41} with $\omega = 3$. Verify your answer by
   direct evaluation modulo 41.

7. Let ω be a primitive nth root of 1 in a field. For $\zeta = \omega^{-1}$, show that

$$
\begin{pmatrix}
1 & 1 & \cdots & 1 \\
1 & \omega & \cdots & \omega^{n-1} \\
\vdots & \vdots & \ddots & \vdots \\
1 & \omega^{n-1} & \cdots & \omega^{(n-1)(n-1)}
\end{pmatrix}
\begin{pmatrix}
1 & 1 & \cdots & 1 \\
1 & \zeta & \cdots & \zeta^{n-1} \\
\vdots & \vdots & \ddots & \vdots \\
1 & \zeta^{n-1} & \cdots & \zeta^{(n-1)(n-1)}
\end{pmatrix}
$$
$$
=
\begin{pmatrix}
n & 0 & \cdots & 0 \\
0 & n & \cdots & 0 \\
\vdots & \vdots & \ddots & \vdots \\
0 & 0 & \cdots & n
\end{pmatrix}
$$

8. Verify the equation in the previous problem directly for ω equal to the
   complex number i.

9. Show that

$$
\det \begin{pmatrix}
1 & \alpha_0 & \alpha_0^2 & \cdots & \alpha_0^{n-1} \\
1 & \alpha_1 & \alpha_1^2 & \cdots & \alpha_1^{n-1} \\
1 & \alpha_2 & \alpha_2^2 & \cdots & \alpha_2^{n-1} \\
\vdots & \vdots & \vdots & \ddots & \vdots \\
1 & \alpha_{n-1} & \alpha_{n-1}^2 & \cdots & \alpha_{n-1}^{n-1}
\end{pmatrix} = \prod_{i>j} (\alpha_i - \alpha_j)
$$

10. Verify that

$$
\det \begin{pmatrix}
1 & 1 & 1 \\
1 & 2 & 4 \\
1 & 3 & 9
\end{pmatrix} = (3 - 2)(3 - 1)(2 - 1)
$$

by evaluating both sides.

11. Let θ be a root of $x^4 + x + 1$ in GF_{16}. Verify that

$$
\det \begin{pmatrix}
1 & 1 & 1 \\
1 & \theta & \theta^2 \\
1 & \theta^2 & \theta^4
\end{pmatrix} = (\theta^2 - \theta)(\theta^2 - 1)(\theta - 1)
$$

by evaluating both sides.

12.6 Fast Fourier Interpolation

The fast Fourier transform is a quick way to transform coefficients into function values. The reverse process, going from function values to coefficients, is called **interpolation** (see page 294). It is a remarkable fact that interpolation can also be done with the fast Fourier transform. This is so because if $\mathbf{y} = V(\omega)\mathbf{x}$, then

$$
\mathbf{x} = n^{-1} V(\omega^{-1}) \mathbf{y}
$$

Note that if ω is a primitive nth root of 1, then so is ω^{-1}.

The fast Fourier interpolation algorithm is simply the FFT with one additional step.

Algorithm 12.5 Fast Fourier interpolation

Input: N a power of two, ω a primitive N-th root of one,
$$\mathbf{y} - (y_0, y_1, \ldots, y_{N-1})$$
Output: $\mathbf{x} = (x_0, x_1, \ldots, x_{N-1})$ such that
$$y_i = x_0 + x_1 \omega^i + x_2 \left(\omega^i\right)^2 + \cdots + x_{N-1} \left(\omega^i\right)^{N-1}$$
Function FFI$(N, \mathbf{y}, \omega)$
Return $\mathbf{x} = N^{-1} \mathrm{FFT}\left(N, \mathbf{y}, \omega^{-1}\right)$

Example 12.14 Consider the polynomial $x(t) = x_0 + x_1t + x_2t^2 + \cdots + x_{N-1}t^{N-1}$ with given function values

$$x\left(2^0\right) = 5 \quad x\left(2^1\right) = 3 \quad x\left(2^2\right) = 7 \quad x\left(2^3\right) = 0$$
$$x\left(2^4\right) = 13 \quad x\left(2^5\right) = 9 \quad x\left(2^6\right) = 2 \quad x\left(2^7\right) = 12$$

We use FFI to find the coefficients of the polynomial $x(t)$. We first apply the permutation phase to get Figure 12.9.

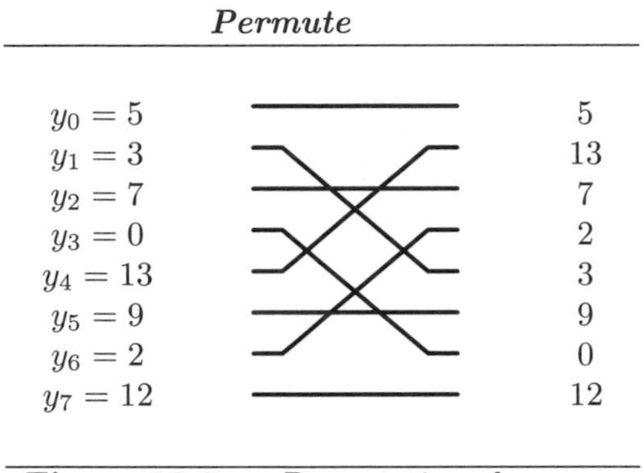

Permute

$$y_0 = 5$$
$$y_1 = 3$$
$$y_2 = 7$$
$$y_3 = 0$$
$$y_4 = 13$$
$$y_5 = 9$$
$$y_6 = 2$$
$$y_7 = 12$$

Figure 12.9   Permutation phase

Since $\omega^{-1} = 2^{-1} \bmod 17 = 9$, we have the combination phase given by Figure 12.10.

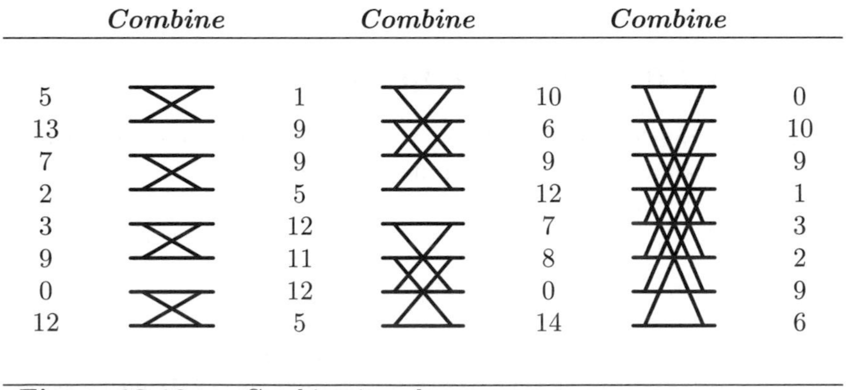

Combine   *Combine*   *Combine*

Figure 12.10   Combination phase

Multiplication of the last column by $8^{-1} \bmod 17 = 15$ yields

$$15 \begin{bmatrix} 0 \\ 10 \\ 9 \\ 1 \\ 3 \\ 2 \\ 9 \\ 6 \end{bmatrix} \bmod 17 = \begin{bmatrix} 0 \\ 14 \\ 16 \\ 15 \\ 11 \\ 13 \\ 16 \\ 5 \end{bmatrix}$$

As a check, we evaluate the polynomial

$$y(t) = 14t + 16t^2 + 15t^3 + 11t^4 + 13t^5 + 16t^6 + 5t^7$$

at powers of 2 and note that the results

$$\begin{array}{llll}
y(2^0) = 5 & y(2^1) = 3 & y(2^2) = 7 & y(2^3) = 0 \\
y(2^4) = 13 & y(2^5) = 9 & y(2^6) = 2 & y(2^7) = 12
\end{array}$$

produce the initial coefficients.

Is the FFT really fast? Let's see how long it takes to do a Fourier transformation without it. Let F be a field,

$$f(t) = y_0 + y_1 t + y_2 t^2 + \cdots + y_{n-1} t^{n-1}$$

a polynomial of degree at most $N - 1$ with coefficients in F, and ω a primitive N-th root of one in F. We need to evaluate $f(\omega^i)$ for $i = 0, 1, \ldots, N - 1$. The most efficient way to evaluate $f(t)$ at an element α in F is by nesting. We illustrate with $N = 8$:

$$f(\alpha) = y_0 + \alpha(y_1 + \alpha(y_2 + \alpha(y_3 + \alpha(y_4 + \alpha(y_5 + \alpha(y_6 + \alpha y_7))))))$$

You can see that this computation requires seven additions and seven multiplications for a polynomial of degree at most seven. For a polynomial of degree at $N - 1$, it requires $N - 1$ additions and $N - 1$ multiplications, a total of $2N - 2$ steps. So evaluating $f(\omega^i)$ for $i = 0, 1, \ldots, N - 1$ in this way requires $N(2N - 2) = 2N(N - 1)$ steps.

How many steps does it take to do these computations with the FFT? Let c_n be the number of steps the FFT algorithm takes for a polynomial of degree at most $n - 1$. Recall that for the FFT, the number n is a power of 2. The basic equation for the recursive algorithm is

$$c_n = 2c_{n/2} + 4\frac{n}{2}$$

The first 2 is because the FFT algorithm for n calls the FFT algorithm for $n/2$ twice. The term $4(n/2)$ is for the four operations, two multiplications, one addition, and one subtraction, each performed $n/2$ times after the two calls to the FFT algorithm for $n/2$. By iterating this equation we get

$$\begin{aligned}
c_n &= 2c_{n/2} + 4\frac{n}{2} = 2c_{n/2} + 2n \\
&= 2\left(2c_{n/4} + 4\frac{n}{4}\right) + 2n = 2^2 c_{n/4} + 4n \\
&= 2^2\left(2c_{n/8} + 4\frac{n}{8}\right) + 4n = 2^3 c_{n/8} + 6n
\end{aligned}$$

The pattern is now fairly clear. For each i from 1 to $\log_2 n$ we have

$$c_n = 2^i c_{n/2^i} + 2in$$

In particular, when $i = \log_2 n$ we get

$$c_n = nc_1 + 2n \log_2 n$$

But $c_1 = 0$ because there are no operations involved in the first step. So

$$c_n = 2n \log_2 n$$

What about the computation of the powers ω^k in the loop? The most efficient thing to do is to calculate all $N/2$ of these powers at the beginning, before entering the recursive FFT algorithm. This will require $N/2 - 1$ more multiplications in addition to the c_N steps of the recursive algorithm, so the total number of steps is

$$\frac{N}{2} - 1 + 2N \log_2 N$$

Compare these two numbers when $N = 1024$. The straightforward Fourier transform takes

$$2N (N - 1) = 2 \cdot 1024 \cdot 1023 = 2,095,104$$

steps, while the fast Fourier transform takes

$$\frac{N}{2} - 1 + 2N \log_2 N = 511 + 2 \cdot 1024 \cdot 10 = 20,991$$

steps.

Suppose F is a field and $\omega \in F$ is a primitive N-th root of 1. For example, the field GF_p has a primitive N-th root of 1 if $p - 1$ is divisible by N. If N is a power of 2, then the FFT and FFI algorithms can be combined to produce a fast multiplication for polynomials $f(t)$ and $g(t)$ in $F[t]$ whose product has degree less than N. Here is how you do it.

Let $\mathbf{f}$ and $\mathbf{g}$ be the lists of coefficients of $f(t)$ and $g(t)$. First form the Fourier transforms

$$\text{FFT}(N, \mathbf{f}, \omega) = \left(f\left(\omega^0\right), f\left(\omega^1\right), \ldots, f\left(\omega^{N-1}\right) \right)$$

and

$$\text{FFT}(N, \mathbf{g}, \omega) = \left(g\left(\omega^0\right), g\left(\omega^1\right), \ldots, g\left(\omega^{N-1}\right) \right)$$

Then take the componentwise product of these two transformations

$$\mathbf{y} = \left(f\left(\omega^0\right) g\left(\omega^0\right), f\left(\omega^1\right) g\left(\omega^1\right), \ldots, f\left(\omega^{N-1}\right) g\left(\omega^{N-1}\right) \right)$$

Finally, $\mathbf{x} = \text{FFI}(N, \mathbf{y}, \omega)$ gives the list of coefficients of the product $f(t) g(t)$ of the two polynomials.

Why is this faster than ordinary multiplication of polynomials? Suppose the degrees of $f(t)$ and $g(t)$ are $N/2$ and $N/2 - 1$. The polynomials have $N/2 + 1$ and $N/2$ coefficients, so ordinary multiplication requires $N^2/4 + N/2$ multiplications in the field F. We have to sum these $N^2/4 + N/2$ products into N sums, so the number of additions required is $N^2/4 + N/2 - N = N^2/4 - N/2$. Thus the total number of operations is $N^2/2$.

With fast multiplication of polynomials, we have three Fourier transforms and N multiplications in the field. The number of steps in a fast Fourier transformation is $N/2 - 1 + 2N \log_2 N$. For $N = 1024$, this last number is $20\,991$, so the number of steps for the fast multiplication is $3 \cdot 20\,991 + 1024 = 63\,997$, while the number of steps for ordinary multiplication is $1024^2/2 = 524\,288$.

Example 12.15 We compute the product of the two polynomials $f(t) = 1 + 5t + 3t^2 + t^4$ and $g(t) = 3 + 5t + 4t^4$ in $GF_{17}[x]$ by using FFT to evaluate each of these polynomials at w^i for $i = 0, 1, 2, 3, 4, 5, 6, 7$, where $w = 2$; then using pointwise multiplication to get the function values of the product at w^i for $i = 0, 1, 2, 3, 4, 5, 6, 7$; and then using FFI to get the coefficients of the product.

We first apply FFT as in Example 12.13 to get function values of $f(t) = 1 + 5t + 3t^2 + t^4$ and $g(t) = 3 + 5t + 4t^4$, then calculate products to generate the following table:

i	$f(2^i)$	$g(2^i)$	$f(2^i)g(2^i)$
0	10	12	1
1	5	9	11
2	2	10	3
3	11	5	4
4	0	2	0
5	2	6	12
6	13	4	1
7	16	10	7

To calculate the coefficients of $f(t) g(t)$, we use FFT with $w^{-1} = 2^{-1} \bmod 17 = 9$ as in Example 12.14 to get the polynomial

$$f(t) g(t) = 3 + 3t + 15t^3 + 7t^4 + 8t^5 + 12t^6 + 4t^8$$

Problems 12.6

1. Find a prime of the form $p = 2^5 k + 1$, and then find an element w in GF_p of order 2^5.

2. Verify that $p = 2^5 29 + 1 = 929$ is a prime. Calculate $a^{29 \cdot 2^4} \bmod 929$ for a few values of a until you find one such that $a^{29 \cdot 2^4} \bmod 929 \neq 1$. Use that to construct an element of order $32 = 2^5$ in GF_{929}.

3. Observe that $2^8 + 1 = 257$ is a prime. Find an element w in GF_{257} of order 2^8.

4. Find a prime of the form $p = 2^n + 1$ for $n > 8$. Find an element w in GF_p of order $p - 1$.

5. Verify that $p = 1 + 2^{30} \cdot 3^7 \cdot 5^8 \cdot 7^3 \cdot 11^2 \cdot 13 \cdot 17^2 \cdot 19 \cdot 23$ is a prime. Find an element of order $p-1$ in GF_p. Use this to find an element of order 2^{30} in GF_p.

6. Show that if ω is a primitive nth root of 1, then ω^{-1} is also a primitive nth root of 1.

7. Let ω be a root of $f(x) = x^5 + 2x + 1$ in GF_{243}. Show that ω is a primitive 242nd root of 1. Find ω^{-1}. What is the minimal polynomial of ω^{-1}?

8. With reference to the discussion on page 303, prove by induction on i that the formula $c_n = 2^i c_{n/2^i} + 2in$ is correct for $i = 1, 2, \ldots, \log_2 n$.

9. Let $f(t)$ and $g(t)$ be polynomials of degrees a and b with coefficients in a field. Show that multiplication of $f(t)$ by $g(t)$ requires $(a+1)(b+1)$ multiplications of field elements, and ab additions of field elements.

10. How many steps does the fast Fourier transform take for $N = 2048$? How does this compare to the number of steps for $N = 1024$?

Appendix A

Topics in Algebra and Number Theory

The application of algebra to codes, ciphers, and discrete algorithms relies on several topics in algebra and number theory. These topics are introduced in the text, but we collect some of the basic definitions and properties together here for easy reference.

A.1 Number Theory

Number theory is the study of the integers, $\ldots, -3, \ 2, -1, 0, 1, 2, 3, \ldots$, and how they behave under addition and multiplication. Often we deal with just the nonnegative integers, $0, 1, 2, 3, \ldots$, or with the positive integers, $1, 2, 3, 4, \ldots$.

We say that an integer n **divides** an integer m, and write $n|m$, if there is an integer a such that $na = m$. We also say that m is a **multiple** of n or that n is a **divisor** of m. If a and b are integers that are not both zero, then the **greatest common divisor** d of a and b is the largest of the common divisors of a and b. We write the greatest common divisor of a and b as $d = \gcd(a, b)$. Euclid gave an algorithm to compute the greatest common divisor over 2000 years ago, and this is still used extensively. The Euclidean algorithm takes only a modest number of steps to compute $\gcd(a, b)$.

An integer $p > 1$ is a **prime** if its only positive divisors are 1 and p. The first five primes are 2, 3, 5, 7, and 11. Primes, especially very large primes, play a very important role in cryptography. An integer $n > 1$ that is not prime is called a **composite**. The first five composites are 4, 6, 8, 9, and 10.

The fundamental theorem of arithmetic says that factorization of positive integers into primes is unique. Of course $2 \cdot 2 \cdot 3$ and $2 \cdot 3 \cdot 2$ are both ways of writing 12 as a product of primes, but they are not essentially different. The product is unique if we arrange the primes in order of size.

Theorem (The Fundamental Theorem of Arithmetic) Every integer greater than 1 is either a prime or can be written uniquely as a product of primes.

Modular arithmetic provides many powerful tools. If n and m are integers with $m > 0$, then we define $n \bmod m = n - \lfloor n/m \rfloor m$, where $\lfloor n/m \rfloor$ is the largest integer that is less than or equal to n/m. Note that $n \bmod m$ is the remainder when dividing n by m. For each positive integer m, this determines a congruence relation, on the integers. We say a **is congruent to** b modulo m and write $a \equiv b$ if $a \bmod m = b \bmod m$, or equivalently, if m divides the difference $a - b$. The Chinese remainder theorem is a statement about simultaneous congruences.

Theorem (Chinese Remainder Theorem)   Let $m_1, m_2, \ldots, m_r$ be pairwise relatively prime positive integers. Then the system $x \equiv a_i \pmod{m_i}$, $i = 1, 2, ..., r$ has a unique solution modulo $m_1 m_2 \cdots m_r$.

Codes, ciphers, and discrete algorithms are often based on the computational complexity of "hard" problems, such as the integer factorization problem in number theory. Among the important theorems and techniques in number theory we present in this text are the Euclidean algorithm, Fermat's little theorem, Euler's theorem, the principle of mathematical induction, Wilson's theorem, and the Chinese remainder theorem. The theorems and techniques of number theory are so pervasive in cryptography that if we gave any reasonably complete summary here of pertinent parts of the field, we would essentially be repeating this entire text.

For further information about number theory and how it relates to codes, ciphers, and discrete algorithms, see the following items in the index: $a^{-1} \bmod m$; $a \equiv b \bmod m$; Absolute value; Carmichael numbers; Ceiling; Chinese Remainder Theorem; Common divisor; Complete residue system; Composite; Congruence class; Congruence relation; Congruent; Disjoint cycles; Division algorithm for integers; Divisor; Euclidean algorithm; Euler phi function; Euler's theorem; Fermat's last theorem; Fermat's little theorem; Floor; Fundamental theorem of arithmetic; Greatest integer; Induction; Integer; Inverse modulo m; Least common multiple; Least nonnegative residue system; Mathematical induction; Method of exhaustion; Miller's test; Modular arithmetic; Mod function; Modular representation; Non-negative integer; Pairwise relatively prime; Pascal's triangle; Positive integer; Primality testing; Prime; Principle of mathematical induction; Probabilistic primes; Pseudo prime; Rabin's probabilistic primality test; Reduced residue system; Relatively prime; Residue system; Sieve of Eratosthenes; Systems of linear equations modulo n; Wilson's theorem.

A.2   Groups

A group is a basic algebraic system. The theory of groups is one of the oldest branches of algebra, and has applications in many areas of science as well as in other parts of mathematics. A **binary operation** on a set is a way of putting two elements of a set together to get a third element of that set; that is, a binary operation on a set S is a mapping $S \times S \to S$. A group is a set with a particular kind of binary operation on it. Here is a formal definition.

Definition   A **group** is a nonempty set G with one binary operation, often indicated simply by juxtaposition of elements of G, such that

1. $a(bc) = (ab)c$ for all a, b, c in G.

2. There is an element 1 in G such that $a1 = 1a = a$ for all a in G.

3. For each a in G, there is an element a^{-1} in G such that $aa^{-1} = a^{-1}a = 1$.

   The group is **commutative** if (4.) holds:

4. $ab = ba$ for all a, b in G.

   The group is **cyclic** if (5.) holds:

5. There is an element $g \in G$ such that every element $a \in G$ can be written as $a = g^m$ for some integer m, positive, negative, or zero. Here the convention is that $g^{-m} = \left(g^{-1}\right)^m$ and $g^0 = 1$.

If G is a finite group, the number of elements of G is called the order of the group G. If a is an element of a finite group G, there is a smallest positive integer t such that $a^t = 1$. This integer is the **order** of the group element a.

If S is a subset of G for which $ab \in S$ and $a^{-1} \in S$ for every pair of elements $a, b \in S$, then S is a group under the operation of G. Such an S is called a subgroup of G. If $a \in G$ has order t, then $\langle a \rangle = \{1, a, a^2, a^3, ..., a^{t-1}\}$ is a subgroup of G. Note that $\langle a \rangle$ is a cyclic group. Also note that the order of a and the order of $\langle a \rangle$ are the same, so the two uses of the word "order" are consistent.

Here are two basic facts relating the orders of groups and of their elements and subgroups:

Theorem (Lagrange)   If S is a subgroup of the finite group G, then the order of S divides the order of G.

Applying Lagrange's theorem to the case $S = \langle a \rangle$ gives the following.

Theorem   If a is an element of a finite group G, then the order of a divides the order of G.

Example   Here are several important examples of groups:

- The integers form a group under addition. This is an infinite group.

- The integers mod m form a finite group under addition.

- The set of permutations of $\{0, 1, 2, ..., n\}$ forms a finite group under the operation of concatenation.

  For further information about groups and how they relate to codes, ciphers, and discrete algorithms, see the following items in the index: Commutative group; Cyclic group; Group; Group of symmetries; Inverse permutation; Lagrange's theorem; Multiplicative group of a field; Order of a group; Order of a group element; Permutation; Product of permutations; Subgroup; Symmetric group.

A.3   Rings and Polynomials

In a ring you can add, subtract, and multiply.

Definition   A **ring** is a set R with an addition and a multiplication satisfying

1. $a + (b + c) = (a + b) + c$ for all a, b, c in R

2. $a + b = b + a$ for all a, b in R

3. there is an element 0 in R such that $a + 0 = a$ for all a in R

4. for each a in R there exists $-a$ in R such that $a + (-a) = 0$

5. $a(bc) = (ab)c$ for all a, b, c in R

6. $a(b + c) = ab + ac$ for all a, b, c in R

   A ring R is a **ring with identity** if

7. there is an element 1 in R such that $1a = a1 = a$ for all a in R

   A ring R is **commutative** if

8. $ab = ba$ for all a, b in R

Example   Here are several important examples of rings:

- The integers are a commutative ring with identity.

- The 2×2 matrices over the integers with the usual matrix addition and multiplication form a noncommutative ring with identity.

- The set of even integers is a commutative ring without an identity.

Consider polynomials with coefficients in an arbitrary ring. Given two polynomials

$$
\begin{aligned}
f(x) &= a_0 + a_1 x + a_2 x^2 + \cdots + a_n x^n \\
g(x) &= b_0 + b_1 x + b_2 x^2 + \cdots + b_m x^m
\end{aligned}
$$

we add them by adding their coefficients. The coefficient of x^i in the sum $f(x) + g(x)$ is $a_i + b_i$, where we set $a_i = 0$ if $i > n$ and $b_i = 0$ if $i > m$. The coefficient d_k of x^k in the product $f(x) \cdot g(x)$ is given by

$$
d_k = \sum_{i+j=k} a_i b_j
$$

that is $d_0 = a_0 b_0$, $d_1 = a_0 b_1 + a_1 b_0$, $d_2 = a_0 b_2 + a_1 b_1 + a_0 b_2$, ..., $d_{m+n-1} = a_{n-1} b_m + a_n b_{m-1}$, and $d_{m+n} = a_n b_m$. This is the product you get if you simply multiply $f(x)$ and $g(x)$ using the usual laws.

- The set of polynomials with coefficients in a commutative ring with identity is itself a commutative ring with identity.

- Of particular interest is the ring $F[x]$ of polynomials over a field F (see next section). Note that a polynomial of degree greater than zero does not have a multiplicative inverse.

  For further information about rings and polynomials and how they relate to codes, ciphers, and discrete algorithms, see the following items in the index: Commutative ring; Division algorithm for polynomials; Extended Euclidean algorithm; Generator polynomial. Greatest common divisor of integers; Greatest common divisor of polynomials; Greatest integer; Irreducible polynomial; Kronecker's algorithm; Lagrange interpolation; Minimal polynomial; Neville's iterated interpolation algorithm; Polynomial division; Polynomial division algorithm; Polynomial interpolation; Polynomial remainder; Polynomial ring; Primitive element; Primitive polynomial; Primitive root; Rational root theorem.

A.4 Fields

In a field you can add, subtract, multiply, and divide.

Definition A **field** is a set F with an addition and a multiplication satisfying

1. $a + (b + c) = (a + b) + c$ for all a, b, c in F

2. $a + b = b + a$ for all a, b in F

3. There is an element 0 in F such that $a + 0 = a$ for all a in F

4. For each a in F there is an element $-a$ in F such that $a + (-a) = 0$

5. $a(bc) = (ab)c$ for all a, b, c in F

6. $ab = ba$ for all a, b in F

7. There is an element 1 in F such that $a1 = a$ for all a in F

8. $a(b + c) = ab + ac$ for all a, b, c in F, and

9. For each nonzero a in F there is an element a^{-1} in F such that $aa^{-1} = 1$.

Example Here are several important examples of fields:

- Take F to be $\mathbb{Z}_p = \{0, 1, 2, \ldots, p - 1\}$ with addition and multiplication modulo p, then $\mathbb{Z}_p$ is a field. This field is also denoted by GF_p, and called the **Galois field** of order p.

- The rational numbers $\mathbb{Q}$ form a field as do the real numbers $\mathbb{R}$, and the complex numbers $\mathbb{C}$. These three fields are infinite. The applications to codes, ciphers, and finite algorithms in this text rely primarily on finite fields.

The following idea of how to construct a field from a field and an irreducible polynomial is due to Kronecker.

- Let F be a field and $h(x)$ an irreducible polynomial of degree m in $F[x]$. Define K to be the set of congruence classes of polynomials in $F[x]$ modulo $h(x)$. Let θ be the congruence class of the polynomial x, so if $f(x) \in F[x]$, then the congruence class of $f(x)$ is $f(\theta)$.

  Define multiplication in K by setting $f(\theta) g(\theta) = r(\theta)$ where $r(x) = (f(x) g(x)) \bmod h(x)$. This turns the set K into a field. If F is a finite field of order q, then the order of K is q^m.

For further information about fields and how they relate to codes, ciphers, and discrete algorithms, see the following items in the index: Binary representation of field element; Cayley table; Fast Fourier interpolation; Fast Fourier transform; Field; Galois field; GFp; GF_{p^n}; Multiplicative group of a field; Primitive element; Primitive polynomial; Primitive root; Splitting field.

A.5   Linear Algebra and Matrices

A **matrix** is a rectangular array of numbers or other algebraic expressions that can be added and multiplied. One application of matrices is to keep track of coefficients of linear equations. They are used in many ways in all of the mathematical sciences.

The horizontal lines of a matrix are called **rows** and the vertical lines are called **columns**. A matrix with m rows and n columns is called an m-by-n matrix. If either $m = 1$ or $n = 1$, the matrix is also called a **vector**.

The **product** of an $m \times k$ matrix with a $k \times n$ matrix

$$
\begin{pmatrix}
a_{11} & a_{12} & \cdots & a_{1k} \\
a_{21} & a_{22} & \cdots & a_{2k} \\
\vdots & \vdots & \ddots & \vdots \\
a_{m1} & a_{m2} & \cdots & a_{mk}
\end{pmatrix}
\begin{pmatrix}
b_{11} & b_{12} & \cdots & b_{1n} \\
b_{21} & b_{22} & \cdots & b_{2n} \\
\vdots & \vdots & \ddots & \vdots \\
b_{k1} & b_{k2} & \cdots & b_{kn}
\end{pmatrix}
$$

is the $m \times n$ matrix

$$
C =
\begin{pmatrix}
c_{11} & c_{12} & \cdots & c_{1n} \\
c_{21} & c_{22} & \cdots & c_{2n} \\
\vdots & \vdots & \ddots & \vdots \\
c_{m1} & c_{m2} & \cdots & c_{mn}
\end{pmatrix}
$$

where the ij^{th} entry of C is given by $c_{ij} = \sum_{r=1}^{k} a_{ir} b_{rj}$.

The $n \times n$ **identity matrix** is

$$I = \begin{pmatrix} 1 & 0 & \cdots & 0 \\ 0 & 1 & \cdots & 0 \\ \vdots & \vdots & \ddots & \vdots \\ 0 & 0 & \cdots & 1 \end{pmatrix}$$

The **inverse** of an $n \times n$ matrix A is an $n \times n$ matrix B such that $AB = I$. The inverse is denoted $B = A^{-1}$.

A system of m linear equations in n variables

$$\begin{aligned} a_{11}x_1 + a_{12}x_2 + \cdots + a_{1n}x_n &= b_1 \\ a_{21}x_1 + a_{22}x_2 + \cdots + a_{2n}x_n &= b_2 \\ &\vdots \\ a_{m1}x_1 + a_{m2}x_2 + \cdots + a_{mn}x_n &= b_m \end{aligned}$$

can be represented by the matrix equation

$$\begin{pmatrix} a_{11} & a_{12} & \cdots & a_{1n} \\ a_{21} & a_{22} & \cdots & a_{2n} \\ \vdots & \vdots & \ddots & \vdots \\ a_{m1} & a_{m2} & \cdots & a_{mn} \end{pmatrix} \begin{pmatrix} x_1 \\ x_2 \\ \vdots \\ x_n \end{pmatrix} = \begin{pmatrix} b_1 \\ b_2 \\ \vdots \\ b_m \end{pmatrix}$$

or by the **augmented matrix**

$$\begin{pmatrix} a_{11} & a_{12} & \cdots & a_{1n} & b_1 \\ a_{21} & a_{22} & \cdots & a_{2n} & b_2 \\ \vdots & \vdots & \ddots & \vdots & \vdots \\ a_{m1} & a_{m2} & \cdots & a_{mn} & b_m \end{pmatrix}$$

A system of linear equations can be solved by a series of steps (**1**) add a multiple of one equation to a second equation, (**2**) interchange two equations, or (**3**) multiply both sides of an equation by a nonzero constant. Applied to a matrix, this method is called **row reduction**. The object of row reduction is to end up with a matrix such that the first nonzero entry in each row is 1 (the leading 1) and the leading 1 in each row is to the right of the leading 1 in the previous row, as in the matrix

$$\begin{pmatrix} 1 & c_{12} & c_{13} & c_{14} & c_{15} & \cdots & d_1 \\ 0 & 0 & 1 & c_{15} & c_{25} & \cdots & d_2 \\ 0 & 0 & 0 & 1 & c_{35} & \cdots & d_3 \\ 0 & 0 & 0 & 0 & 1 & \cdots & d_4 \\ \vdots & \vdots & \ddots & \vdots & \vdots & \ddots & \vdots \\ 0 & 0 & 0 & 0 & 0 & \cdots & d_n \end{pmatrix}$$

Such a matrix is said to be in **row echelon form**. To get the solution of the system of linear equations, reinterpret each row of the matrix as an equation and use back substitution. If the last nonzero row is

$$\begin{pmatrix} 0 & 0 & \cdots & 1 & c_{ij} & \cdots & d_i \end{pmatrix}$$

it follows that

$$x_{j-1} = d_i - \sum_{k=j}^{n} c_{ik} x_k$$

where $x_j, \ldots, x_n$ can be assigned arbitrary values. You get a similar equation for each variable corresponding to a column with a leading 1, which you can evaluate working from the last nonzero row to the first. The variables corresponding to columns without a leading 1 can be assigned arbitrary values.

A system of n linear equations in n variables can be solved by using an inverse matrix, if you have one. If

$$\begin{pmatrix} a_{11} & a_{12} & \cdots & a_{1n} \\ a_{21} & a_{22} & \cdots & a_{2n} \\ \vdots & \vdots & \ddots & \vdots \\ a_{n1} & a_{n2} & \cdots & a_{nn} \end{pmatrix} \begin{pmatrix} x_1 \\ x_2 \\ \vdots \\ x_n \end{pmatrix} = \begin{pmatrix} b_1 \\ b_2 \\ \vdots \\ b_n \end{pmatrix}$$

then

$$\begin{pmatrix} x_1 \\ x_2 \\ \vdots \\ x_n \end{pmatrix} = \begin{pmatrix} a_{11} & a_{12} & \cdots & a_{1n} \\ a_{21} & a_{22} & \cdots & a_{2n} \\ \vdots & \vdots & \ddots & \vdots \\ a_{n1} & a_{n2} & \cdots & a_{nn} \end{pmatrix}^{-1} \begin{pmatrix} b_1 \\ b_2 \\ \vdots \\ b_n \end{pmatrix}$$

provided there is an inverse matrix.

The **affine transformation** given by a matrix A and a vector $\mathbf{b}$ is the map taking $\mathbf{x}$ to $\mathbf{y} = A\mathbf{x} + \mathbf{b}$. When such a transformation is used for encryption, we need to be able to reverse it. This can be done when A is invertible, in which case $\mathbf{x} = A^{-1}(\mathbf{y} - \mathbf{b})$.

The **determinant** function associates with every square matrix A a unique number $\det A$ obtained as the sum and difference of certain products of the entries. The **determinant** of the $n \times n$ matrix (a_{ij}) is

$$\det(a_{ij}) = \sum_{\sigma} (-1)^{sgn(\sigma)} a_{1\sigma(1)} a_{2\sigma(2)} \cdots a_{n\sigma(n)}$$

where σ ranges over all the permutations of $\{1, 2, \ldots, n\}$ and $(-1)^{sgn(\sigma)} = \pm 1$, depending on whether σ permutes an even or odd number of digits. Here are some examples:

$$\det \begin{pmatrix} a & b \\ c & d \end{pmatrix} = ad - bc$$

$$\det \begin{pmatrix} a_{11} & a_{12} & a_{13} \\ a_{21} & a_{22} & a_{23} \\ a_{31} & a_{32} & a_{33} \end{pmatrix} = a_{11}a_{22}a_{33} - a_{11}a_{23}a_{32} - a_{12}a_{21}a_{33}$$

$$+ a_{12}a_{31}a_{23} + a_{21}a_{13}a_{32} - a_{13}a_{22}a_{31}$$

$$\det \begin{pmatrix} 824 & -65 & -814 \\ -741 & -979 & -764 \\ 216 & 663 & 880 \end{pmatrix} = -96\,396\,086$$

One important use of the determinant is that it gives a test for invertibility: a square matrix A is invertible if and only if $\det A \neq 0$.

For further information about linear algebra and matrices and how they relate to codes, ciphers, and discrete algorithms, see the following items in the index: Boolean matrix, Hilbert matrix, Inverse matrix, Linear combination, Permutation matrix, Row reduction, Vandermonde matrix.

Solutions to Odd Problems

Chapter 1   Integers and Computer Algebra

Problems 1.1

1. Let $n = ab$ with $a > 1$ and $b > 1$. If $a^2 \leq n$, then a is a divisor of the right kind. Otherwise $a^2 > n$ and

$$b^2 = \frac{n^2}{a^2} < \frac{n^2}{n} = n$$

is a divisor of the right kind.

3. Note that $15\,(-3) + 24\,(2) = 3$, where $3|15$ and $3|24$.

5. A sum of two cubes factors as $a^3 + b^3 = (a \mid b)\,(a^2 - ab + b^2)$, so

$$
\begin{aligned}
2^9 + 5^{12} &= \left(2^3\right)^3 + \left(5^4\right)^3 \\
&= \left(2^3 + 5^4\right)\left(\left(2^3\right)^2 - \left(2^3\right)\left(5^4\right) + \left(5^4\right)^2\right) \\
&= (633)\,(385\,689) \\
&= (3 \times 211)\,(3 \times 128\,563) \\
&= 3^2 \times 211 \times 128\,563
\end{aligned}
$$

7. Let a, b, c, and d be arbitrary integers.

   (a) This is false. Indeed $4|12$ and $6|12$, but $4 \cdot 6 = 24$ does not divide 12. Even simpler, but maybe too simple: $2|2$ and $2|2$ but $2 \cdot 2 = 4$ does not divide 2.

   (b) If $a|b$, then $b = sa$ for some integer s. If $a|c$, then $c = ta$ for some integer t. So $b + c = sa + ta = (s + t)a$, which means that $a|(b + c)$. See also part iii of the theorem on properties of divisors.

   (c) This is false. Take $a = 2$, $b = 4$, and $c = 6$.

   (d) If $a|b$, then $b = at$ for some t, so $b^2 = a^2t^2$, which means that $a^2|b^2$.

(e) This is false. Take $a = 2$, $b = 4$, $c = 3$, and $d = 9$. Note that $2|4$ and $3|9$, but $5 \nmid 13$.

(f) Let $b = as$ and $d = ct$. Then $bd = asct = ac \cdot st$, which means that $ac|bd$.

(g) Let $b = as$. Then $bc = asc$, which means that $a|bc$.

9. No. The number 13 is not divisible by 3.

11. Let p be a prime divisor of $p_1 p_2 \cdots p_k + 1$. If $p = p_i$ for some i, then $p_1 p_2 \cdots p_k + 1 = p_i n$ for some integer n. So

$$1 = p_i n - p_1 p_2 \cdots p_k$$
$$= p_i \left(n - p_1 p_2 \cdots p_{i-1} p_{i+1} \cdots p_k \right)$$

making p_i a factor of 1. That can't be, so p must be different from each prime p_i.

This shows how, given any finite set of primes, you can find another prime not in that set. So there are infinitely many primes.

13.

$4 = 2 + 2$	$24 = 19 + 5$	$44 = 37 + 7$	$64 = 61 + 3$	$84 = 79 + 5$
$6 = 3 + 3$	$26 = 19 + 7$	$46 = 43 + 3$	$66 = 61 + 5$	$86 = 79 + 7$
$8 = 5 + 3$	$28 = 23 + 5$	$48 = 43 + 5$	$68 = 61 + 7$	$88 = 83 + 5$
$10 = 7 + 3$	$30 = 23 + 7$	$50 = 43 + 7$	$70 = 67 + 3$	$90 = 83 + 7$
$12 = 7 + 5$	$32 = 29 + 3$	$52 = 47 + 5$	$72 = 67 + 5$	$92 = 89 + 3$
$14 = 11 + 3$	$34 = 29 + 5$	$54 = 47 + 7$	$74 = 67 + 7$	$94 = 89 + 5$
$16 = 13 + 3$	$36 = 29 + 7$	$56 = 53 + 3$	$76 = 73 + 3$	$96 = 89 + 7$
$18 = 13 + 5$	$38 = 31 + 7$	$58 = 53 + 5$	$78 = 73 + 5$	$98 = 79 + 19$
$20 = 13 + 7$	$40 = 37 + 3$	$60 = 53 + 7$	$80 = 73 + 7$	
$22 = 19 + 3$	$42 = 37 + 5$	$62 = 59 + 3$	$82 = 79 + 3$	

15. Note that each factorization of $n! + 1$ involves a prime or primes larger than n.

(a) $2! + 1 = 3$ is prime

(b) $3! + 1 = 7$ prime

(c) $4! + 1 = 5^2$ composite

(d) $5! + 1 = 11^2$ composite

(e) $6! + 1 = 7 \times 103$ composite

(f) $7! + 1 = 71^2$ composite

(g) $8! + 1 = 61 \times 661$ composite

(h) $9! + 1 = 19 \times 71 \times 269$ composite

17. If three numbers are consecutive odd numbers, they must look like n, $n+2$, $n+4$, where n is odd. But if n is odd, then $n = 2m + 1$, where m is an integer. Thus, the three numbers are $2m + 1$, $2m + 3$, $2m + 5$. If $m - 1$, we get the triplet 3, 5, 7.

Suppose $m > 1$. If $2m + 1$ is not divisible by 3, then dividing $2m + 1$ by 3 gives a remainder of 1 or 2. If the remainder is 1, then $2m + 1 = 3k + 1$ for some integer k, and $2m + 3 = 3k + 3$ which is divisible by 3. If the remainder is 2, then $2m + 1 = 3k + 2$, and $2m + 5 = 3k + 6$ which is divisible by 3. Thus one of the three numbers is divisible by 3, and all are greater than 3, one of them is composite.

19. If $n = 1$, then $n^2 + 1 = 2$, a prime. But if n is an odd number greater than 1, then its square is also odd, and adding 1 gives an even number greater than 2, which is thus composite. So the answer to the first question is, if n is odd, $n^2 + 1$ is a prime if and only if $n = 1$.

If n is even, then $n^2 + 1$ can be prime, for example $2^2 + 1 = 5$ is a prime. But $n^2 + 1$ is not a prime for $n = 8$ because $8^2 + 1 = 65 = 5 \times 13$.

21. Yes. If n is composite, then $n = ab$, where $a > 1$ and $b > 1$, so

$$2^n - 1 = 2^{ab} - 1$$
$$= (2^a - 1)\left(2^{ab-a} + 2^{ab-2a} + 2^{ab-3a} + \cdots + 1\right)$$

which shows that $2^n - 1$ is composite. Thus if $2^n - 1$ is prime, then n is also prime.

Problems 1.2

1. Note that $|3.14 - \pi| = 1.5927 \times 10^{-3}$ and $\left|\frac{22}{7} - \pi\right| = 1.2645 \times 10^{-3}$, so $22/7$ is the closer approximation to π.

3. A computer algebra system produces $\pi = 3.141\,592\,654$ to ten decimal places

5. Note that $(-1)^{2/4} = \sqrt[4]{(-1)^2} = 1$, whereas $(-1)^{1/2} = \sqrt{1} = i$.

7. Let $x = 0.33333333333\ldots$. Then $10x = 3.33333333333\ldots$ so subtracting gives $9x = 3.0$, which implies $x = 3/9 = 1/3$.

9. Note that $x = 3.4\overline{89}$ and $100x = 348.9\overline{89}$, so $99x = 348.9\overline{89} - 3.4\overline{89} = 345.5 = 691/2$. This gives

$$x = \frac{691}{2 \cdot 99} = \frac{691}{198}$$

As a check, compute $691/198 \approx 3.489\,898\,990$.

11. Here is one solution:

$$\frac{a}{b} \approx 0.469\,387\,755 \approx \frac{1}{0.469\,387\,755^{-1}} \approx \frac{1}{2 + 0.\,130\,434\,783}$$

$$\approx \frac{1}{2 + \frac{1}{0.\,130\,434\,783^{-1}}} \approx \frac{1}{2 + \frac{1}{7 + 0.\,666\,666\,644}} \approx \frac{1}{2 + \frac{1}{7 + \frac{1}{0.\,666\,666\,644^{-1}}}}$$

$$\approx \frac{1}{2 + \frac{1}{7 + \frac{1}{1 + .\,500\,000\,051}}} \approx \frac{1}{2 + \frac{1}{7 + \frac{1}{1 + \frac{1}{.\,500\,000\,051^{-1}}}}} \approx \frac{1}{2 + \frac{1}{7 + \frac{1}{1 + \frac{1}{1 + \frac{1}{.2}}}}} = \frac{23}{49}$$

Checking,

$$\frac{23}{49} \approx 0.469\,387\,755\,1$$

Problems 1.3

1. $\displaystyle\sum_{j=1}^{10} 2 = \overbrace{2 + 2 + \cdots + 2}^{10} = 10 \cdot 2 = 20$

3. $\displaystyle\sum_{i=0}^{10}\sum_{j=0}^{10} (i+1)(j+1) = 4356$. To see why this is correct, note that

$$\sum_{i=0}^{10}\sum_{j=0}^{10}(i+1)(j+1) = \left(\sum_{i=0}^{10}(i+1)\right) \cdot \left(\sum_{j=0}^{10}(j+1)\right)$$

$$= \left(\frac{11 \cdot 12}{2}\right)^2 = 4356$$

5. $\displaystyle\sum_{k=2}^{n} k^3 = \frac{1}{4}(n+1)^4 - \frac{1}{2}(n+1)^3 + \frac{1}{4}(n+1)^2 - 1$

7. $\displaystyle\sum_{k=n}^{m} 1 = \overbrace{1 + 1 + \cdots + 1}^{m-n+1 \text{ times}} = m - n + 1$

9. $\displaystyle\sum_{k=0}^{5}(1+k^3) = (1+0^3) + (1+1^3) + (1+2^3) + (1+3^3) + (1+4^3) + (1+5^3) = 1 + 2 + 9 + 28 + 65 + 126 = 231$

11. $\displaystyle\prod_{i=1}^{10} i^2 = 1^2 \cdot 2^2 \cdot 3^2 \cdot 4^2 \cdot 5^2 \cdot 6^2 \cdot 7^2 \cdot 8^2 \cdot 9^2 \cdot 10^2 = 1^2 \cdot 2^2 \cdot 3^2 \cdot 2^4 \cdot 5^2 \cdot 2^2 \cdot 3^2 \cdot 7^2 \cdot 2^6 \cdot 3^4 \cdot 2^2 \cdot 5^2 = 2^{16}3^85^47^2 = 13\,168189\,440\,000$

13. $\prod\limits_{n=1}^{5} (n^2 + n) = 2^7 3^3 5^2 = 86\,400$

15. $\sum\limits_{i=n}^{m} (a_i + b_i) = a_n + b_n + a_{n+1} + b_{n+1} + \cdots + a_m + b_m = a_n + a_{n+1} + \cdots +$

$a_m + b_n + b_{n+1} + \cdots + b_m = \sum\limits_{i=n}^{m} a_i + \sum\limits_{i=n}^{m} b_i$

Problems 1.4

1. $\sum\limits_{i=1}^{n} i^2 = \frac{1}{3}(n+1)^3 - \frac{1}{2}(n+1)^2 + \frac{1}{6}n + \frac{1}{6} = \frac{1}{6}n(n+1)(2n+1)$. To verify this result by induction on n, first note that with $n = 1$ we have

$$\sum_{i=1}^{1} i^2 = 1^2 = 1 = \frac{1}{6}(1)(1+1)(2(1)+1)$$

Then assume that

$$\sum_{i=1}^{k} i^2 = \frac{1}{6}k(k+1)(2k+1)$$

for some $k \geq 1$. Then

$$\sum_{i=1}^{k+1} i^2 = (k+1)^2 + \sum_{i=1}^{k} i^2$$

$$= (k+1)^2 + \frac{1}{6}k(k+1)(2k+1)$$

$$= \frac{1}{6}(k+1)((k+1)+1)(2(k+1)+1)$$

so, by the principle of mathematical induction, for all $n \geq 1$ we have

$$\sum_{i=1}^{n} i^2 = \frac{1}{6}n(n+1)(2n+1)$$

3. A computer algebra system yields

$$\sum_{k=1}^{n} k2^k = (n+1)\,2^{n+1} - 2 \times 2^{n+1} + 2$$

which further simplifies to

$$\sum_{k=1}^{n} k2^k = 2^{n+1}(n+1-2) + 2 = 2^{n+1}(n-1) + 2$$

See the disc for the proof by mathematical induction.

5. The sum is given by

$$\sum_{r=0}^{n} \binom{n}{r} x^r = (1+x)^n$$

To verify this by induction, first observe that

$$\sum_{r=0}^{0} \binom{0}{r} x^r = \binom{0}{0} x^0 = 1 \text{ and } (1+x)^0 = 1$$

Assume that for some $n \geq 1$, $\sum_{r=0}^{n} \binom{n}{r} x^r = (1+x)^n$. Then,

$$\sum_{r=0}^{n+1} \binom{n+1}{r} x^r = \binom{n+1}{0} x^0 + \binom{n+1}{n+1} x^{n+1} + \sum_{r=1}^{n} \binom{n+1}{r} x^r$$

$$= 1 + x^{n+1} + \sum_{r=1}^{n} \left(\binom{n}{r} + \binom{n}{r-1} \right) x^r$$

$$= 1 + x^{n+1} + \sum_{r=1}^{n} \binom{n}{r} x^r + \sum_{r=1}^{n} \binom{n}{r-1} x^r$$

$$= 1 + x^{n+1} + \sum_{r=1}^{n} \binom{n}{r} x^r + \sum_{r=0}^{n-1} \binom{n}{r} x^{r+1}$$

$$= \left(1 + \sum_{r=1}^{n} \binom{n}{r} x^r \right) + \left(x^{n+1} + x \sum_{r=0}^{n-1} \binom{n}{r} x^r \right)$$

$$= \sum_{r=0}^{n} \binom{n}{r} x^r + x \sum_{r=0}^{n} \binom{n}{r} x^r = (1+x)^n + x(1+x)^n$$

$$= (1+x)(1+x)^n = (1+x)^{n+1}$$

7. For $n = 1$, $\sum_{k=1}^{1} \frac{1}{1(1+1)} = \frac{1}{2} = \frac{1}{1+1}$. Assume that $\sum_{k=1}^{n} \frac{1}{k(k+1)} = \frac{n}{n+1}$ holds for some n. Then

$$\sum_{k=1}^{n+1} \frac{1}{k(k+1)} = \frac{n}{n+1} + \frac{1}{(n+1)((n+1)+1)}$$

$$= \frac{n(n+2)}{(n+1)(n+2)} + \frac{1}{(n+1)(n+2)}$$

$$= \frac{n^2 + 2n + 1}{(n+1)(n+2)} = \frac{(n+1)^2}{(n+1)(n+2)} = \frac{n+1}{n+2}$$

9. For $n = 1$, we have $\sum_{r=0}^{1} \binom{1}{r} = \binom{1}{0} + \binom{1}{1} = 1 + 1 = 2^1$. Assume that $\sum_{r=0}^{n} \binom{n}{r} = 2^n$ holds for some n. In the next problem you are asked to

prove that

$$\binom{n+1}{r+1} = \binom{n}{r} + \binom{n}{r+1}$$

if $0 < r \le n$, and we use that here. So

$$\sum_{r=0}^{n+1}\binom{n+1}{r} = \sum_{r=-1}^{n}\binom{n+1}{r+1}$$

$$= \binom{n+1}{0} + \sum_{r=0}^{n-1}\binom{n}{r} + \sum_{r=0}^{n-1}\binom{n}{r+1} + \binom{n+1}{n+1}$$

$$= \binom{n+1}{0} + \left(\sum_{r=0}^{n}\binom{n}{r} - \binom{n}{n}\right)$$

$$+ \left(\sum_{r=0}^{n}\binom{n}{r} - \binom{n}{0}\right) + \binom{n+1}{n+1}$$

$$= 1 + (2^n - 1) + (2^n - 1) + 1 = 2 \cdot 2^n = 2^{n+1}$$

11. For $n = 1$, we have $(a + b)^1 = a + b = \binom{1}{0}a^1 b^0 + \binom{1}{1}a^{1-1}b^1$ since $\binom{1}{0} = \binom{1}{1} = 1$. Assume that $(a + b)^n = \sum_{r=0}^{n}\binom{n}{r}a^{n-r}b^r$. Then

$$(a + b)^{n+1} = (a + b)(a + b)^n$$

$$= (a + b)\sum_{r=0}^{n}\binom{n}{r}a^{n-r}b^r$$

$$= a\sum_{r=0}^{n}\binom{n}{r}a^{n-r}b^r + b\sum_{r=0}^{n}\binom{n}{r}a^{n-r}b^r$$

$$= \cdots = \sum_{r=0}^{n+1}\binom{n+1}{r}a^{n+1-r}b^r$$

which shows that the formula also holds with n replaced with $n + 1$. (See the disc for details of this simplification.)

13. For $n = 1$ we have

$$(x + y)^1 = x + y = \binom{1}{0}x^1 y^0 + \binom{1}{1}x^0 y^1$$

since $\binom{1}{0} - \binom{1}{1} = 1$. Assume that

$$(x + y)^n = \sum_{r=0}^{n}\binom{n}{r}x^{n-r}y^r$$

Then

$$(x+y)^{n+1} = (x+y)(x+y)^n = (x+y)\sum_{r=0}^{n}\binom{n}{r}x^{n-r}y^r$$

$$= x\sum_{r=0}^{n}\binom{n}{r}x^{n-r}y^r + y\sum_{r=0}^{n}\binom{n}{r}x^{n-r}y^r$$

$$= \sum_{r=0}^{n}\binom{n}{r}x^{n+1-r}y^r + \sum_{r=0}^{n}\binom{n}{r}x^{n-r}y^{r+1}$$

$$= \sum_{r=0}^{n}\binom{n}{r}x^{n+1-r}y^r + \sum_{r=1}^{n+1}\binom{n}{r-1}x^{n-(r-1)}y^r$$

$$= \sum_{r=0}^{n}\binom{n}{r}x^{n+1-r}y^r + \sum_{r=1}^{n+1}\binom{n}{r-1}x^{n-r+1}y^r$$

$$= \binom{n}{0}x^{n+1}y^0 + \sum_{r=1}^{n+1}\binom{n}{r}x^{n+1-r}y^r + \sum_{r=1}^{n+1}\binom{n}{r-1}x^{n-r+1}y^r$$

$$= \binom{n}{0}x^{n+1}y^0 + \sum_{r=1}^{n+1}\left[\binom{n}{r}+\binom{n}{r-1}\right]x^{n+1-r}y^r$$

$$= \binom{n}{0}x^{n+1}y^0 + \sum_{r=1}^{n+1}\binom{n+1}{r}x^{n+1-r}y^r$$

$$= \sum_{r=0}^{n+1}\binom{n+1}{r}x^{n+1-r}y^r$$

which shows that the formula also holds with n replaced with $n+1$.

15. Direct evaluation yields

$$16! = 20\,922\,789\,888\,000$$
$$14!5!2! = 20\,922\,789\,888\,000$$

Note that

$$14! \times 5! \times 2! = 14! \times 2^4 \times 3 \times 5 = 14! \times 16 \times 15 = 16!$$

17. For $n = 1$ we have

$$\frac{1}{\sqrt{5}}\left(\frac{1+\sqrt{5}}{2}\right)^1 - \frac{1}{\sqrt{5}}\left(\frac{1-\sqrt{5}}{2}\right)^1 = 1 = F_1$$

$$\frac{1}{\sqrt{5}}\left(\frac{1+\sqrt{5}}{2}\right)^0 - \frac{1}{\sqrt{5}}\left(\frac{1-\sqrt{5}}{2}\right)^0 = 0 = F_0$$

Suppose these two equations hold for $n = k$, so

$$F_k = \frac{1}{\sqrt{5}} \left(\frac{1+\sqrt{5}}{2} \right)^k - \frac{1}{\sqrt{5}} \left(\frac{1-\sqrt{5}}{2} \right)^k$$

$$F_{k-1} = \frac{1}{\sqrt{5}} \left(\frac{1+\sqrt{5}}{2} \right)^{k-1} - \frac{1}{\sqrt{5}} \left(\frac{1-\sqrt{5}}{2} \right)^{k-1}$$

Then

$$F_{k+1} = F_k + F_{k-1}$$

$$= \frac{1}{\sqrt{5}} \left(\frac{1+\sqrt{5}}{2} \right)^k - \frac{1}{\sqrt{5}} \left(\frac{1-\sqrt{5}}{2} \right)^k$$

$$+ \frac{1}{\sqrt{5}} \left(\frac{1+\sqrt{5}}{2} \right)^{k-1} - \frac{1}{\sqrt{5}} \left(\frac{1-\sqrt{5}}{2} \right)^{k-1}$$

$$= \frac{1}{\sqrt{5}} \left(\frac{1+\sqrt{5}}{2} \right)^{k-1} \left(1 + \frac{1+\sqrt{5}}{2} \right)$$

$$- \frac{1}{\sqrt{5}} \left(\frac{1-\sqrt{5}}{2} \right)^{k-1} \left(1 + \frac{1-\sqrt{5}}{2} \right)$$

$$= \frac{1}{\sqrt{5}} \left(\frac{1+\sqrt{5}}{2} \right)^{k-1} \left(\frac{1+\sqrt{5}}{2} \right)^2$$

$$- \frac{1}{\sqrt{5}} \left(\frac{1-\sqrt{5}}{2} \right)^{k-1} \left(\frac{1-\sqrt{5}}{2} \right)^2$$

$$= \frac{1}{\sqrt{5}} \left(\frac{1+\sqrt{5}}{2} \right)^{k+1} - \frac{1}{\sqrt{5}} \left(\frac{1-\sqrt{5}}{2} \right)^{k+1}$$

so the two equations hold for $n = k + 1$.

19. Notice that if $x = 2$, then

$$\sum_{k=1}^{n} k 2^k = \sum_{k=1}^{n} k x^k = x \sum_{k=1}^{n} k x^{k-1} = x \sum_{k=0}^{n-1} (k+1) x^k$$

If we define

$$f(x) = \sum_{k=0}^{n-1} (k+1) x^k$$

then

$$\int f(x)\, dx = \sum_{k=0}^{n-1} x^{k+1} = x \sum_{k=0}^{n-1} x^k = x \frac{x^n - 1}{x - 1} = \frac{x^{n+1} - x}{x - 1}$$

and hence

$$xf(x) = x\frac{d}{dx}\left(\frac{x^{n+1} - x}{x - 1}\right)$$

$$= x\frac{((n+1)x^n - 1)(x-1) - (x^{n+1} - x)}{(x-1)^2}$$

$$= \frac{-x^{n+1} + x^{n+2}n - x^{n+1}n + x}{(x-1)^2}$$

$$= \frac{x^{n+1}(nx - n - 1) + x}{(x-1)^2}$$

Therefore,

$$\sum_{k=1}^{n} k2^k = 2f(2)$$

$$= \frac{2^{n+1}(2n - n - 1) + 2}{(2-1)^2}$$

$$= 2^{n+1}(n-1) + 2$$

Chapter 2   Codes

Problems 2.1

1. Note that

$$\underbrace{101}_{5}\ \underbrace{1010}_{A}\ \underbrace{1011}_{B}\ \underbrace{1001}_{9}\ \underbrace{0010}_{2} = (1011010101110010010)_2$$

3. Evaluation produces

$$((((5 \cdot 16) + 10) \cdot 16 + 11) \cdot 16 + 9) \cdot 16 + 2 = 371\,602$$

The definition, in terms of the decimal representation, produces

$$(5AB92)_{16} = 5 \times 16^4 + 10 \times 16^3 + 11 \times 16^2 + 9 \times 16 + 2$$
$$= (371\,602)_{10}$$

5. Note that $(10101011100001110101001101010100)_2 =$

$$\underbrace{1010}_{A}\ \underbrace{1011}_{B}\ \underbrace{1000}_{8}\ \underbrace{0111}_{7}\ \underbrace{0101}_{5}\ \underbrace{0011}_{3}\ \underbrace{0101}_{5}\ \underbrace{0100}_{4} = (AB875354)_{16}$$

7. We have $(10101011100001110101001101010100)_2 = 2^{31} + 2^{29} + 2^{27} + 2^{25}$
$+ 2^{24} + 2^{23} + 2^{18} + 2^{17} + 2^{16} + 2^{14} + 2^{12} + 2^9 + 2^8 + 2^6 + 2^4 + 2^2 = 2877\,772\,628$

9. Conversion yields

$$50927341 = (11000010010001011011101101)_2$$

11 Note that

$50927341 \bmod 16$	=	13	$(50927341 - 13)/16$	=	3182958
$3182958 \bmod 16$	=	14	$(3182958 - 14)/16$	=	198934
$198934 \bmod 16$	=	6	$(198934 - 6)/16$	=	12433
$12433 \bmod 16$	=	1	$(12433 - 1)/16$	=	777
$777 \bmod 16$	=	9	$(777 - 9)/16$	=	48
$48 \bmod 16$	=	0	$(48 - 0)/16$	=	3
$3 \bmod 16$	=	3			

and hence

$$(50927341)_{10} = (30916ED)_{16}$$

13. Multiplication yields

$$
\begin{array}{rccccc}
 & & C & 1 & F \\
 & \times & 2 & B \\
\hline
 & & & A & 5 \\
 & & & B & 0 \\
 & 8 & 4_2 & 0 & 0 \\
 & & 1 & E & 0 \\
 & & 2 & 0 & 0 \\
 1_1 & 8 & 0 & 0 & 0 \\
\hline
 2 & 0 & 9 & 3 & 5 \\
\end{array}
$$

and hence

$$(2B)_{16} \times (C1F)_{16} = (20935)_{16}$$

15. Multiplication yields

$$
\begin{array}{cccccccccc}
 & & & 1 & 0 & 1 & 1 & 0 & 1 \\
 & & & \times & 1 & 1 & 0 & 1 \\
\hline
 & & & 1_1 & 0_1 & 1_1 & 1 & 0 & 1 \\
 & & 1_1 & 0_{10} & 1 & 1 & 0 & 1 \\
 & 1_1 & 0 & | & 1 & 0 & 1 \\
\hline
 1 & 0 & 0 & 1 & 0 & 0 & 1 & 0 & 0 & 1 \\
\end{array}
$$

and hence

$$(101101)_2 \times (1101)_2 = (1001001001)_2$$

Check by converting to decimal:

$$
\begin{aligned}
(101101)_2 &= 2^5 + 2^3 + 2^2 + 2^0 = 45 \\
(1101)_2 &= 2^3 + 2^2 + 2^0 = 13 \\
45 \cdot 13 &= 585 \\
(1001001001)_2 &= 2^9 + 2^6 + 2^3 + 1 = 585
\end{aligned}
$$

17.

$$
\begin{array}{r}
4\quad 8\\
2\quad B\,\big|\,\overline{C_2\quad 1\quad F}\\
A\quad C\\
\hline
1\quad 5_5\quad F\\
1\quad 5\quad 8\\
\hline
7
\end{array}
$$

and hence

$$
\frac{(C1F)_{16}}{(2B)_{16}} = (47)_{16} + \frac{(22)_{16}}{(2B)_{16}}
$$

19. Long division yields

$$
\begin{array}{r}
1\quad 1\\
1\quad 1\quad 0\quad 1\,\big|\,\overline{1\quad 0\quad 1\quad 1\quad 0\quad 1}\\
1\quad 1\quad 0\quad 1\\
\hline
1\quad 0\quad 0\quad 1\quad 1\\
1\quad 1\quad 0\quad 1\\
\hline
1\quad 1\quad 0
\end{array}
$$

and hence

$$
\frac{(101101)_2}{(1101)_2} = (11)_2 + \frac{(110)_2}{(1101)_2}
$$

Checking,

$$
\begin{aligned}
(1101)_2 \times (11)_2 + (110)_2 &= \left(2^3 + 2^2 + 1\right)(2+1) + 6 = 45\\
(101101)_2 &= 2^5 + 2^3 + 2^2 + 1 = 45
\end{aligned}
$$

Problems 2.2

1. *Number theory is the queen of mathematics.*

3.

```
4D 61 74 68 65 6D 61 74 69 63 73 20 69 73
20 74 68 65 20 71 75 65 65 6E 20 6F 66 20
74 68 65 20 73 63 69 65 6E 63 65 73 2E
```

5. *ASCII is pronounced ask-ee.*

7. *Unicode contains over thirty thousand distinct coded characters.*

Problems 2.3

1. H E L P M Y F L O

    A

    T I N G P O I N T N U M

    B E R S A R E S I N K I N

    G

Help, my floating point numbers are sinking.

3. A T Y A L E C O L L

    E G E M O R S E D E L I

    G H T E D I N P A I N T I

    N G M I N I A T U R E P

    O R T R A I T S

At Yale College Morse delighted in painting miniature portraits.

5. Using the numbers of units from Problem 4, we have

Letter	A	B	C	D	E	F	G	H	I
%	7.3	0.9	3.0	4.4	13.0	2.8	1.6	3.5	7.4
Units	5	9	11	7	1	9	9	7	3
Exp.#	0.37	0.08	0.33	0.31	0.13	0.25	0.14	0.24	0.22
Letter	J	K	L	M	N	O	P	Q	R
%	0.2	0.3	3.5	2.5	7.8	7.4	2.7	0.3	7.7
Units	13	9	9	7	5	11	11	13	7
Exp.#	0.03	0.03	0.32	0.18	0.39	0.81	0.30	0.04	0.54
Letter	S	T	U	V	W	X	Y	Z	
%	6.3	9.3	2.7	1.3	1.6	0.5	1.9	0.1	
Units	5	3	7	9	9	11	13	11	
Exp.#	0.32	0.28	0.19	0.12	0.14	0.06	0.25	0.01	

By adding together the contributions from each letter, the expected number of time units per letter is 6.06. Notice that the most frequently occurring letter (E) has the shortest code.

7. *Suddenly, Samuel had an idea. Maybe electricity could transmit messages.*

Problems 2.4

1. *At the age of ten he attended a school for blind boys in Paris.*

    A   T   T   H   E   A   G   E

3. *Braille is used in virtually every language.*

5. *Braille became a talented cellist.*

7. *Braille was admired as a teacher.*

Problems 2.5

1. The combinations are 1100, 1010, 1001, 0110, 0101, and 0011, so there are six arrangements that can be made by choosing 2 from 4. Note also that this is the value of the binomial coefficient

$$\binom{4}{2} = 6$$

3.

$$
\begin{array}{ccccc}
000111 & 001011 & 001101 & 001110 & 010011 \\
010101 & 010110 & 011001 & 011010 & 011100 \\
100011 & 100101 & 100110 & 101001 & 101010 \\
101100 & 110001 & 110010 & 110100 & 111000 \\
\end{array}
$$

5. From Problem 4 we see that $\binom{10}{3} = 120$ and $\binom{9}{4} = 126$ are reasonable choices. A single error would be easier to recognize than for a 7-bit binary code. With a 7-bit binary code, a single error would yield a different ASCII value. With a 4-out-of-9 code, a single error would lead to 3 out of 9 or 5 out of 9, which could immediately be recognized as an error.

7. Use induction. If a set has 1 element x, then the subsets are $\{x\}$ and the empty set, so there are $2^1 = 2$ subsets. Assume that a set with n elements has a total of 2^n subsets. Consider a set with $n+1$ elements formed by adding a new element z. There are 2^n subsets that do not contain z and there are 2^n subsets that do contain z, so there are

$$2^n + 2^n = 2^{n+1}$$

subsets of the set with $n+1$ elements.

The formula

$$\sum_{k=0}^{m} \binom{m}{k} = 2^m$$

simply states that every subset has a certain number k of elements.

Problems 2.6

1. Lining up the slant cuts in a deck of IBM cards assures that all the cards are right side up and facing in the correct direction.

3. Using 12 locations 2 at a time yields $\binom{12}{2} = 66$ combinations, plus $\binom{12}{1} = 12$ single punches for a total of 78.

5. There are $\binom{12}{4} = 495$ patterns with four holes.

Chapter 3   Euclidean Algorithm

Problems 3.1

1. Find remainders modulo 5:

$$
\begin{array}{llll}
9374 \bmod 5 & = & 4 & \quad (9374 - 4)/5 & = & 1874 \\
1874 \bmod 5 & = & 4 & \quad (1874 - 4)/5 & = & 374 \\
374 \bmod 5 & = & 4 & \quad (374 - 4)/5 & = & 74 \\
74 \bmod 5 & = & 4 & \quad (74 - 4)/5 & = & 14 \\
14 \bmod 5 & = & 4 & \quad (14 - 4)/5 & = & 2 \\
2 \bmod 5 & = & 2 & &
\end{array}
$$

so $9374 = (244444)_5$. As a check, note that

$$((((2 \cdot 5 + 4) \cdot 5 + 4) \cdot 5 + 4) \cdot 5 + 4) \cdot 5 + 4 = 9374.$$

3. The addition and multiplication tables are given by

$\oplus$	0	1	2	3
0	0	1	2	3
1	1	2	3	0
2	2	3	0	1
3	3	0	1	2

$\otimes$	0	1	2	3
0	0	0	0	0
1	0	1	2	3
2	0	2	0	2
3	0	3	2	1

We see from the addition table that $2 \oplus 3 = 1$. However, $2 \otimes x = 1$ has no solution because 1 does not appear in the row 2 0 2 0 2.

5. The addition and multiplication tables are given by

$\oplus$	0	1	2	3	4	5	6
0	0	1	2	3	4	5	6
1	1	2	3	4	5	6	0
2	2	3	4	5	6	0	1
3	3	4	5	6	0	1	2
4	4	5	6	0	1	2	3
5	5	6	0	1	2	3	4
6	6	0	1	2	3	4	5

$\otimes$	0	1	2	3	4	5	6
0	0	0	0	0	0	0	0
1	0	1	2	3	4	5	6
2	0	2	4	6	1	3	5
3	0	3	6	2	5	1	4
4	0	4	1	5	2	6	3
5	0	5	3	1	6	4	2
6	0	6	5	4	3	2	1

If a and b are in the set $\{0, 1, 2, 3, 4, 5, 6\}$, then you can always solve the equation $a + x = b$ because b appears in every row. If $a = 0$, then $a \otimes x = b$ can be solved only if $b = 0$. Otherwise, b appears in the row headed by a and hence $a \otimes x = b$ has a solution.

7. There are many possibilities. For example, given a letter that corresponds to the number x, associate the letter that corresponds to

$$y = x + a \bmod 26$$

for any fixed integer u strictly between 0 and 26. Another possibility would be to associate the letter that corresponds with x, the letter that corresponds with the number

$$y = ax \bmod 26$$

where a is a particular fixed integer such as 3 or 5. For a choose an odd integer other than 13 between 0 and 26. A third possibility would be to do a combination of these two schemes. Given a letter that corresponds to the number x, associate the letter that corresponds to the number

$$y - ax + b \bmod 26$$

where $b \neq 0$ and a is an odd integer between 0 and 26 other than 13.

9. Add 3 to both sides of the equation $5x + 8 = 4$ to get $5x + 11 = 7$, which is the same as $5x = 7$ in the integers modulo 11. From the multiplication table

×	1	2	3	4	5	6	7	8	9	10
1	1	2	3	4	5	6	7	8	9	10
2	2	4	6	8	10	1	3	5	7	9
3	3	6	9	1	4	7	10	2	5	8
4	4	8	1	5	9	2	6	10	3	7
5	5	10	4	9	3	8	2	7	1	6
6	6	1	7	2	8	3	9	4	10	5
7	7	3	10	6	2	9	5	1	8	4
8	8	5	2	10	7	4	1	9	6	3
9	9	7	5	3	1	10	8	6	4	2
10	10	9	8	7	6	5	4	3	2	1

locate 7 in row 5. The column header is 8 and $5 \cdot 8 \bmod 11 = 7$. Checking, $5 \cdot 8 + 8 \bmod 11 = 4$, so $x = 8$ is indeed a solution.

11. There are only 11 possible choices for x, so just try them. Let $f(x) = x^2 + 9x + 9 \bmod 11$. Then

$$
\begin{array}{llll}
f(0) = 9 & f(1) = 8 & f(2) = 9 & f(3) = 1 \\
f(4) = 6 & f(5) = 2 & f(6) = 0 & f(7) = 0 \\
f(8) = 2 & f(9) = 6 & f(10) = 1 &
\end{array}
$$

Thus $x = 6$ or $x = 7$. Now that the solution is known, it is easy to see that factorization would also work. That is, $(x - 6)(x - 7) \bmod 11 = x^2 + 9x + 9$.

Problems 3.2

1. The positive divisors of 48 and of 72 are

$$d(48) = \{1, 2, 3, 4, 6, 8, 12, 16, 24, 48\}$$
$$d(72) = \{1, 2, 3, 4, 6, 8, 9, 12, 18, 24, 36, 72\}$$

The intersection of these two sets is $\{1, 2, 3, 4, 6, 8, 12, 24\}$ and $\max\{1, 2, 3, 4, 6, 8, 12, 24\} = 24$.

3. Factoring gives $55\,440 = 2^4 3^2 5 \times 7 \times 11$ and $48\,000 = 2^7 3 \times 5^3$, so

$$\gcd{(55\,440, 48\,000)} = 2^{\min\{4,7\}} 3^{\min\{1,2\}} 5^{\min\{1,3\}} 7^{\min\{0,1\}} 11^{\min\{0,1\}}$$
$$= 2^4 3^1 5^1 7^0 11^0 = 240$$

5. From the table

n	q_n	r_n	n	q_n	r_n
0		29 432 403	4	9	677 131
1		22 254 869	5	1	45 136
2	1	7177 534	6	15	91
3	3	722 267	7	496	0

we see that $\gcd{(29\,432\,403, 22\,254\,869)} = 91$

7. Let a and $a + 5$ denote the two integers. If $k \mid a$ and $k \mid a + 5$ then also $k \mid (a + 5) - a = 5$ and hence $k \mid 5$. Thus $\gcd{(a, a + 5)} = 1$ or $\gcd{(a, a + 5)} = 5$.

9. Note that

$$\begin{aligned} \gcd{(a, b)} &= \gcd{(b, a \bmod b)} = \gcd{(b, r_0)} = \gcd{(r_0, b \bmod r_0)} \\ &= \gcd{(r_0, r_1)} = \gcd{(r_1, r_0 \bmod r_1)} = \gcd{(r_1, r_2)} \end{aligned}$$

and, in general,

$$\gcd{(r_{k-1}, r_k)} = \gcd{(r_k, r_{k-1} \bmod r_k)} = \gcd{(r_k, r_{k+1})}$$

11. The remainders, starting with 233 and 144, are

$$89, \ 55, \ 34, \ 21, \ 13, \ 8, \ 5, \ 3, \ 2, \ 1, \ 0$$

so the number of steps is 11 (or maybe it's 10). If we start with F_{21} and F_{20}, then the remainders are

$$F_{19}, \ F_{18}, \ \ldots, \ F_2, \ 0$$

so the number of steps is 19. You can calculate F_{20} and F_{21}. They are 6765 and 10946.

Problems 3.3

1. Let $a = 43$ and $b = 56$. Then successive applications of the division algorithm give the equations

$$56 = 1 \cdot 43 + 13 \text{ and } 43 = 3 \cdot 13 + 4 \text{ and } 13 = 3 \cdot 4 + 1$$

Solve the first equation for 13, the second for 4, and the third for 1

$$13 = 56 - 1 \cdot 43 \text{ and } 4 = 43 - 3 \cdot 13 \text{ and } 1 = 13 - 3 \cdot 4$$

Start with the third of these equations and substitute the expression for 4 from the second equation

$$1 = 13 - 3 \cdot 4 = 13 - 3 \cdot (43 - 3 \cdot 13)$$
$$= 13 - 3 \cdot 43 + 9 \cdot 13 = 10 \cdot 13 - 3 \cdot 43$$

Now substitute the expression for 13 from the first equation into this

$$1 = 10 \cdot 13 - 3 \cdot 43 = 10 \cdot (56 - 1 \cdot 43) - 3 \cdot 43$$
$$= 10 \cdot 56 - 10 \cdot 43 - 3 \cdot 43 = 10 \cdot 56 - 13 \cdot 43$$

3. $1 = \gcd(43, 56) = (-13) \cdot 43 + 10 \cdot 56$. (See disc for details.)

5. The solution is

$$3 = \gcd(742789479, 9587374758)$$
$$= 742789479 \, (-1357\,468\,553) + 9587374758 \, (105\,170\,955)$$

7. Consider the array

$x \backslash y$	−5	−4	−3	−2	−1	0	1	2	3	4	5	6
−5	−50	−44	−38	−32	−26	−20	−14	−8	−2	4	10	16
−4	−46	−40	−34	−28	−22	−16	−10	−4	2	8	14	20
−3	−42	−36	−30	−24	−18	−12	−6	0	6	12	18	24
−2	−38	−32	−26	−20	−14	−8	−2	4	10	16	22	28
−1	−34	−28	−22	−16	−10	−4	2	8	14	20	26	32
0	−30	−24	−18	−12	−6	0	6	12	18	24	30	36
1	−26	−20	−14	−8	−2	4	10	16	22	28	34	40
2	−22	−16	−10	−4	2	8	14	20	26	32	38	44
3	−18	−12	−6	0	6	12	18	24	30	36	42	48
4	−14	−8	−2	4	10	16	22	28	34	40	46	52
5	−10	−4	2	8	14	20	26	32	38	44	50	56
6	−6	0	6	12	18	24	30	36	42	48	54	60

Note that 2 is the smallest positive integer that appears in the array. By reading the row and column headers, we see that $2 = 5 \cdot 4 + (-3) \cdot 6 = 2 \cdot 4 + (-1) \cdot 6 = (-1) \cdot 4 + 1 \cdot 6 = (-4) \cdot 4 + 3 \cdot 6$. Thus $d = 2$ and (x, y) could be any of the pairs $(5, -3)$, $(2, -1)$, $(-1, 1)$, or $(-4, 3)$.

9. Assume $a | bc$ and $a \perp b$, say $1 = ax + by$. Then multiplication by c yields $c = acx + bcy$. Since $a | a$ and $a | bc$, it follows that $a | (acx + bcy) = c$.

Problems 3.4

In problems 1-7, $\mathbb{M} = \{1, 4, 7, 10, 13, 16, 19, 22, 25, \ldots\}$.

1. The next six elements after 25 in the set $\mathbb{M}$ are $28, 31, 34, 37, 40, 43$.

3. The first six composites in $\mathbb{M}$ are

$$
\begin{array}{lll}
16 = 4 \cdot 4 & 28 = 4 \cdot 7 & 40 = 4 \cdot 10 \\
49 = 7 \cdot 7 & 52 = 4 \cdot 13 & 64 = 4 \cdot 4 \cdot 4
\end{array}
$$

5. Since p and q are ordinary primes, the only factorization of pq in the positive integers is $p \cdot q$, but p and q are not elements of $\mathbb{M}$ so pq must be prime in $\mathbb{M}$.

7. Each of ab, cd, ac, bd, ad, and bc is prime in $\mathbb{M}$ because each is of the form $(3m + 2)(3n + 2)$, where $3m + 2$ and $3n + 2$ are ordinary primes.

9. It takes the mutual cooperation of a factor of 2 and a factor of 5 to generate a multiple of 10 and thus generate a trailing zero. There are 994 factors of 2 available. The number of factors of 5 is given by

$$
\lfloor 1000/5 \rfloor + \lfloor 1000/25 \rfloor + \lfloor 1000/125 \rfloor + \lfloor 1000/625 \rfloor = 200 + 40 + 8 + 1
$$
$$
= 249
$$

and hence there are exactly 249 factors of the form $2 \cdot 5$, each of which contributes 1 zero at the end of $1000!$.

11. Let

$$
n = \prod_{i=1}^{k} p_i^{e_i}, \quad a = \prod_{i=1}^{k} p_i^{f_i}, \quad \text{and} \quad b = \prod_{i=1}^{k} p_i^{g_i}
$$

where $f_i \le e_i$ and $g_i \le e_i$ for $i = 1, 2, \ldots, k$. Then

$$
\operatorname{lcm}(a, b) = \prod_{i=1}^{k} p_i^{\max\{f_i, g_i\}} \mid \prod_{i=1}^{k} p_i^{e_i} = n
$$

Problems 3.5

1. Finding a solution to the congruence $10x \equiv 14 \pmod{8}$ is equivalent to finding an integer solution to the equation $10x = 14 + 8y$, which is equivalent to finding an integer solution to the equation $5x = 7 + 4y$, which is equivalent to finding a solution to the congruence $5x \equiv 7 \pmod{4}$. The integer solutions to $10x = 14 + 8y$ are given by $x = 3 + 4N_1$ and $y = 2 + 5N_1$ and the integer solutions to $5x = 7 + 4y$ are given by $x = 3 + 4N_1$ and $y = 2 + 5N_1$.

3. The inverse of 8 modulo 13 is 5 and $3 \cdot 5 \bmod 13$ is 2. So $3/8 \bmod 13 = 2$. The inverse of 3 modulo 17 is 6, $-14 \cdot 6 \bmod 17 = 1$, so $-14/3 \bmod 17 = 1$.

5. We have

$$\begin{pmatrix} 0 & 1 \\ 1 & -\lfloor 34/113 \rfloor \end{pmatrix} \begin{pmatrix} 34 & 1 \\ 113 & 0 \end{pmatrix} = \begin{pmatrix} 113 & 0 \\ 34 & 1 \end{pmatrix}$$

$$\begin{pmatrix} 0 & 1 \\ 1 & -\lfloor 113/34 \rfloor \end{pmatrix} \begin{pmatrix} 113 & 0 \\ 34 & 1 \end{pmatrix} = \begin{pmatrix} 34 & 1 \\ 11 & -3 \end{pmatrix}$$

$$\begin{pmatrix} 0 & 1 \\ 1 & -\lfloor 34/11 \rfloor \end{pmatrix} \begin{pmatrix} 34 & 1 \\ 11 & -3 \end{pmatrix} = \begin{pmatrix} 11 & -3 \\ 1 & 10 \end{pmatrix}$$

$$\begin{pmatrix} 0 & 1 \\ 1 & -\lfloor 11/1 \rfloor \end{pmatrix} \begin{pmatrix} 11 & -3 \\ 1 & 10 \end{pmatrix} = \begin{pmatrix} 1 & 10 \\ 0 & -113 \end{pmatrix}$$

so that $34^{-1} \bmod 113 = 10$. As a check, note that $10 \cdot 34 \bmod 113 = 1$.

7. We get $7382784739^{-1} \cdot 1727372727 \bmod 2783479827 = 213\,764\,595$. As a
check, $7382784739 \cdot 213\,764\,595 \bmod 2783479827 = 1727\,372\,727$.

9. The multiplication table for the integers modulo 5 using $-2, -1, 0, 1, 2$ is

$\times$	-2	-1	0	1	2
-2	-1	2	0	-2	1
-1	2	1	0	-1	-2
0	0	0	0	0	0
1	-2	-2	0	1	2
2	1	-2	0	2	-1

Note that $(-2)^2 = -1$ and $(2^2) = -1$ in this multiplication table.

11. Multiply the first equation by $37^{-1} \bmod 101 = 71$ and the second by
$17^{-1} \bmod 101 = 6$ to get $17x + y = 85$ and $15x + y = 49$. Then subtract the second equation from the first to get $2x = 36$ so that $x =
2^{-1} 36 \bmod 101 = 18$; and from the equation $15x + y = 49$ get $y =
49 - 15 \cdot 18 \bmod 101 = 82$. Checking, we see that $23 \cdot 18 + 37 \cdot 82 \bmod 101 =
14$ and $53 \cdot 18 + 17 \cdot 82 \bmod 101 = 25$.

13. If -1 has a square root modulo p, then $x^2 + 1$ should factor modulo p. Factoring with a computer algebra system produces results like $x^2 + 1 \bmod 2 =
(x + 1)^2$, $x^2 + 1 \bmod 3 = x^2 + 1$, and $x^2 + 1 \bmod 5 = (x - 2)(x + 2)$. The
prime $p = 2$ is a special case since $-1 = 1$ and $x^2 + 1$ has 1 as a double
root. Try some primes on your own and you will see that, for the odd
primes, it appears that $x^2 + 1$ factors exactly when $p \bmod 4 = 1$. To give
you a start, the solution on the disc factors the first 30 primes.

Chapter 4   Ciphers

Problems 4.1

1. KHOOR

3. KTIXEVZOUT VXUJAIZY COZN RKYY ZNGT YODZE LUAX HOZY GXK
    LXKKRE KDVUXZGHRK

5. *Goodbye.*

7. *Caesar is considered to be one of the first persons to have ever employed encryption for the sake of securing messages.*

9. Drive for show; putt for dough.

11. GHXSG NOOMN NHKFS UEKCE FPJSG WEUYT ADBOS DYFEM HUSFS UVITN
     RCODP PNENG CAESH YDGFH HSTY

13. $y = 15\,(3x + 5) + 4\,\mathrm{mod}\,26 = 45x + 79\,\mathrm{mod}\,26 = 19x + 1\,\mathrm{mod}\,26$

Problems 4.2

1. The ciphertext was generated using the shift $y = x + 6\,\mathrm{mod}\,26$, so the inverse is given by the shift $x = y - 6\,\mathrm{mod}\,26$. The plaintext is

    The oldest known encryption device is the scytale.

3. The ciphertext was generated using $y = 5x\,\mathrm{mod}\,26$. Since $5^{-1}\,\mathrm{mod}\,26 = 21$ the inverse affine cipher is given by $x = 21y\,\mathrm{mod}\,26$. The plaintext is

    The scytale consisted of two round pieces of wood which had exactly the same dimensions.

5. The ciphertext was generated using $y = 11x + 4\,\mathrm{mod}\,26$, so the inverse is given by $x = (y - 4)11^{-1}\,\mathrm{mod}\,26$. Since $11^{-1}\,\mathrm{mod}\,26 = 19$ the inverse is the affine character cipher $x = (y - 4)\,19\,\mathrm{mod}\,26 = 19y + 2\,\mathrm{mod}\,26$. The plaintext is

    Messages were encoded by winding a piece of parchment around the cylinder in a spiral and writing the message along the length of the cylinder.

7. The shift ciphers $y = x + 7\,\mathrm{mod}\,26$ and $y = x + 11\,\mathrm{mod}\,26$ were used to encrypt the odd and even letters. Use the inverse shift ciphers $y = x + 19\,\mathrm{mod}\,26$ and $y = x + 15\,\mathrm{mod}\,26$ to obtain the plaintext

    Thirty-five code talkers attended the dedication of the Navajo code talker exhibit. The exhibit includes a display of photographs, equipment, and the original code, along with an explanation of how the code worked. Dedication ceremonies included speeches by the then Deputy Secretary of Defense Donald Atwood, U.S. Senator John McCain of Arizona, and Navajo President Peterson Zah. The Navajo veterans and their families traveled to the ceremony from their homes on the Navajo Reservation, which includes parts of Arizona, New Mexico, and Utah.

Problems 4.3

1. CPCRW CEPST UYVOW T

3. EVIHAA ERSETC

5. VNLPL HZQBW QJPZL CQXHB BHWJD VTHCP LWQFC HEHCF VNRVR
   WHINL PHZVP RCZIQ ZHVHQ CRCXC QVZJK ZVHVJ VHQC

7. If $\sigma = (0\ 4\ 7)(1\ 6\ 5\ 3)(2\ 9\ 8)$ and $\tau = (0)(1\ 3\ 5\ 9\ 7)(2\ 8\ 6\ 4)$, then

$$\sigma\tau = (0\ 2\ 7)\,(1\ 4)\,(3)\,(5)\,(6\ 9)\,(8)$$
$$\tau\sigma = (0\ 4\ 9)\,(1)\,(2)\,(3)\,(5\ 8)\,(6\ 7)$$

9. Let σ be a permutation on $\{1, 2, \ldots, n\}$. If $x^\sigma = x$ for all x, then $\sigma = (1)\,(2)\cdots(n)$ is a product of disjoint cycles. We say that an integer x is **moved** by σ if $x^\sigma \neq x$. We induct on the number k of integers moved by a permutation σ. If $k = 0$, then $x^\sigma = x$ for all x and hence $\sigma = (1)\,(2)\cdots(n)$ is a product of disjoint cycles, so the theorem holds.

   For $k > 0$, assume that all permutations τ that move fewer than k integers can be written as products of disjoint cycles. Let a be an integer moved by σ and consider the finite sequence $a_1 = a$, $a_2 = a_1^\sigma$, $a_3 = a_2^\sigma$, and so forth. Since $a_i \in \{1, 2, \ldots, n\}$, there exist integers $i < j$ such that $a_i = a_j$. Let m be the least integer such that $a_i = a_m$ with $1 \leq i < m$. If $i > 1$, then $a_i = a_{i-1}^\sigma = a_m = a_{m-1}^\sigma$, contrary to the minimality of m.

   Consider the permutation $(a_1 a_2 \cdots a_m)^{-1}\sigma = \tau$. Note that $a_i' = a_i$ for $i = 1\ldots m$ and $x^\sigma = x^t$ for all other $x \in \{1, 2, \ldots, n\}$. Thus τ moves fewer than k integers and hence τ is a product of disjoint cycles. Thus $\sigma = (a_1 a_2 \cdots a_m)\tau$ is also a product of disjoint cycles.

Problems 4.4

1. IS $\longrightarrow (\ 8\quad 18\ )\begin{pmatrix} 2 & 1 \\ 5 & 3 \end{pmatrix} \bmod 26 = (\ 2\quad 10\ ) \longrightarrow$ CK

   PY $\longrightarrow (\ 15\quad 24\ )\begin{pmatrix} 2 & 1 \\ 5 & 3 \end{pmatrix} \bmod 26 = (\ 20\quad 9\ ) \longrightarrow$ UJ

3. MC TX BX HN FZ IY AN TI OH WN BC PD RT XN SV DH
   DV IE ZW IR MP CV QB RF IY PB VO VR JN YM IN KA
   IG KT WC RS AH CL YO KS UD UQ DY KM BH HX IC CJ
   OU RS HN KU

5. The inverse cipher is

$$X\begin{pmatrix} 0 & 7 & 3 & 6 & 8 \\ 5 & 8 & 1 & 9 & 5 \\ 3 & 7 & 3 & 4 & 5 \\ 6 & 1 & 7 & 9 & 6 \\ 8 & 6 & 9 & 3 & 1 \end{pmatrix}^{-1} \bmod 26 = X\begin{pmatrix} 20 & 11 & 12 & 16 & 19 \\ 3 & 15 & 11 & 11 & 14 \\ 18 & 5 & 13 & 17 & 2 \\ 9 & 0 & 24 & 19 & 6 \\ 23 & 11 & 13 & 12 & 15 \end{pmatrix} \bmod 26$$

7. Assuming the fictitious Social Security number is 555-55-5555, the affine transformation yields the encryption 5 0 3 8 6 5 9 8 5 . See the solution on the disc for details.

9. The plaintext is

> *If sensitive data falls into the wrong hands,*
> *it can lead to fraud or identity theft.*

See the solution on the disc for details.

Problems 4.5

1. The Playfair square is given by

C	I	R	S	L
P	E	A	K	M
H	U	G	N	W
B	F	O	V	X
D	Q	T	Y	Z

The plaintext and ciphertext are given by

WH	EA	TS	TO	NE	NA	ME	DT	HE	PL
↓	↓	↓	↓	↓	↓	↓	↓	↓	↓
HU	AK	YR	RT	UK	GK	PA	QY	UP	MC

AY	FA	IR	CI	PH	ER	AF	TE	RH	IS
↓	↓	↓	↓	↓	↓	↓	↓	↓	↓
KT	OE	RS	IR	HB	AI	EO	QA	OI	RL

FR	EI	ND	LY	ON	PL	AY	FA	IR
↓	↓	↓	↓	↓	↓	↓	↓	↓
OI	UE	HY	SZ	VG	MC	KT	OE	RS

3. Use the Playfair square

R	H	E	O	S
D	C	B	A	T
F	G	IJ	K	L
U	Q	P	N	M
V	W	X	Y	Z

to obtain the message

> *The primary use of a Wheatstone bridge is the measurement of resistance.*

5. *The Playfair cipher may be improved by seriating its input text.*

7. *Who? Not me!*

Problems 4.6

1. LOSCH POTYX YXQBL EEYWT NIWAJ MIOMY IPINP PPKLH B

3. *Who is the weakest link?*

5. *Why Playfair (play fair?) when the other guys are using RSA?*

7. Since $1 + 1 \bmod 2 - 0$, it follows that adding the same binary vector twice gives back the original vector. In particular, if

$$\begin{pmatrix} 1 & 0 & 0 & 0 & 1 & 0 & 1 & 1 \end{pmatrix} + \begin{pmatrix} 1 & 1 & 0 & 1 & 1 & 0 & 0 & 1 \end{pmatrix} \bmod 2$$
$$= \begin{pmatrix} 0 & 1 & 0 & 1 & 0 & 0 & 1 & 0 \end{pmatrix}$$

then

$$\begin{pmatrix} 0 & 1 & 0 & 1 & 0 & 0 & 1 & 0 \end{pmatrix} + \begin{pmatrix} 1 & 1 & 0 & 1 & 1 & 0 & 0 & 1 \end{pmatrix} \bmod 2$$
$$= \begin{pmatrix} 1 & 0 & 0 & 0 & 1 & 0 & 1 & 1 \end{pmatrix}$$

Problems 4.7

1. We take $x = 7 \leftarrow$ 'H' and compute

$$
\begin{aligned}
y &= (x + s_1 \bmod 26)^{\sigma_1} = (7 + 25 \bmod 26)^{\sigma_1} = 6^{\sigma_1} = 9 \\
z &= (y + s_2 \bmod 26)^{\sigma_2} = (9 + 13 \bmod 26)^{\sigma_2} = 22^{\sigma_2} = 21 \\
w &= (z + s_3 \bmod 26)^{\sigma_3} = (21 + 4 \bmod 26)^{\sigma_3} = 25^{\sigma_3} = 1 \\
u &= (w)^\rho = 1^\rho = 9 \\
z &= \left(u^{\sigma_3^{-1}} - s_3 \bmod 26 \right) = 9^{\sigma_3^{-1}} - 4 \bmod 26 = 18 - 4 \bmod 26 = 14 \\
y &= (z^{\sigma_2^{-1}} - s_2 \bmod 26) = 14^{\sigma_2^{-1}} - 13 \bmod 26 = 13 - 13 \bmod 26 = 0 \\
x &= (y^{\sigma_1^{-1}} - s_3 \bmod 26) = 0^{\sigma_1^{-1}} - 25 \bmod 26 = 23 - 25 \bmod 26 = 24
\end{aligned}
$$

and hence the first ciphertext character is $24 \rightarrow$ 'Y'. Finally, we update the shift constants:

$$
\begin{aligned}
s_1 &\leftarrow s_1 + 1 \bmod 26 - 0 \\
s_2 &\leftarrow s_2 + 1 \bmod 26 = 14
\end{aligned}
$$

Continuing in this way, we see that the ciphertext is YQ.

3. *Attack now!*

5. The number is $\dfrac{\binom{26}{2}\binom{24}{2}\binom{22}{2} \cdots \binom{8}{2}}{10!}$. In the example, there are $\binom{26}{2}$ ways to choose the first pair $\{A, Z\}$, then $\binom{24}{2}$ ways to choose the second pair $\{B, Y\}$, and so on. But the 10! ways of listing these 10 pairs all give the same pairings, so we must divide by 10! to get the number of distinct plugboard settings.

7. $\sigma^{-1} = \begin{pmatrix} 1 & 2 & 3 & 4 & 5 & 6 & 7 \\ 6 & 7 & 2 & 5 & 3 & 1 & 4 \end{pmatrix}$ and $\tau^{-1} = \begin{pmatrix} 1 & 2 & 3 & 4 & 5 & 6 & 7 \\ 2 & 3 & 7 & 1 & 5 & 4 & 6 \end{pmatrix}$

9. Let $\rho = (a_1, a_2, \ldots, a_k)$ and set $\tau = \sigma^{-1}$. Then $(a_i^\tau)^{\sigma\rho\tau} = (a_i)^{\tau\sigma\rho\tau} = a_i^{\rho\tau} = (a_{i+1})^\tau$ for $i = 1, 2, \ldots, k-1$, and $(a_k^\tau)^{\sigma\rho\tau} = (a_k)^{\tau\sigma\rho\tau} = a_k^{\rho\tau} = (a_1)^\tau$. If $x \notin \{a_1, a_2, \ldots, a_k\}$, then $x^{\tau\sigma\rho\tau} = x^{\rho\tau} = x\tau$ so x^τ is fixed by the permutation $\sigma\rho\tau$. Thus $\sigma\rho\tau = (a_1^\tau, a_2^\tau, \ldots, a_k^\tau)$ is also a k-cycle.

Chapter 5   Error-Control Codes

Problems 5.1

1. The probability of getting 10 out of 10 on a true false exam is $\left(\frac{1}{2}\right)^{10} = \frac{1}{1024} \approx 9.7656 \times 10^{-4}$.

3. The codewords 000 and 111 are a distance 3 apart.

5. There are 200 places where the error could occur; so the probability of a single error is $\binom{200}{1}(1 - 0.0001)^{199}(0.0001) = 1.9606 \times 10^{-2}$.

7. The expected number of errors is the sum of $n \times$ (the probability of exactly n errors), which is given by

$$\sum_{n=0}^{1000} n \times \binom{1000}{n}(1 - 0.001)^{1000-n}(0.001)^n = 1.0$$

The contribution by the nth bit is $1 \times (0.001) = 0.001$. Since there are 1000 bits, the expected number of bit errors is $1000 \times (0.001) = 1.0$.

9. If the probability of a letter being changed is $1/10$, then you can expect about a tenth of the letters to be changed. For two consecutive sentences, about one fifth of the letters can be expected to be different. The length of the sentence is 65, and $65/5 = 13$, so on average the distance between two sentences should be roughly 13. Indeed, the distances between pairs of sentences range from 11 to 14, so an average of 13 is reasonable.

Problems 5.2

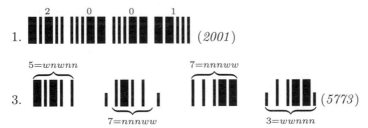

1. (2001)

3. (5773)

(Note that the short bars were added to indicate the width of beginning and ending spaces.)

5. The 2 of 5 bar code  decodes
as (*9112001; September 11, 2001*)

7. The Code 39 bar code

decodes as (*Code 39*).

9. The Interleaved 2 of 5 bar code

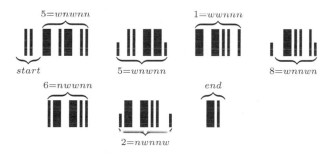

decodes as (*551862; Cinco de Mayo*).

(Note that the short bars were added to indicate the width of beginning and
ending spaces.)

11. The Code 39 bar code

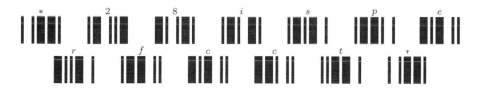

decodes as *28 is perfect*. (A **perfect number** is a number that is equal to
the sum of its proper divisors. In this case, $1 + 2 + 4 + 7 + 14 = 28$.)

Problems 5.3

1.

3. Let $a_1 - a_2 a_3 a_4 - a_5 a_6 a_7 a_8 a_9 - a_{10}$ be a valid ISBN number so that

$$\sum_{i=1}^{10} i a_i \bmod 11 = 0$$

Assume that a_k and a_n are interchanged, $a_k \neq a_n$, and $1 \leq k < n \leq 9$. Set $b_k = a_n$, $b_n = a_k$, and otherwise $b_i = a_i$. Then

$$\sum_{i=1}^{10} i b_i \bmod 11 = \left(\sum_{i=1}^{10} i b_i - \sum_{i=1}^{10} i a_i \right) \bmod 11$$

$$= (k(b_k - a_k) + n(b_n - a_n)) \bmod 11$$

$$= (k(a_n - a_k) + n(a_k - a_n)) \bmod 11$$

$$= (k - n)(a_n - a_k) \bmod 11 \neq 0$$

because 11 divides neither of the factors $k - n$ or $a_n - a_k$. (Why?) Thus

$$b_1 - b_2 b_3 b_4 - b_5 b_6 b_7 b_8 b_9 - b_{10}$$

cannot be a valid ISBN number.

5.

7. The first digit (0) represents the country and the second part represents the publisher. The last digit is the check sum

$$0 \cdot 1 + 4 \cdot 2 + 7 \cdot 3 + 1 \cdot 4 + 8 \cdot 5 + 9 \cdot 6 + 6 \cdot 7 + 1 \cdot 8 + 4 \cdot 9 \bmod 11 = 4$$

9. Do a web search on a bar code such as code 93.

11. A short element could be added to any of

There are $\frac{5!}{2!} = 60$ rearrangements of

$\frac{5!}{3!2!} = 10$ rearrangements of

$$\left\{ \begin{matrix} | \\ | \end{matrix} \middle| \begin{matrix} | \\ | \end{matrix}, |, |, | \right\}$$

and $\frac{5!}{2!2!} = 30$ rearrangements of

$$\left\{ \begin{matrix} | \\ | \end{matrix}, |, \begin{matrix} | \\ | \end{matrix}, | \right\}$$

for a total of $60 + 10 + 30 = 100$ symbols.

Problems 5.4

In the following problems,

$$H = \begin{pmatrix} 1 & 1 & 1 & 0 & 0 & 0 & 0 \\ 1 & 0 & 0 & 1 & 1 & 0 & 0 \\ 0 & 1 & 0 & 1 & 0 & 1 & 0 \\ 1 & 1 & 0 & 1 & 0 & 0 & 1 \end{pmatrix}, P = \begin{pmatrix} 0 & 0 & 1 \\ 0 & 1 & 0 \\ 0 & 1 & 1 \\ 1 & 0 & 0 \\ 1 & 0 & 1 \\ 1 & 1 & 0 \\ 1 & 1 & 1 \end{pmatrix}, \text{ and}$$

$$D = \begin{pmatrix} 0 & 0 & 0 & 0 \\ 0 & 0 & 0 & 0 \\ 1 & 0 & 0 & 0 \\ 0 & 0 & 0 & 0 \\ 0 & 1 & 0 & 0 \\ 0 & 0 & 1 & 0 \\ 0 & 0 & 0 & 1 \end{pmatrix}$$

1. The codeword is

$$\begin{pmatrix} 1 & 0 & 1 & 1 \end{pmatrix} H \bmod 2 = \begin{pmatrix} 0 & 1 & 1 & 0 & 0 & 1 & 1 \end{pmatrix}$$

3. The parity is even, so we assume there are either 0 or 2 errors. An error check yields

$$\begin{pmatrix} 1 & 0 & 1 & 1 & 0 & 1 & 0 \end{pmatrix} P \bmod 2 = \begin{pmatrix} 0 & 0 & 0 \end{pmatrix}$$

which indicates no errors, so the plaintext is given by

$$\begin{pmatrix} 1 & 0 & 1 & 1 & 0 & 1 & 0 \end{pmatrix} D \bmod 2 = \begin{pmatrix} 1 & 0 & 1 & 0 \end{pmatrix}$$

5. The parity is odd, so we assume a single error occurred. An error check yields

$$\begin{pmatrix} 1 & 0 & 1 & 1 & 0 & 1 & 0 \end{pmatrix} P \bmod 2 = \begin{pmatrix} 0 & 0 & 0 \end{pmatrix}$$

so we assume the error occurred in the 8th bit. The plaintext is given by

$$\begin{pmatrix} 1 & 0 & 1 & 1 & 0 & 1 & 0 \end{pmatrix} D \bmod 2 = \begin{pmatrix} 1 & 0 & 1 & 0 \end{pmatrix}$$

7. The codewords are given by

$$(0 \quad 0 \quad 0 \quad 1) \, H \bmod 2 = (1 \quad 1 \quad 0 \quad 1 \quad 0 \quad 0 \quad 1) =$$

$$(0 \quad 0 \quad 1 \quad 0) \, H \bmod 2 = (0 \quad 1 \quad 0 \quad 1 \quad 0 \quad 1 \quad 0) =$$

$$(0 \quad 0 \quad 1 \quad 1) \, H \bmod 2 = (1 \quad 0 \quad 0 \quad 0 \quad 0 \quad 1 \quad 1) =$$

$$(0 \quad 1 \quad 0 \quad 0) \, H \bmod 2 = (1 \quad 0 \quad 0 \quad 1 \quad 1 \quad 0 \quad 0) =$$

$$(0 \quad 1 \quad 0 \quad 1) \, H \bmod 2 = (0 \quad 1 \quad 0 \quad 0 \quad 1 \quad 0 \quad 1) =$$

$$(0 \quad 1 \quad 1 \quad 0) \, H \bmod 2 = (1 \quad 1 \quad 0 \quad 0 \quad 1 \quad 1 \quad 0) =$$

$$(0 \quad 1 \quad 1 \quad 1) \, H \bmod 2 = (0 \quad 0 \quad 0 \quad 1 \quad 1 \quad 1 \quad 1) =$$

$$(1 \quad 0 \quad 0 \quad 0) \, H \bmod 2 = (1 \quad 1 \quad 1 \quad 0 \quad 0 \quad 0 \quad 0) =$$

$$(1 \quad 0 \quad 0 \quad 1) \, H \bmod 2 = (0 \quad 0 \quad 1 \quad 1 \quad 0 \quad 0 \quad 1) =$$

$$(1 \quad 0 \quad 1 \quad 0) \, H \bmod 2 = (1 \quad 0 \quad 1 \quad 1 \quad 0 \quad 1 \quad 0) =$$

$$(1 \quad 0 \quad 1 \quad 1) \, H \bmod 2 = (0 \quad 1 \quad 1 \quad 0 \quad 0 \quad 1 \quad 1) =$$

$$(1 \quad 1 \quad 0 \quad 0) \, H \bmod 2 = (0 \quad 1 \quad 1 \quad 1 \quad 1 \quad 0 \quad 0) =$$

$$(1 \quad 1 \quad 0 \quad 1) \, H \bmod 2 = (1 \quad 0 \quad 1 \quad 0 \quad 1 \quad 0 \quad 1) =$$

$$(1 \quad 1 \quad 1 \quad 0) \, H \bmod 2 = (0 \quad 0 \quad 1 \quad 0 \quad 1 \quad 1 \quad 0) =$$

Omitting patterns with 1 or 4 bars, there are 10 remaining two- and three-bar codes that could be used to encode the digits 0–9. This bar code symbology would give single error correction.

9. An error check yields

$$(0 \ \ 0 \ \ 0 \ \ 0 \ \ 0 \ \ 1 \ \ 1) \, P \bmod 2 = (0 \ \ 1 \ \ 1)$$

which indicates an error in bit $(0 \ \ 1 \ \ 1)_2 = 3$. Thus

$$(0 \ \ 0 \ \ 1 \ \ 0 \ \ 0 \ \ 1 \ \ 1) \begin{pmatrix} 0 & 0 & 0 & 0 \\ 0 & 0 & 0 & 0 \\ 1 & 0 & 0 & 0 \\ 0 & 0 & 0 & 0 \\ 0 & 1 & 0 & 0 \\ 0 & 0 & 1 & 0 \\ 0 & 0 & 0 & 1 \end{pmatrix} \bmod 2 = (1 \ \ 0 \ \ 1 \ \ 1)$$

11. The parity is even, so we assume there are either 0 or 2 errors. An error check yields

$$(0 \ \ 1 \ \ 0 \ \ 0 \ \ 1 \ \ 0 \ \ 1) \, P \bmod 2 = (0 \ \ 0 \ \ 0)$$

so assume there are no errors. Thus the plaintext is

$$(0 \ \ 1 \ \ 0 \ \ 0 \ \ 1 \ \ 0 \ \ 1) \, D \bmod 2 = (0 \ \ 1 \ \ 0 \ \ 1)$$

13. The parity is odd, so we assume a single error occurred. An error check yields

$$(1 \ \ 1 \ \ 1 \ \ 1 \ \ 1 \ \ 1 \ \ 1) \, P \bmod 2 = (0 \ \ 0 \ \ 0)$$

so a single error must have occurred at bit 8. The plaintext is

$$(1 \ \ 1 \ \ 1 \ \ 1 \ \ 1 \ \ 1 \ \ 1) \, D \bmod 2 = (1 \ \ 1 \ \ 1 \ \ 1)$$

Chapter 6   Chinese Remainder Theorem

Problems 6.1

1. Both equations are satisfied by $x = 3$ and $y = 4$; that is, $2{\cdot}3 + 3{\cdot}4 \bmod 11 = 7$ and $3 \cdot 3 + 4 \bmod 11 = 2$. See the disc for details of the solution.

3. Use row reduction to get

$$\begin{pmatrix} 3 & 5 & 1 & 0 \\ 4 & 8 & 0 & 1 \end{pmatrix} \longrightarrow \begin{pmatrix} 1 & 0 & 2 & -\frac{5}{4} \\ 0 & 1 & -1 & \frac{3}{4} \end{pmatrix}$$

and conclude that

$$\begin{pmatrix} 3 & 5 \\ 4 & 8 \end{pmatrix}^{-1} = \begin{pmatrix} 2 & -\frac{5}{4} \\ -1 & \frac{3}{4} \end{pmatrix}$$

Reduce modulo 11 to get

$$\begin{pmatrix} 3 & 5 \\ 4 & 8 \end{pmatrix}^{-1} \bmod 11 = \begin{pmatrix} 2 & 7 \\ 10 & 9 \end{pmatrix}$$

Check your answer by showing that

$$\begin{pmatrix} 3 & 5 \\ 4 & 8 \end{pmatrix}\begin{pmatrix} 2 & 7 \\ 10 & 9 \end{pmatrix} \bmod 11 = \begin{pmatrix} 1 & 0 \\ 0 & 1 \end{pmatrix}$$

See the disc for further details.

5. Note that

$$\begin{pmatrix} 2 & 3 & 1 \\ 0 & 1 & 5 \\ 3 & 7 & 2 \end{pmatrix}^{-1} = \begin{pmatrix} \frac{11}{8} & -\frac{1}{24} & -\frac{7}{12} \\ -\frac{5}{8} & -\frac{1}{24} & \frac{5}{12} \\ \frac{1}{8} & \frac{5}{24} & -\frac{1}{12} \end{pmatrix}$$

Reduce each entry modulo 17 to get

$$\begin{pmatrix} 2 & 3 & 1 \\ 0 & 1 & 5 \\ 3 & 7 & 2 \end{pmatrix}^{-1} \bmod 17 = \begin{pmatrix} 12 & 12 & 15 \\ 10 & 12 & 16 \\ 15 & 8 & 7 \end{pmatrix}$$

See the disc for further details.

7. The answer is $x = 1$, $y = 12$, and $z = 3$. See the disc for details of the solution.

9. A computer algebra system gives the solution

$$w = \frac{121}{2}, x = -89, y = \frac{139}{2}, z = -18$$

Reduce modulo 91 to find the solution $\frac{121}{2} \bmod 91 = 15$, $-89 \bmod 91 = 2$, $\frac{139}{2} \bmod 91 = 24$, and $-18 \bmod 91 = 73$. See the disc for further details.

Problems 6.2

1. The solutions to the congruence $x \equiv 5 \pmod 9$ are $\ldots, 5, 14, 23, 32, 41,$ $50, 59, 68, 77, 86, 94, \ldots$ and the solutions of $x \equiv 4 \pmod{11}$ are $\ldots, 4, 15,$ $26, 37, 48, 59, 70, 81, 92, 103, 114, \ldots$ and hence the common solutions are $59 + 99k$ for $k \in \mathbb{Z}$.

3. There should be a unique solution x such that $0 \le x < 3 \cdot 4 \cdot 5 = 60$. The solutions in the interval $0 \le x < 60$ to three congruences are

$$A = \{2, 5, 8, 11, 14, 17, 20, 23, 26, 29, 32, 35, 38, 41, 44, 47, 50, 53, 56, 59\}$$
$$B = \{3, 7, 11, 15, 19, 23, 27, 31, 35, 39, 43, 47, 51, 55, 59\}$$
$$C = \{4, 9, 14, 19, 24, 29, 34, 39, 44, 49, 54, 59\}$$

Note that $A \cap B \cap C = \{59\}$.

5. The Chinese remainder algorithm yields 353.

7. The Chinese remainder algorithm yields 419.

9. The congruence $2x + 3 \equiv 7 \pmod{11}$ is equivalent to $2x \equiv 7 - 3 \equiv 4 \pmod{11}$, and hence $x \equiv 4 \cdot 2^{-1} \equiv 2 \pmod{11}$. The congruence $3x + 4 \equiv 5 \pmod{13}$ is equivalent to $3x \equiv 5 - 4 \equiv 1 \pmod{13}$, and hence $x \equiv 3^{-1} \equiv 9 \pmod{13}$. Thus the original pair of congruences is equivalent to the system $x \equiv 2 \pmod{11}$ and $x \equiv 9 \pmod{13}$. Thus x must satisfy

$$x = 2 + 11k - 9 + 13r$$
$$11k \equiv 9 - 2 \pmod{13}$$
$$k = 7 \cdot 11^{-1} \bmod 13 = 3$$

so $x \equiv 2 + 11 \cdot 3 \equiv 35 \pmod{143}$. Checking, note that $2 \cdot 35 + 3 \bmod 11 = 7$ and $3 \cdot 35 + 4 \bmod 13 = 5$.

Problems 6.3

1. The product is $37759097376 \cdot 116389305648 = 4394\,755\,125\,487\,858\,779\,648$. See the disc for details of the computation using base $b = 1000$ arithmetic.

3. The product is $37759097376 \cdot 116389305648 = 4394\,755\,125\,487\,858\,779\,648$. See the disc for details of the computation using the modular basis (997, 999, 1000, 1001, 1003, 1007, 1009, 1013) and the Chinese remainder algorithm.

5. We have $2^{40} \approx 1.099\,511\,628 \times 10^{12}$. Modulo 1000, we have $2^{40} \bmod 1000 = 2^{10} \times 2^{10} \times 2^{10} \times 2^{10} \bmod 1000 = 24 \times 24 \times 24 \times 24 \bmod 1000 = 776$. Modulo 100 000, have $2^{40} \bmod 100\,000 = 2^{10} \times 2^{10} \times 2^{10} \times 2^{10} \bmod 100\,000 = \cdots = 27\,776$, and hence the last five digits are 27 776. It follows that $2^{40} = 1099\,511\,627\,776$. See the disc for more detail.

7. Because $100! = 2^{97} 3^{48} 5^{24} 7^{16} 11^9 13^7 17^5 19^5 23^4 29^3 31^3 37^2 41^2 43^2 47^2 53 \times 59 \times 61 \times 67 \times 71 \times 73 \times 79 \times 83 \times 89 \times 97$, the modular representation is

$$100! = (0,0,0,0,0,0,0,0,0,0,0,0,0,0,0,0,0,0,0,0,0,0,0,0,0)$$

Note that the elements of the modular basis are all powers of distinct primes, and are thus pairwise relatively prime.

Problems 6.4

1. Note that $2x^3 + 3x^2 + 3x + 1 = (2x + 1)(x^2 + x + 1)$ and $2x^2 + 5x + 2 = (2x + 1)(x + 2)$, so $\gcd(2x^3 + 3x^2 + 3x + 1, 2x^2 + 5x + 2) = 2x + 1$.

3. Modulo 5 we have

$$\frac{2x^3 + 3x^2 + 3x + 1}{2x^2 + 5x + 2} = x - 1 + \frac{6x + 3}{2x^2 + 5x + 2}$$

$$2x^3 + 3x^2 + 3x + 1 = (x + 4)\left(2x^2 + 5x + 2\right) + x + 3$$

$$\frac{2x^2 + 5x + 2}{x + 3} = 2x - 1$$

and modulo 7 we have

$$\frac{2x^3 + 3x^2 + 3x + 1}{2x^2 + 5x + 2} = x - 1 + \frac{6x + 3}{2x^2 + 5x + 2}$$

$$2x^3 + 3x^2 + 3x + 1 = (x + 6)\left(2x^2 + 5x + 2\right) + 6x + 3$$

$$\frac{2x^2 + 5x + 2}{6x + 3} = \frac{1}{3}x + \frac{2}{3}$$

The leading coefficient must be 1 or 2. If the leading coefficient is 1, then modulo 5 we have $x + 3$ and modulo 7 we have $6^{-1}(6x + 3) \bmod 7 = x + 4$ and $x + 18$ does not divide $2x^3 + 3x^2 + 3x + 1$. Assume the leading coefficient is 2. Then modulo 5 we have $2(x + 3) = 2x + 1$, and modulo 7 we have $2(x + 4) = 2x + 1$. Note that $\frac{2x^3 + 3x^2 + 3x + 1}{2x + 1} = x^2 + x + 1$ and $\frac{2x^2 + 5x + 2}{2x + 1} = x + 2$, so $2x + 1$ is a common divisor.

5. The greatest common divisor is $1 + 2x^2$. As a check, note that

$$\frac{6x^5 - 7x^3 + 8x^2 - 5x + 4}{2x^2 + 1} = 3x^3 - 5x + 4$$

$$\frac{8x^4 + 6x^3 + 8x^2 + 3x + 2}{2x^2 + 1} = 4x^2 + 3x + 2$$

See the disc for details of the polynomial division.

7. *Scientific Notebook* produces

$$\gcd\left(6x^5 - 7x^3 + 8x^2 - 5x + 4, 8x^4 + 6x^3 + 8x^2 + 3x + 2\right) = 1 + 2x^2$$

9. Use long division to get

$$\frac{6x^6 + 4x^5 + 9x^4 + 19x^3 + 2x^2 + 15x + 5}{3x^3 + 2x^2 + 5} = 2x^3 + 3x + 1$$

$$\frac{12x^5 + 11x^4 + 11x^3 + 26x^2 + 5x + 15}{3x^3 + 2x^2 + 5} = 4x^2 + x + 3$$

which shows $3x^3 + 2x^2 + 5$ is a common divisor. Note that $\frac{2x^3 + 3x + 1}{4x^2 + x + 3} = \frac{1}{2}x - \frac{1}{8} + \frac{\frac{13}{8}x + \frac{11}{8}}{4x^2 + x + 3}$ implies $2x^3 + 3x + 1 = \left(\frac{1}{2}x - \frac{1}{8}\right)\left(4x^2 + x + 3\right) + \frac{13}{8}x + \frac{11}{8}$ and $\frac{4x^2 + x + 3}{\frac{13}{8}x + \frac{11}{8}} = \frac{32}{13}x - \frac{248}{169} + \frac{848}{169\left(\frac{13}{8}x + \frac{11}{8}\right)}$ which indicates that $2x^3 + 3x + 1$ and $4x^2 + x + 3$ are relatively prime.

Problems 6.5

1. See the answer to problem 3. See the disc for details of the solution using elementary row operations and rational arithmetic.

3. Direct calculation with a computer algebra system yields

$$
\begin{pmatrix}
1 & \frac{1}{2} & \frac{1}{3} & \frac{1}{4} \\
\frac{1}{2} & \frac{1}{3} & \frac{1}{4} & \frac{1}{5} \\
\frac{1}{3} & \frac{1}{4} & \frac{1}{5} & \frac{1}{6} \\
\frac{1}{4} & \frac{1}{5} & \frac{1}{6} & \frac{1}{7}
\end{pmatrix}^{-1}
=
\begin{pmatrix}
16 & -120 & 240 & -140 \\
-120 & 1200 & -2700 & 1680 \\
240 & -2700 & 6480 & -4200 \\
-140 & 1680 & -4200 & 2800
\end{pmatrix}
$$

5. Note that

$$
\det H_2 = \frac{1}{12} = \frac{1}{2^2 3}
$$

$$
\det H_3 = \frac{1}{2160} = \frac{1}{2^4 3^3 5}
$$

$$
\det H_4 = \frac{1}{6048\,000} = \frac{1}{2^8 3^3 5^3 7}
$$

$$
\det H_5 = \frac{1}{266\,716\,800\,000} = \frac{1}{2^{10} 3^5 5^5 7^3}
$$

$$
\det H_6 = \frac{1}{186\,313\,420\,339\,200\,000} = \frac{1}{2^{14} 3^9 5^5 7^5 11}
$$

7. The row operations

$$
\begin{pmatrix}
100 & 31 & 1 & 0 \\
29 & 9 & 0 & 1
\end{pmatrix}
\rightarrow
\begin{pmatrix}
1 & \frac{31}{100} & \frac{1}{100} & 0 \\
1 & \frac{9}{29} & 0 & \frac{1}{29}
\end{pmatrix}
\rightarrow
\begin{pmatrix}
1 & \frac{31}{100} & \frac{1}{100} & 0 \\
0 & \frac{1}{2900} & -\frac{1}{100} & \frac{1}{29}
\end{pmatrix}
$$

$$
\rightarrow
\begin{pmatrix}
1 & \frac{31}{100} & \frac{1}{100} & 0 \\
0 & 1 & -29 & 100
\end{pmatrix}
\rightarrow
\begin{pmatrix}
1 & 0 & 9 & -31 \\
0 & 1 & -29 & 100
\end{pmatrix}
$$

give
$$
\begin{pmatrix}
100 & 31 \\
29 & 9
\end{pmatrix}^{-1}
=
\begin{pmatrix}
9 & -31 \\
-29 & 100
\end{pmatrix}.
$$

9. Note that indeed $\det \begin{pmatrix} 127 & -24 \\ -37 & 7 \end{pmatrix} = 1$. Modulo 11, the inverse is $\begin{pmatrix} 7 & 2 \\ 4 & 6 \end{pmatrix}$, and modulo 13, the inverse is $\begin{pmatrix} 7 & 11 \\ 11 & 10 \end{pmatrix}$. The Chinese remainder theorem yields

$$
\begin{pmatrix}
127 & -24 \\
-37 & 7
\end{pmatrix}^{-1}
=
\begin{pmatrix}
c\,([7,7],[11,13]) & c\,([2,11],[11,13]) \\
c\,([4,11],[11,13]) & c\,([6,10],[11,13])
\end{pmatrix}
$$
$$
=
\begin{pmatrix}
7 & 24 \\
37 & 127
\end{pmatrix}
$$

Checking, note that

$$\begin{pmatrix} 127 & -24 \\ -37 & 7 \end{pmatrix} \begin{pmatrix} 7 & 24 \\ 37 & 127 \end{pmatrix} = \begin{pmatrix} 1 & 0 \\ 0 & 1 \end{pmatrix}$$

Chapter 7   Theorems of Fermat and Euler

Problems 7.1

1. Note that $(101 - 1)! \bmod 101 = 100 \equiv -1 \pmod{101}$, $(103 - 1)! \bmod 103 = 102 \equiv -1 \pmod{103}$, $(105 - 1)! \bmod 105 = 0$, $(107 - 1)! \bmod 107 = 106 \equiv -1 \pmod{107}$, and $(109 - 1)! \bmod 109 = 108 \equiv -1 \pmod{109}$, so 101, 103, 107, and 109 are primes and 105 is not.

3. $1^2 \bmod 11 = 1$ $\qquad$ $2^2 \bmod 11 = 4$
    $3^2 \bmod 11 = 9$ $\qquad$ $4^2 \bmod 11 = 5$
    $5^2 \bmod 11 = 3$ $\qquad$ $6^2 \bmod 11 = 3$
    $7^2 \bmod 11 = 5$ $\qquad$ $8^2 \bmod 11 = 9$
    $9^2 \bmod 11 = 4$ $\qquad$ $10^2 \bmod 11 = 1$

5. Computing $x^2 \bmod 15$ for $x = 1, 2, 3, ..., 14$ shows that the solutions are

$$1^2 \bmod 15 = 4^2 \bmod 15 = 11^2 \bmod 15 = 14^2 \bmod 15 = 1$$

7. We know that $(p - 1)! \bmod p = p - 1$. Since $p - 1 \perp p$, it follows that $(p - 2)! \bmod p = 1$. Since $p - 2 \pmod{p} \equiv -2$, it follows that

$$\begin{aligned}
(p - 3)! \, (p - 2) \bmod p &= 1 \\
-2 \, (p - 3)! \bmod p &= 1 \\
2 \, (p - 3)! \bmod p &= p - 1.
\end{aligned}$$

9. Rearrange the product as $2 \cdot 3 \cdot 4 \cdot 5 \cdot 6 \cdot 7 \cdot 8 \cdot 9 \cdot 10 \cdot 11 \cdot 12 \cdot 13 \cdot 14 \cdot 15 = (2 \cdot 9) \cdot (3 \cdot 6) \cdot (4 \cdot 13) \cdot (5 \cdot 7) \cdot (8 \cdot 15) \cdot (10 \cdot 12) \cdot (11 \cdot 14)$ and observe that $2 \cdot 9 \bmod 17 = 1$, $3 \cdot 6 \bmod 17 = 1$, $4 \cdot 13 \bmod 17 = 1$, $5 \cdot 7 \bmod 17 = 1$, $8 \cdot 15 \bmod 17 = 1$, $10 \cdot 12 \bmod 17 = 1$, and $11 \cdot 14 \bmod 17 = 1$. Hence

$$16! \bmod 17 = 1 \cdot 16 \cdot (2 \cdot 9) \cdot (3 \cdot 6) \cdot (4 \cdot 13) \cdot$$
$$(5 \cdot 7) \cdot (8 \cdot 15) \cdot (10 \cdot 12) \cdot (11 \cdot 14) \bmod 17$$
$$= 1 \cdot 16 \cdot 1 \cdot 1 \cdot 1 \cdot 1 \cdot 1 \cdot 1 \bmod 17 = 16 \bmod 17 \equiv -1 \pmod{17}$$

Problems 7.2

1. Using Algorithm 7.1, gets

k	p
	1
1	$1 \cdot 11 \bmod 15 = 11$
2	$11 \cdot 11 \bmod 15 = 1$
3	$1 \cdot 11 \bmod 15 = 11$
4	$11 \cdot 11 \bmod 15 = 1$

so $11^4 \bmod 15 = 1$.

k	p
	1

3. Using Algorithm 7.1, gets

1	$1 \cdot 16 \bmod 29 = 16$
2	$16 \cdot 16 \bmod 29 = 24$
3	$24 \cdot 16 \bmod 29 = 7$
4	$7 \cdot 16 \bmod 29 = 25$

so $16^4 \bmod 29 = 25$.

5. Using Algorithm 7.2, we get $5^{97} \bmod 127 = 80$. See the disc for details.

7. Using Algorithm 7.2, we get $4^{63} \bmod 127 = 1$. See the disc for details.

9. We get $12^{72387894339363242} \bmod 243682743764 = 17\,298\,641\,040$.

11. The second variation computes x^n and then reduces the result modulo m. For x and n integers, x^n is feasible to calculate exactly only if x and n are relatively small integers, say at most 3 or 4 digits each. The first variation makes it easy to calculate $x^n \bmod m$ for numbers x, n, and m that are each perhaps a hundred decimal digits long.

Problems 7.3

1.

$x^2 \bmod 11 \longrightarrow x$	$\lfloor n/2 \rfloor \longrightarrow n$	$n \bmod 2$	$p \cdot x \bmod 11 \longrightarrow p$
2	10	0	1
$2^2 \bmod 11 = 4$	$\lfloor 10/2 \rfloor = 5$	1	$1 \cdot 4 \bmod 11 = 4$
$4^2 \bmod 11 = 5$	$\lfloor 5/2 \rfloor = 2$	0	
$5^2 \bmod 11 = 3$	$\lfloor 2/2 \rfloor = 1$	1	$3 \cdot 4 \bmod 11 = 1$

and we see that $2^{10} \bmod 11 = 1$.

3. $2^{1000} \bmod 13 = \left(2^{12}\right)^{83} 2^4 \bmod 13 = 1^{83} 2^4 \bmod 13 = 16 \bmod 13 = 3$

5. $5^{2000} \bmod 17 = \left(5^{16}\right)^{125} \bmod 17 = 1^{125} \bmod 17 = 1$

7. $11^{1234} \bmod 29 = \left(11^{28}\right)^{44} 11^2 \bmod 29 = 1^{44} 11^2 \bmod 29 = 121 \bmod 29 = 5$

9. $5^{217} \bmod 217 = 5$ although $217 = 7 \times 31$

11. To verify this by hand, write 217 as a sum of powers of 2:

$$217 = 128 + 64 + 16 + 8 + 1$$
$$= 2^8 + 2^7 + 2^4 + 2^3 + 1$$

then compute $5^n \pmod{217}$ for $n = 1, 2, 2^2, 2^3, \ldots, 2^8$ using the fact that each number $5^n \pmod{217}$ is the square of the preceding one:

$$5, 25, 191, 125, 191, 125, 191, 125, 191$$

Finally, compute

$$
\begin{aligned}
5^{217} &= 5^{128+64+16+8+1} \\
&= 5^{128}5^{64}5^{16}5^{8}5 \\
&\equiv 191 \cdot 125 \cdot 191 \cdot 125 \cdot 5 \pmod{217} \\
&\equiv 5 \pmod{217}
\end{aligned}
$$

Trying to divide 217 by $3, 5, 7, 11, \ldots$ results quickly in

$$217 = 7 \times 31$$

13. Note that $1729 = 7 \times 13 \times 19$ and $1728 = 2^6 3^3$. Since 7, 13, and 19 are primes, it follows that $x^7 \bmod 7 = x \bmod 7$ and $x^{13} \bmod 13 = x \bmod 13$ and $x^{19} \bmod 19 = x \bmod 19$ for every integer x. Suppose $x \perp 7$, $x \perp 13$, and $x \perp 19$. Then

$$
x^{1728} \bmod 7 = \left(x^6\right)^{288} \bmod 7 = 1^{288} \bmod 7 = 1
$$

$$
x^{1728} \bmod 13 = \left(x^{12}\right)^{144} \bmod 13 = 1^{144} \bmod 13 = 1
$$

$$
x^{1728} \bmod 19 = \left(x^{18}\right)^{96} \bmod 19 = 1^{96} \bmod 19 = 1
$$

and hence by the Chinese remainder theorem, $x^{1728} \bmod 1729 = 1$, which means that $x^{1729} \bmod 1729 = x \bmod 1729$. Suppose that $x \bmod 7 = 0$, $x \perp 23$, and $x \perp 41$. Then

$$
x^{1729} \bmod 7 = 0 = x \bmod 7
$$

$$
x^{1728} \bmod 13 = \left(x^{12}\right)^{144} \bmod 13 = 1^{144} \bmod 13 = 1
$$

$$
x^{1728} \bmod 19 = \left(x^{18}\right)^{96} \bmod 19 = 1^{96} \bmod 19 = 1
$$

and hence $x^{1729} \bmod 7 = 0 = x \bmod 7$ and $x^{1729} \bmod 13 = x \bmod 13$ and $x^{1729} \bmod 19 = x \bmod 19$ so that $x^{1729} \bmod 1729 = x \bmod 1729$ by the Chinese remainder theorem. The remaining cases can be handled in a similar manner. Observe that $1729 = 10^3 + 9^3 = 12^3 + 1^3$.

15. Let x be an integer and p a prime dividing n. If p divides x, then

$$
x^n \bmod p = 0 = x \bmod p
$$

If p does not divide x, then

$$
x^{n-1} = \left(x^{p-1}\right)^{(n-1)/(p-1)} \equiv 1^{(n-1)/(p-1)} \equiv 1 \pmod{p}
$$

so $x^n \bmod p = x \bmod p$. Thus this equation holds for any prime p dividing n. By the Chinese remainder theorem,

$$
x^n \bmod n = x \bmod n
$$

17. Set $m = 1$ and do a loop such as

> **for k from 1 to 100 do;**
>> if isprime(6*k+1) = true and isprime(12*k+1) = true
>>> and isprime(18*k+1) = true then
>>
>> c[m]:=(6*k+1)*(12*k+1)*(18*k+1);
>>
>> m:=m+1;
>
> **end if;**
>
> **end for**

and observe that

$$
\begin{aligned}
c(1) &= 1729 \\
c(2) &= 294409 \\
c(3) &= 56052361 \\
c(4) &= 118901521 \\
c(5) &= 172947529 \\
c(6) &= 216821881 \\
c(7) &= 228842209 \\
c(8) &= 1299963601 \\
c(9) &= 2301745249
\end{aligned}
$$

Problems 7.4

1. Since $\lfloor \sqrt{899} \rfloor = 29$, we check

$$
\begin{array}{lll}
\frac{899}{2} = 449\tfrac{1}{2} & \frac{899}{3} = 299\tfrac{2}{3} & \frac{899}{5} = 179\tfrac{4}{5} \\
\frac{899}{7} = 128\tfrac{3}{7} & \frac{899}{11} = 81\tfrac{8}{11} & \frac{899}{13} = 69\tfrac{2}{13} \\
\frac{899}{17} = 52\tfrac{15}{17} & \frac{899}{19} = 47\tfrac{6}{19} & \frac{899}{23} = 39\tfrac{2}{23} \\
\frac{899}{29} = 31 & &
\end{array}
$$

and hence $899 = 29 \cdot 31$.

3. Note that $\gcd(2, 899) = 1$. Factoring, we have $898 = 2 \times 449$. The calculations $2^{449} \bmod 899 = 698$ and $698^2 \bmod 899 = 845$ show that 899 is composite.

5. Since $2^{p-1} \bmod p = 26\,747$, it follows from Fermat's little theorem that p is composite. Factorization shows that $p = 449 \times 457$.

7. We have $187\,736\,503 - 1 = 2 \times 93\,868\,251$. Note that

$$2^{93\,868\,251} \bmod 187\,736\,503 = 1$$
$$3^{93\,868\,251} \bmod 187\,736\,503 = -1$$
$$5^{93\,868\,251} \bmod 187\,736\,503 = -1$$
$$7^{93\,868\,251} \bmod 187\,736\,503 = 1$$
$$11^{93\,868\,251} \bmod 187\,736\,503 = -1$$
$$13^{93\,868\,251} \bmod 187\,736\,503 = -1$$
$$17^{93\,868\,251} \bmod 187\,736\,503 = -1$$
$$19^{93\,868\,251} \bmod 187\,736\,503 = -1$$
$$23^{93\,868\,251} \bmod 187\,736\,503 = 1$$
$$29^{93\,868\,251} \bmod 187\,736\,503 = -1$$

and hence $187\,736\,503$ passes Miller's test with $n = 10$ trials.

9. Let $f(x) = \frac{x}{\ln x}$. Then $f'(x) = \frac{\ln x - x \cdot \frac{1}{x}}{(\ln x)^2} = \frac{\ln x - 1}{\ln^2 x} \approx \frac{1}{\ln x}$ and hence $f'\left(10^{100}\right) = 4.324\,083\,649 \times 10^{-3}$, so the average gap between primes is $\frac{1}{4.324\,083\,649 \times 10^{-3}} \approx 231$. Assuming that on average you start halfway between two primes, and you only check even numbers, you will expect to test about $231/4 \approx 58$ numbers.

11. The gaps are

$$949 - 268 = 681 \qquad\qquad 1243 - 949 = 294$$
$$1293 - 1243 = 50 \qquad\qquad 1983 - 1293 = 690$$
$$2773 - 1983 = 790 \qquad\qquad 2809 - 2773 = 36$$
$$2911 - 2809 = 102 \qquad\qquad 2967 - 2911 = 56$$
$$3469 - 2967 = 502 \qquad\qquad 3501 - 3469 = 32$$

so the average gap is

$$\frac{681 + 294 + 50 + 690 + 790 + 36 + 102 + 56 + 502 + 32}{10} = 323\frac{3}{10}$$

which is somewhat larger than the expected gap given in problem 9.

Problems 7.5

1. Note that $\gcd(7, 100) = 1$. We have $73^7 \bmod 101 = 40$ and $7^{-1} \bmod 100 = 43$ and $40^{43} \bmod 101 = 73$, and hence 73 is recovered.

3. For example, let

$$p = 3391\,691\,164\,919\,859\,649\,719\,340\,532\,627\,567\,207$$
$$607\,656\,859\,034\,356\,995\,566\,589\,707\,894\,210\,757$$
$$866\,827\,613\,621\,721\,127\,496\,191\,413$$

be the smallest prime $> 5^{150}$ mod 10^{100}

5. The numbers

$$2\,047\,679\,804\,982\,929\,369\,090\,035\,623\,\mathbf{502}\,\mathbf{251}\,062\,360\,\mathbf{672}\,859\,815\,645$$
$$1\,503\,714\,396\,228\,131\,674\,104\,263\,081\,904\,\mathbf{252}\,158\,592\,322\,776\,941\,888$$

$$958\,411\,\mathbf{586}\,724\,601\,781\,\mathbf{693}\,914\,292\,292\,892\,305\,522\,540\,318\,192$$
$$363\,442\,\mathbf{582}\,571\,176\,226\,\mathbf{543}\,020\,134\,304\,037\,361\,597\,914\,795\,908$$

agree at 10 spots (see the bold digits). This is not a surprise because the process does a good job of mixing, and random 100-digit numbers would be expected to agree in about 10 spots (one-tenth of the time).

Problems 7.6

1. Since $24 = 2^3 3$, the integers $\{1, 5, 7, 11, 13, 17, 19, 23\}$ are relatively prime to 24 and $\varphi(24) = 8$.

3. We have $\varphi(27) = \varphi(3^3) = 3^3 - 3^2 = 18$. There are 18 integers

$$1, 2, 4, 5, 7, 8, 10, 11, 13, 14, 16, 17, 19, 20, 22, 23, 25, 26$$

less than or equal to 27 that are relatively prime to 27.

5. Since $1001 = 7 \times 11 \times 13$, we have

$$\begin{aligned} \varphi(1001) &= \varphi(7)\varphi(11)\varphi(13) \\ &= 6 \cdot 10 \cdot 12 \\ &= 720 \end{aligned}$$

Note that
$$5^{720} \bmod 1001 = 1$$

7. We start with the least nonnegative residue system $R = \{0, 1, 2, 3, 4, 5, 6, 7, 8, 9\}$ and construct

$$\{7r + 4 \mid r \in R\} = \{4, 11, 18, 25, 32, 39, 46, 53, 60, 67\}$$

Note that

$4 \bmod 10$	=	4	$39 \bmod 10$	=	9
$11 \bmod 10$	=	1	$46 \bmod 10$	=	6
$18 \bmod 10$	=	8	$53 \bmod 10$	=	3
$25 \bmod 10$	=	5	$60 \bmod 10$	=	0
$32 \bmod 10$	=	2	$67 \bmod 10$	=	7

and hence each element of $\{7r + 4 \mid r \in R\}$ is congruent modulo 10 to exactly one element of R.

9. Let p be a prime that divides n. If p does not divide a, then $a^{p-1} \equiv 1 \pmod{p}$ so
$$a^{\varphi(n)} \equiv 1 \pmod{p}$$
because $p - 1$ divides $\varphi(n)$. Multiplying by a we get
$$a^{\varphi(n)+1} \equiv a \pmod{p}$$

On the other hand, if p does divide a, then $a \equiv 0 \equiv a^{\varphi(n)+1} \pmod{p}$ so, trivially,
$$a^{\varphi(n)+1} \equiv a \pmod{p}$$

Thus $a^{\varphi(n)+1} \equiv a \pmod{p}$ for each prime dividing n. But this says that each prime that divides n also divides $a^{\varphi(n)+1} - a$, so if n is a product of distinct primes, then n must divide $a^{\varphi(n)+1} - a$, that is, $a^{\varphi(n)+1} \equiv a \pmod{n}$.

Chapter 8   Public Key Ciphers

Problems 8.1

1. As $m = 5 \cdot 7$, we have $\varphi(35) = \varphi(5)\varphi(7) = 4 \cdot 6 = 24$. Then $d = e^{-1} \bmod \varphi(m) = 11^{-1} \bmod 24 = 11$, so

$$\begin{aligned} y &= x^e \bmod m = 22^{11} \bmod 35 = 8 \\ z &= y^d \bmod m = 8^{11} \bmod 35 = 22 \end{aligned}$$

See the disc for details of these computations.

3. As $m = 29 \cdot 31$, we have $\varphi(m) = \varphi(29)\varphi(31) = 28 \cdot 30 = 840$ and $d = e^{-1} \bmod \varphi(m) = 101^{-1} \bmod 840 = 341$. So

$$y = x^e \bmod m = 555^{101} \bmod 899 = 731$$
$$z = y^d \bmod m = 731^{341} \bmod 899 = 555$$

5. Let $x = 99999999999999$, $m = 25972641171898723$, $e = 997$, and $\varphi = 25972640809676568$ and note that

$$y = x^e \bmod m = 4815\,828\,410\,330\,867$$
$$d = e^{-1} \bmod \varphi = 11\,514\,450\,589\,645\,981$$

yields $y^d \bmod m = 99\,999\,999\,999\,999$.

7. The numbers p and q are found using $p+q = m - \varphi(m) + 1 = 15\,481\,643\,766$ $690\,322\,656\,570\,354\,930\,733\,675\,907\,594\,344\,877\,898\,320\,382\,154\,352\,656\,607\,014$ and $p - q = \sqrt{(p+q)^2 - 4m} = 12\,326\,200\,145\,806\,231\,064\,298\,751\,424\,387$ $459\,058\,535\,358\,406\,473\,593\,017\,996\,013\,777\,132\,860$.

Then $p = \frac{(p+q)+(p-q)}{2} = 1\,577\,721\,810\,442\,045\,796\,135\,801\,753\,173\,108\,424$
$529\,493\,235\,712\,363\,682\,079\,169\,439\,737\,077$ and $q = \frac{(p+q)-(p-q)}{2} = 13\,903$
$921\,956\,248\,276\,860\,434\,553\,177\,560\,567\,483\,064\,851\,642\,185\,956\,700\,075\,183$
$216\,869\,937.$

As a check, $pq = 21\,936\,520\,921\,056\,942\,428\,185\,744\,321\,881\,874\,204\,790\,829$
$920\,570\,235\,226\,904\,516\,467\,385\,564\,406\,736\,567\,597\,367\,535\,979\,699\,930\,859$
$170\,667\,289\,061\,009\,756\,151\,158\,068\,196\,185\,554\,149$

9. There are $\varphi(m) = (p-1)(q-1)$ numbers between 1 and m that are relatively prime to m. So there are $m-(p-1)(q-1)-p+q-1$ numbers x in that range with $\gcd(x,m) \neq 1$. Thus the probability of choosing a number x with $\gcd(x,m) \neq 1$ is

$$\frac{p+q-1}{m} = \frac{1}{q} + \frac{1}{p} - \frac{1}{m}$$

Problems 8.2

1. We have $12^3 \bmod 15 = 3$ and $3^5 \bmod 21 = 12$ and hence Kyle sends the message 12 to Sarah.

3. The product $p_b q_b$ is given by $n_b = 2748\,401 \cdot 157\,849\,763 = 433\,834$
$446\,478\,963$ so the public key is $(e_b, n_b) = (9587, 433\,834\,446\,478\,963)$. Since $\varphi(n_b) = (2748\,401 - 1)(157\,849\,763 - 1) = 433\,834\,285\,880\,800$, it follows that $d_b = 9587^{-1} \bmod 433\,834\,285\,880\,800 = 394\,147\,974\,342\,523$ and hence the private key is

$$(d_s, n_s) = (394\,147\,974\,342\,523, 433\,834\,446\,478\,963)$$

5. Since $n_s = 37\,847\,755\,706\,513 < n_b = 433\,834\,446\,478\,963$, Sean decrypts the message by first applying Brendon's public key, then Sean's private key to get

$$\left(386\,686\,175\,803\,129^{9587} \bmod n_b\right)^{28\,526\,126\,032\,457} \bmod n_s$$
$$= 5112\,397\,193\,243^{28\,526\,126\,032\,457} \bmod n_s = 1234\,567\,890$$

7. Since $n_j < n_b$, the plaintext message is encrypted by first using Janet's public key, then using Brendon's private key to get

$$\left(99\,999\,999^{7853} \bmod 15\,092\,177\right)^{164165} \bmod 29\,143\,171$$
$$= 6136\,937^{164165} \bmod 29\,143\,171 = 24\,506\,902$$

9. Encryption yields

$$\left(2000\,000^{7907} \bmod 49\,839\,739\right)^{22\,913\,165} \bmod 26\,513\,567$$
$$= 45\,574\,626^{22\,913\,165} \bmod 26\,513\,567 = 6765\,427$$

and decryption yields

$$\left(6765\,427^{4637} \bmod 26\,513\,567\right)^{45\,679\,243} \bmod 49\,839\,739$$
$$= 19\,061\,059^{45\,679\,243} \bmod 49\,839\,739 = 44\,781\,288$$

Information can be lost when modular arithmetic with a large modulus is followed by modular arithmetic with a smaller modulus.

11. Since $n_j > n_b$, the plaintext message is encrypted by first using Brendon's private key, then using Janet's public key to get

$$\left(99\,999\,999^{260\,733\,157\,830\,321\,636\,619} \bmod 986\,577\,727\,411\,807\,628\,569\right)^{177\,264\,463}$$
$$\bmod 7029\,757\,456\,346\,007\,993\,017$$
$$= 889\,504\,247\,126\,301\,481\,385^{177\,264\,463} \bmod 7029\,757\,456\,346\,007\,993\,017$$
$$= 6165\,005\,940\,876\,305\,088\,312$$

13. Encryption yields $70\,849\,375\,393\,993\,065\,812$ and decryption yields $333\,146$ $600\,281\,273\,156\,600\,517$. Information can be lost when modular arithmetic with a large modulus is followed by modular arithmetic with a smaller modulus. By first applying Sarah's private key followed by Preston's public key, the encrypted message is $30\,564\,624\,919\,862\,526\,996\,750$ and decryption yields $101\,010\,101\,010\,101\,010\,101$. See the disc for details of these computations.

Problems 8.3

1. The ASCII values are given by

2	^	4	=	1	6
50	94	52	61	49	54

and hence the equivalent positive integer is

$$50 + 94 \cdot 256 + 52 \cdot 256^2 + 61 \cdot 256^3 + 49 \cdot 256^4 + 54 \cdot 256^5 = 59\,585\,108\,139\,570$$

3. Sarah first translates the plaintext to large integers using the table

R	o	n		R	i	v	e	s	t	
82	111	110	32	82	105	118	101	115	116	32
i	s		a		p	r	o	f	e	s
105	115	32	97	32	112	114	111	102	102	115
s	o	r		a	t		M	I	T	.
115	111	114	32	97	116	32	77	73	84	46

and the calculations

$$x = 82 + 111c + 110c^2 + 32c^3 + 82c^4 + 105c^5 + 118c^6 + 101c^7 + 115c^8$$
$$+ 116c^9 + 32c^{10} + 105c^{11} + 115c^{12} + 32c^{13} + 97c^{14} + 32c^{15}$$
$$+ 112c^{16} + 114c^{17} + 111c^{18} + 102c^{19}$$
$$= 584\,802\,410\,296\,329\,453\,294\,993\,117\,073\,566\,607\,047\,018\,835\,794$$
$$y = 102 + 115c + 115c^2 + 111c^3 + 114c^4 + 32c^5 + 97c^6 + 116c^7$$
$$+ 32c^8 + 77c^9 + 73c^{10} + 84c^{11} + 46c^{12}$$
$$= 3670\,580\,832\,286\,885\,393\,984\,052\,753\,254$$

Since $n_k < n_s$, Sarah first uses Kyle's public key, then her private key to encrypt the message as

$$(x^{e_k} \bmod n_k)^{d_s} \bmod n_s = 6642\,642\,701\,546\,739\,554\,009\,203$$
$$231\,203\,625\,130\,908\,152\,661\,584\,571$$

$$(y^{e_k} \bmod n_k)^{d_s} \bmod n_s = 5781\,828\,020\,001\,339\,664\,110\,566$$
$$141\,724\,658\,323\,575\,128\,107\,486\,053$$

5. The message is

> *Multiplication is easy, but factorization is hard.*

See the disc for details.

7. Since $n_k < n_s$, Kyle first uses Sarah's public key then his own private key to calculate the numbers $\left(x^{997} \bmod n_s\right)^{d_k} \bmod n_k$, $\left(y^{997} \bmod n_s\right)^{d_k} \bmod n_k$, and $\left(z^{997} \bmod n_s\right)^{d_k} \bmod n_k$. The message is

> *RSA keys are typically 1024 to 2048 bits long*

See the disc for details.

Problems 8.4

1. Note that

$$2 = 2 < 3$$
$$2 + 3 = 5 < 6$$
$$2 + 3 + 6 = 11 < 12$$
$$2 + 3 + 6 + 12 = 23 < 24$$
$$2 + 3 + 6 + 12 + 24 = 47 < 48$$
$$2 + 3 + 6 + 12 + 24 + 48 = 95 < 96$$
$$2 + 3 + 6 + 12 + 24 + 48 + 96 = 191 < 200$$

so the sequence $\{2, 3, 6, 12, 24, 48, 96, 200\}$ is super increasing.

3. Since $2 + 3 + 6 + 12 + 24 + 48 + 96 + 200 = 391 < 453$, it follows that $m = 453$ is an appropriate modulus. Also, $\gcd(453, 61) = 1$ and hence $k = 61$ has an inverse modulo 453. The public key is given by

$$2 \cdot 61 \bmod 453 = 122$$
$$3 \cdot 61 \bmod 453 = 183$$
$$6 \cdot 61 \bmod 453 = 366$$
$$12 \cdot 61 \bmod 453 = 279$$
$$24 \cdot 61 \bmod 453 = 105$$
$$48 \cdot 61 \bmod 453 = 210$$
$$96 \cdot 61 \bmod 453 = 420$$
$$200 \cdot 61 \bmod 453 = 422$$

The encrypted message is given by

$$(122, 183, 366, 279, 105, 210, 420, 422) \cdot (1, 0, 0, 1, 0, 1, 1, 0) = 1031$$

To decrypt the message, note first that $61^{-1} \bmod 453 = 52$ and hence the original superincreasing sequence is given by

$$122 \cdot 52 \bmod 453 = 2$$
$$183 \cdot 52 \bmod 453 = 3$$
$$366 \cdot 52 \bmod 453 = 6$$
$$279 \cdot 52 \bmod 453 = 12$$
$$105 \cdot 52 \bmod 453 = 24$$
$$210 \cdot 52 \bmod 453 = 48$$
$$420 \cdot 52 \bmod 453 = 96$$
$$422 \cdot 52 \bmod 453 = 200$$

and

$$1031 \cdot 52 \bmod 453 = 158$$

We now deduce that

$$158 = 96 + 62$$
$$62 = 48 + 14$$
$$14 = 12 + 2$$

so

$$158 = 96 + 48 + 12 + 2$$

and the plaintext is $\left(\begin{array}{cccccccc} 1 & 0 & 0 & 1 & 0 & 1 & 1 & 0 \end{array}\right)_2$.

5. Note that

$$1 = 1 < 3$$
$$1 + 3 = 4 < 5$$
$$4 + 5 = 9 < 10$$
$$9 + 10 = 19 < 20$$
$$19 + 20 = 39 < 40$$
$$39 + 40 = 79 < 80$$
$$79 + 80 = 159 < 160$$
$$159 + 160 = 319 < 320$$
$$319 + 320 = 639 < 640$$
$$639 + 640 = 1279 < 1280$$
$$1279 + 1280 = 2559 < 2560$$
$$2559 + 2560 = 5119 < 5120$$
$$5119 + 5120 = 10\,239 < 10240$$
$$10239 + 10240 = 20\,479 < 20480$$
$$20\,479 + 20480 = 40\,959 < 40960$$
$$40\,959 + 40960 = 81\,919 < 81920$$
$$81\,919 + 81920 = 163\,839 < 163840$$

so the sequence $1, 3, 5, 10, 20, 40, 80, 160, 320, 640, 1280, 2560, 5120, 10\,240,$
$20\,480, 40\,960, 81\,920, 163\,840$ is superincreasing.

7. The public key is $997, 2991, 4985, 9970, 19940, 39880, 79760, 159520, 319040,$
$310\,261, 292\,703, 257\,587, 187\,355, 46\,891, 93\,782, 187\,564, 47\,309, 94\,618$. The
ciphertext is $860\,321 \bmod 327819 = 204\,683$.

The private key is $997^{-1} \bmod 327819 = 133\,495$. To decrypt, multiply
by 133495 and reduce modulo 327819 to get the superincreasing sequence
$1,3,5,10,20,40,80,160,320,640,1280,2560,5120,10240,20480,40960,81920,$
163840 and the sum $133495 \cdot 860\,321 \bmod 327819 = 115\,616$. Now

$$115\,616 = 81\,920 + 33\,696$$
$$33\,696 = 20\,480 + 13\,216$$
$$13\,216 = 10\,240 + 2976$$
$$2976 = 2560 + 416$$
$$416 = 320 + 96$$
$$96 = 80 + 16$$
$$16 = 10 + 6$$
$$6 = 5 + 1$$

so the plaintext is 101100101001011010.

Problems 8.5

1. Use a computer algebra system to verify that $q = 1\,091\,381\,946\,026\,415\,813\,817$ $094\,712\,889\,364\,130\,135\,176\,179\,079$ is prime. The evaluations

$$q - 2^{159} = 360\,631\,127\,360\,964\,354\,715\,252$$
$$296\,531\,222\,620\,307\,209\,907\,591$$
$$q - 2^{160} = -370\,119\,691\,304\,487\,104\,386\,590$$
$$119\,826\,918\,889\,520\,756\,363\,897$$

show that $2^{159} < q < 2^{160}$.

3. Evaluation of $p - 1 \bmod q$ should produce 0.

5. The order of g must divide the prime q. Since the order of g is not 1, it must be exactly q.

7. The generalized prime number theorem implies that

$$\pi_{q,m}(x) \approx \frac{1}{\varphi(q)} \cdot \frac{x}{\ln x}$$

Let $f(x) = \frac{1}{\varphi(q)} \cdot \frac{x}{\ln x} = \frac{1}{q-1} \cdot \frac{x}{\ln x}$. The derivative is given by

$$f'(x) = \frac{1}{q-1} \cdot \frac{\ln x - x\frac{1}{x}}{(\ln x)^2} \approx \frac{1}{q \ln x}$$

and hence the average gap between primes in the arithmetic sequence $kq+1$ is roughly $q \ln x$. Since $q \approx 2^{160}$ and $x \approx 2^{512}$ and we only check every $2q$, and on average we start halfway between such primes, the average number of trips through the loop should be about

$$\frac{\ln 2^{512}}{4} \approx 354.\,891\,356\,4/4 \approx 89$$

Chapter 9   Finite Fields

Problems 9.1

1. From the distributive law, $a0 + a0 = a(0 + 0)$. So $a0 + a0 = a0$. Adding $-a0$ to both sides of this equation gives $a0 = 0$.

3. The system

$$3x + 7y + 6z = 2$$
$$4x + 5y + 3z = 2$$
$$2x + 4y + 5z = 4$$

is equivalent to the matrix equation

$$\begin{pmatrix} 3 & 7 & 6 \\ 4 & 5 & 3 \\ 2 & 4 & 5 \end{pmatrix} \begin{pmatrix} x \\ y \\ z \end{pmatrix} = \begin{pmatrix} 2 \\ 2 \\ 4 \end{pmatrix}$$

Calculate the inverse modulo 5 and multiply to get

$$\begin{pmatrix} 3 & 7 & 6 \\ 4 & 5 & 3 \\ 2 & 4 & 5 \end{pmatrix}^{-1} \mod 5 = \begin{pmatrix} 4 & 2 & 3 \\ 3 & 4 & 0 \\ 3 & 1 & 1 \end{pmatrix}$$

$$\begin{pmatrix} 4 & 2 & 3 \\ 3 & 4 & 0 \\ 3 & 1 & 1 \end{pmatrix} \begin{pmatrix} 2 \\ 2 \\ 4 \end{pmatrix} \mod 5 = \begin{pmatrix} 4 \\ 4 \\ 2 \end{pmatrix}$$

Thus $x = 1$, $y = 4$, $z = 1$. As a check, note that

$$3 \cdot 4 + 7 \cdot 4 + 6 \cdot 2 \mod 5 = 2$$
$$4 \cdot 4 + 5 \cdot 4 + 3 \cdot 2 \mod 5 = 2$$
$$2 \cdot 4 + 4 \cdot 4 + 5 \cdot 2 \mod 5 = 4$$

5. Solve the system

$$3x + 9y + 4z = 8$$
$$2x + 3y + 5z = 2$$
$$5x + 4y + 9z = 10$$

over the rational numbers to get

$$\left\{ z = -\frac{10}{7}, y = 0, x = \frac{32}{7} \right\}$$

Reduce modulo 11 to produce

$$\frac{32}{7} \mod 11 = 3$$
$$0 \mod 11 = 0$$
$$-\frac{10}{7} \mod 11 = 8$$

As a check, note that indeed

$$3 \cdot 3 + 9 \cdot 0 + 4 \cdot 8 \bmod 11 = 8$$
$$2 \cdot 3 + 3 \cdot 0 + 5 \cdot 8 \bmod 11 = 2$$
$$5 \cdot 3 + 4 \cdot 0 + 9 \cdot 8 \bmod 11 = 10$$

7.

+	0	1	2	3	4	5	6	7	8	9	10	11	12	13	14	15	16
0	0	1	2	3	4	5	6	7	8	9	10	11	12	13	14	15	16
1	1	2	3	4	5	6	7	8	9	10	11	12	13	14	15	16	0
2	2	3	4	5	6	7	8	9	10	11	12	13	14	15	16	0	1
3	3	4	5	6	7	8	9	10	11	12	13	14	15	16	0	1	2
4	4	5	6	7	8	9	10	11	12	13	14	15	16	0	1	2	3
5	5	6	7	8	9	10	11	12	13	14	15	16	0	1	2	3	4
6	6	7	8	9	10	11	12	13	14	15	16	0	1	2	3	4	5
7	7	8	9	10	11	12	13	14	15	16	0	1	2	3	4	5	6
8	8	9	10	11	12	13	14	15	16	0	1	2	3	4	5	6	7
9	9	10	11	12	13	14	15	16	0	1	2	3	4	5	6	7	8
10	10	11	12	13	14	15	16	0	1	2	3	4	5	6	7	8	9
11	11	12	13	14	15	16	0	1	2	3	4	5	6	7	8	9	10
12	12	13	14	15	16	0	1	2	3	4	5	6	7	8	9	10	11
13	13	14	15	16	0	1	2	3	4	5	6	7	8	9	10	11	12
14	14	15	16	0	1	2	3	4	5	6	7	8	9	10	11	12	13
15	15	16	0	1	2	3	4	5	6	7	8	9	10	11	12	13	14
16	16	0	1	2	3	4	5	6	7	8	9	10	11	12	13	14	15

×	1	2	3	4	5	6	7	8	9	10	11	12	13	14	15	16
1	1	2	3	4	5	6	7	8	9	10	11	12	13	14	15	16
2	2	4	6	8	10	12	14	16	1	3	5	7	9	11	13	15
3	3	6	9	12	15	1	4	7	10	13	16	2	5	8	11	14
4	4	8	12	16	3	7	11	15	2	6	10	14	1	5	9	13
5	5	10	15	3	8	13	1	6	11	16	4	9	14	2	7	12
6	6	12	1	7	13	2	8	14	3	9	15	4	10	16	5	11
7	7	14	4	11	1	8	15	5	12	2	9	16	6	13	3	10
8	8	16	7	15	6	14	5	13	4	12	3	11	2	10	1	9
9	9	1	10	2	11	3	12	4	13	5	14	6	15	7	16	8
10	10	3	13	6	16	9	2	12	5	15	8	1	11	4	14	7
11	11	5	16	10	4	15	9	3	14	8	2	13	7	1	12	6
12	12	7	2	14	9	4	16	11	6	1	13	8	3	15	10	5
13	13	9	5	1	14	10	6	2	15	11	7	3	16	12	8	4
14	14	11	8	5	2	16	13	10	7	4	1	15	12	9	6	3
15	15	13	11	9	7	5	3	1	16	14	12	10	8	6	4	2
16	16	15	14	13	12	11	10	9	8	7	6	5	4	3	2	1

9. The definition of a field contains the following two properties: $a1 = a$ and, if $a \neq 0$, then $ax = 1$ for some x. The element in the set of integers modulo 10 under addition and multiplication such that $a1 = a$ for all a in the set is the congruence class of the number 1. The multiplication table for the set of integers modulo 10 is

·	1	2	3	4	5	6	7	8	9
1	1	2	3	4	5	6	7	8	9
2	2	4	6	8	0	2	4	6	8
3	3	6	9	2	5	8	1	4	7
4	4	8	2	6	0	4	8	2	6
5	5	0	5	0	5	0	5	0	5
6	6	2	8	4	0	6	2	8	4
7	7	4	1	8	5	2	9	6	3
8	8	6	4	2	0	8	6	4	2
9	9	8	7	6	5	4	3	2	1

and you can see that there are several rows that do not contain 1. For example, there is no element in the set for which $2n = 1$. If this were a field, every element would have to have a multiplicative inverse.

11. To factor $x^2 + 7x + 10 = (x - a)(x - b)$ over GF_{11}, look for a and b in the Cayley tables for GF_{11} so that $a + b = -7 \bmod 11 = 4$ and $ab = 10 \bmod 11$. The solution is $x = 6$ or $x = 9$

Problems 9.2

1. The sum $(5x^3 + 4x^2 + 3) + (4x^4 + 6x^2 + 3x + 5) \bmod 7$ is equal to

$$4x^4 + 5x^3 + 3x^2 + 3x + 1$$

3. The product $(5x^3 + 4x^2 + 3)(4x^4 + 6x^2 + 3x + 5) \bmod 7$ is

$$6x^7 + 2x^6 + 2x^5 + 2x^4 + 2x^3 + 3x^2 + 2x + 1$$

5. The division algorithm in $GF_7[x]$ writes $4x^4 + x^2 + 3x + 1$ as

$$(5x + 3)(5x^3 + 4x^2 + 3) + 3x^2 + 2x + 6$$

so the answer is $3x^2 + 2x + 6$.

7. Factoring the polynomial in $\mathbb{Q}[x]$ gives $(x - 3)^2(x + 3)(3x + x^2 - 2)$ which is

$$(x + 4)^2(x + 3)(3x + x^2 + 5) \bmod 7$$

9. The codeword $x^4 p(x) + (x^4 p(x) \bmod g(x)) \bmod 2$ is equal to

$$x^{15} + x^{13} + x^6 + x^3$$

11. Let n be a positive integer. If n is 1 or a prime, then we are done. Otherwise $n = ab$ where a and b are positive integers greater than 1. So $a < n$ and $b < n$. By the principle of complete induction, we may assume that each of a and b is a prime or a product of primes. So $n = ab$ is a product of primes.

13. Computing $(r(x) \bmod g(x)) \bmod 2$ gives $x^{15} + x^6 + x^3 + x + 1$, so at least one error has occurred. The plaintext must be retransmitted.

Problems 9.3

1. The multiplication table is given by

$$
\begin{pmatrix} 0 \\ 1 \\ \theta \\ \theta+1 \end{pmatrix} \begin{pmatrix} 0 & 1 & \theta & \theta+1 \end{pmatrix} \bmod 2 = \begin{pmatrix} 0 & 0 & 0 & 0 \\ 0 & 1 & \theta & \theta+1 \\ 0 & \theta & \theta^2 & \theta(\theta+1) \\ 0 & \theta+1 & \theta(\theta+1) & (\theta+1)^2 \end{pmatrix}
$$

where $(\theta^2 \bmod \theta^2 + \theta + 1) \bmod 2 = \theta + 1$, $(\theta(\theta+1) \bmod \theta^2 + \theta + 1) \bmod 2 = 1$, and $\left((\theta+1)^2 \bmod \theta^2 + \theta + 1\right) \bmod 2 = \theta$, so the multiplication table is given by

$\cdot$	0	1	θ	$\theta+1$
0	0	0	0	0
1	0	1	θ	$\theta+1$
θ	0	θ	$\theta+1$	1
$\theta+1$	0	$\theta+1$	1	θ

Addition is vector addition modulo 2. In particular, $0+0 = 1+1 = \theta+\theta = (\theta+1)+(\theta+1) = 0$, so the addition table is given by

$+$	0	1	θ	$\theta+1$
0	$0+0$	$0+1$	$0+\theta$	$0+\theta+1$
1	$1+0$	$1+1$	$1+\theta$	$1+\theta+1$
θ	$\theta+0$	$\theta+1$	$\theta+1$	$\theta+\theta+1$
$\theta+1$	$\theta+1+0$	$\theta+1+1$	$\theta+1+\theta$	$\theta+1+\theta+1$

which reduces to

$+$	0	1	θ	$\theta+1$
0	0	1	θ	$\theta+1$
1	1	0	$1+\theta$	θ
θ	θ	$\theta+1$	0	1
$\theta+1$	$\theta+1$	θ	1	0

3. Let $p(x), q(x), r(x) \in \mathbb{Z}[x]$ and set $s(x) = x^2 + x + 1$. Then

$$(p(x) + q(x)) + r(x) \bmod s(x) = p(x) + (q(x) + r(x)) \bmod s(x)$$
$$(p(x) q(x)) r(x) \bmod s(x) = p(x) (q(x) r(x)) \bmod s(x)$$

so

$$[[p(x) + q(x)] + r(x)] = [p(x) + [q(x) + r(x)]]$$
$$[p(x) q(x)] [r(x)] = [p(x)] [q(x) r(x)]$$

5. Suppose

$$f_1(x) \equiv f_2(x) \pmod{h(x)}$$

and

$$g_1(x) \equiv g_2(x) \pmod{h(x)}$$

Then

$$f_1(x) = f_2(x) + p(x) h(x)$$
$$g_1(x) = g_2(x) + q(x) h(x)$$

for some $p(x), q(x) \in \mathbb{Z}[x]$. Thus

$$f_1(x) + g_1(x) = f_2(x) + g_2(x) + (p(x) + q(x)) h(x)$$
$$f_1(x) g_1(x) = f_2(x) g_2(x)$$
$$+ (f_2(x) q(x) + g_2(x) p(x) + p(x) q(x) h(x)) h(x)$$

so

$$f_1(x) + g_1(x) \equiv f_2(x) + g_2(x) \pmod{h(x)}$$
$$f_1(x) g_1(x) \equiv f_2(x) g_2(x) \pmod{h(x)}$$

7. Note that f transforms the addition table

+	0	1	θ	$\theta + 1$
0	0	1	θ	$\theta + 1$
1	1	0	$1 + \theta$	θ
θ	θ	$0 + 1$	0	1
$\theta + 1$	$0 + 1$	θ	1	0

into the addition table

+	0	1	β	$\beta + 1$
0	0	1	β	$\beta + 1$
1	1	0	$1 + \beta$	β
β	β	$\beta + 1$	0	1
$\beta + 1$	$\beta + 1$	β	1	0

and the multiplication table

$\times$	0	1	θ	$\theta + 1$
0	0	0	0	0
1	0	1	θ	$\theta + 1$
θ	0	θ	$\theta + 1$	1
$\theta + 1$	0	$\theta + 1$	1	θ

into the multiplication table

$\times$	0	1	β	$\beta + 1$
0	0	0	0	0
1	0	1	β	$\beta + 1$
β	0	β	$\beta + 1$	1
$\beta + 1$	0	$\beta + 1$	1	β

so $f(s+t) = f(s) + f(t)$ and $f(st) = f(s)f(t)$ for all $s, t \in GF_4$.

9. Let $r(x) = f(x) \bmod x^2 + x + 1$. Then

$$f(x) - r(x) = k(x)\left(x^2 + x + 1\right)$$

for some $k(x) \in GF_2[x]$, and $\deg(r(x)) < 2$ or $r(x) = 0$.
If $s(x) \in [f(x)]$ then

$$f(x) - s(x) = h(x)\left(x^2 + x + 1\right)$$

for some $h(x) \in GF_2[x]$.

Suppose that $\deg(s(x)) < 2$ or $s(x) = 0$. We have

$$r(x) - s(x) = \left(f(x) - k(x)\left(x^2 + x + 1\right)\right) - \left(f(x) - h(x)\left(x^2 + x + 1\right)\right)$$
$$= (h(x) - k(x))\left(x^2 + x + 1\right)$$

so if $(h(x) - k(x)) \neq 0$, then

$$\deg(r(x) - s(x)) = \deg(h(x) - k(x)) + 2 \geq 2$$

But this is not possible. Thus $h(x) - k(x) = 0$, implying that $r(x) = s(x)$.

Problems 9.4

1. Note that

$$(\theta^2 + 1)(\theta^2 + \theta) = \theta^6 \cdot \theta^4 = \theta^{10}$$
$$= \theta^7 \cdot \theta^3 = 1 \cdot \theta^3 = \theta + 1$$

3. Note that

$$E \cdot E = (1110) \cdot (1110)$$
$$= (\theta^3 + \theta^2 + \theta) \cdot (\theta^3 + \theta^2 + \theta)$$
$$= \theta^6 + 2\theta^5 + 3\theta^4 + 2\theta^3 + \theta^2 \bmod 2$$
$$= (\theta^6 + \theta^4 + \theta^2 \bmod \theta^4 + \theta + 1) \bmod 2$$
$$= \theta^3 + \theta + 1 = (1011) = B$$

and

$$8 \cdot A = (1000) \cdot (1010)$$
$$= \theta^3 \cdot (\theta^3 + \theta)$$
$$= (\theta^6 + \theta^4 \bmod \theta^4 + \theta + 1) \bmod 2$$
$$= \theta^3 + \theta^2 + \theta + 1 = (1111) = F$$

5. These polynomials are irreducible over GF_2:

$$x^5 + x^4 + x^3 + x + 1$$
$$x^5 + x^4 + x^3 + x^2 + 1$$
$$x^5 + x^3 + 1$$
$$x^5 + x^4 + x^2 + x + 1$$
$$x^5 + x^3 + x^2 + x + 1$$
$$x^5 + x^2 + 1$$

7. Note that in GF_2,

$$x^5 + x + 1 \bmod 2 = (x^3 + x^2 + 1)(x^2 + x + 1)$$

9. The given polynomials are irreducible because they have no roots in GF_2. That is,

$$[x^3 + x^2 + 1]_{x=0} = 1$$
$$[x^3 + x^2 + 1]_{x=1} = 1$$
$$[x^3 + x + 1]_{x=0} = 1$$
$$[x^3 + x + 1]_{x=1} = 1$$

modulo 2. Here is a complete list of the remaining polynomials of degree 3 together with an evaluation that shows that each polynomial has a root:

$$\left[x^3\right]_{x=0} = 0$$
$$\left[x^3 + 1\right]_{x=1} = 0$$
$$\left[x^3 + x\right]_{x=0} = 0$$
$$\left[x^3 + x^2\right]_{x=0} = 0$$
$$\left[x^3 + x^2 + x\right]_{x=0} = 0$$
$$\left[x^3 + x^2 + x + 1\right]_{x=1} = 0$$

This exhausts all eight possible polynomials of degree 3.

Problems 9.5

1. The polynomials

$$
\begin{array}{lll}
x^2 + 4x + 1 & x^2 + x + 2 & x^2 + 4x + 2 \\
x^2 + x + 1 & x^2 + 3x + 4 & x^2 + 3x + 3 \\
x^2 + 2 & x^2 + 2x + 4 & x^2 + 3 \\
x^2 + 2x + 3 & &
\end{array}
$$

are all irreducible.

3. Let k be the least positive integer such that $a^k = 1$ and assume that $a^n = 1$. Write $n = kx + b$, where $0 \le b < k$. Then $a^n = 1 = a^{kx+b} = \left(a^k\right)^x a^b = 1 \cdot a^b$ implies $a^b = 1$. Since k is the least positive integer with this property, it follows that $b = 0$, and hence $k \mid n$. If $k \ne n$, then $k \mid \frac{n}{p}$ for some prime divisor p of n. Thus $a^{n/p} = \left(a^k\right)^r = 1$.

5. The three polynomials are irreducible because

$$
\begin{array}{lll}
\left[x^2 + 1\right]_0 \bmod 3 = 1 & \left[x^2 + x + 2\right]_0 \bmod 3 = 2 & \left[x^2 + 2x + 2\right]_0 \bmod 3 = 2 \\
\left[x^2 + 1\right]_1 \bmod 3 = 2 & \left[x^2 + x + 2\right]_1 \bmod 3 = 1 & \left[x^2 + 2x + 2\right]_1 \bmod 3 = 2 \\
\left[x^2 + 1\right]_2 \bmod 3 = 2 & \left[x^2 + x + 2\right]_2 \bmod 3 = 2 & \left[x^2 + 2x + 2\right]_2 \bmod 3 = 1
\end{array}
$$

shows that none of the polynomials has a root in GF_3.

7. Let θ be a root of $x^3 + 2x^2 + x + 1$ in GF_{27}. The order of θ divides 26 and hence is 2, 13, or 26. Note that

$$\left(\theta^{26} \bmod \theta^3 + 2\theta^2 + \theta + 1\right) \bmod 3 = 1$$
$$\left(\theta^{13} \bmod \theta^3 + 2\theta^2 + \theta + 1\right) \bmod 3 = 2$$
$$\left(\theta^2 \bmod \theta^3 + 2\theta^2 + \theta + 1\right) \bmod 3 = \theta^2$$

and hence θ is an element of order 26.

9 Note that

$$x^{27} - x \bmod 3 = x\,(x+1)\,(x+2)\,(x^3 + 2x^2 + x + 1)$$
$$(x^3 + x^2 + 2)\,(x^3 + 2x + 1)\,(x^3 + 2x + 2)$$
$$(x^3 + 2x^2 + 2x + 2)\,(x^3 + 2x^2 + 1)$$
$$(x^3 + x^2 + 2x + 1)\,(x^3 + x^2 + x + 2)$$

is the product of all monic irreducible polynomials of degrees 1 and 3 in $GF_3\,[x]$.

11. Consider the field GF_9.

a.

Power	Product	Expand	$\bmod x^2 + 1$	$\bmod 3$
y	$x+1$	$x+1$	$x+1$	$x+1$
y^2	$(x+1)(x+1)$	$x^2 + 2x + 1$	$2x$	$2x$
y^3	$(x+1)2x$	$2x^2 + 2x$	$2x - 2$	$2x + 1$
y^4	$(x+1)(2x+1)$	$2x^2 + 3x + 1$	$-1 + 3x$	2
y^5	$(x+1)2$	$2x + 2$	$2x + 2$	$2x + 2$
y^6	$(x+1)(2x+2)$	$2x^2 + 4x + 2$	$4x$	x
y^7	$(x+1)x$	$x^2 + x$	$x - 1$	$x + 2$
y^8	$(x+1)(x+2)$	$x^2 + 3x + 2$	$1 + 3x$	1

b.

×	00	01	02	10	11	12	20	21	22
00	00	00	00	00	00	00	00	00	00
01	00	01	02	10	11	12	20	21	22
02	00	02	01	20	22	21	10	12	11
10	00	10	20	02	12	22	01	11	21
11	00	11	22	12	20	01	21	02	10
12	00	12	21	22	01	10	11	20	02
20	00	20	10	01	21	11	02	22	12
21	00	21	12	11	02	20	22	10	01
22	00	22	11	21	10	02	12	01	20

See the disc for the addition table.

c.

+	0	1	2	3	4	5	6	7	8		×	0	1	2	3	4	5	6	7	8
0	0	1	2	3	4	5	6	7	8		0	0	0	0	0	0	0	0	0	0
1	1	2	0	4	5	3	7	8	6		1	0	1	2	3	4	5	6	7	8
2	2	0	1	5	3	4	8	6	7		2	0	2	1	6	8	7	3	5	4
3	3	4	5	6	7	8	0	1	2		3	0	3	6	2	5	8	1	4	7
4	4	5	3	7	8	6	1	2	0		4	0	4	8	5	6	1	7	2	3
5	5	3	4	8	6	7	2	0	1		5	0	5	7	8	1	3	4	6	2
6	6	7	8	0	1	2	3	4	5		6	0	6	3	1	7	4	2	8	5
7	7	8	6	1	2	0	4	5	3		7	0	7	5	4	2	6	8	3	1
8	8	6	7	2	0	1	5	3	4		8	0	8	4	7	3	2	5	1	6

Problems 9.6

1. The divisors of 12 are $1, 2, 3, 4, 6,$ and 12. The following table shows that there are exactly $\varphi(d)$ elements of order d for each divisor d.

Order	φ(Order)	Elements
1	1	1
2	1	12
3	2	3,9
4	2	5,8
6	2	4,10
12	4	2,6,7,11

3. In any group, $a^n a^m = a^{n+m}$ for any integers n and m. In particular, $a^0 a^n = a^{0+n} = a^n = a^{n+0} = a^n a^0$ so $a^0 = 1$, the identity. The inverse of a^n is a^{-n} because $a^n a^{-n} = a^{n-n} = a^0 = 1$. The product of two elements of (a) is in (a) because $a^n a^m = a^{n+m} \in (a)$. Thus (a) is a subgroup.

5. Let $a, b \in G$.

   (a) Assume $a^n = 1$ and let $n = k \cdot o(a) + r$, where $0 \le r < o(a)$. Then $1 = a^n = a^{k \cdot o(a) + r} = \left(a^{o(a)}\right)^k a^r = a^r$ implies $r = 0$ [since $o(a)$ is the smallest positive integer i such that $a^i = 1$], so that $o(a) \mid n$.

   (b) Since $ab = ba$ it follows that $a^n b^n = (ab)^n$ for every integer n. Let $o(a) = s$ and $o(b) = t$. Note that $(ab)^{st} = a^{st} b^{st} = (a^s)^t (b^t)^s = 1$ implies $o(ab) \mid st$. If $(ab)^n = 1$, then $a^n b^n = 1$ implies $a^n = b^{-n}$ and hence $1 = (a^s)^n = (a^n)^s = (b^{-n})^s = b^{-ns}$ and hence $t \mid ns$, but $s \perp t$ implies $t \mid n$. Similarly, $s \mid n$. Thus $st \mid n$. In particular, if $n = o(ab)$, then $st \mid o(ab)$. Hence $st = o(ab)$.

7. Suppose $m - 1 = nt$. Then $x^{m-1} - 1 = x^{nt} - 1$ is equal to $x^n - 1$ times the polynomial $x^{n(t-1)} + x^{n(t-2)} + x^{n(t-2)} + \cdots + x^n + 1$.

9. The divisors of 30 are 1, 2, 3, 5, 6, 10, 15, and 30. The following table shows that there are exactly $\varphi(d)$ elements of order d for each divisor d.

Order	φ(Order)	Elements
1	1	1
2	1	16
4	2	4,13
8	4	2,8,9,15
16	8	3,5,6,7,10,11,12,14

Problems 9.7

1. Since factor $(997) = 997$ it follows that 997 is the largest three-digit prime. Note that factor $(996) = 2^2 3 \times 83$. Since

$$525^{996/2} \bmod 997 = 996$$
$$525^{996/3} \bmod 997 = 304$$
$$525^{996/83} \bmod 997 = 701$$

it follows that 525 is primitive. The first 10 terms of the pseudorandom number sequence are given by

$$789 \cdot 525 \bmod 997 = 470$$
$$470 \cdot 525 \bmod 997 = 491$$
$$491 \cdot 525 \bmod 997 = 549$$
$$549 \cdot 525 \bmod 997 = 92$$
$$92 \cdot 525 \bmod 997 = 444$$
$$444 \cdot 525 \bmod 997 = 799$$
$$799 \cdot 525 \bmod 997 = 735$$
$$735 \cdot 525 \bmod 997 = 36$$
$$36 \cdot 525 \bmod 997 = 954$$
$$954 \cdot 525 \bmod 997 = 356$$

3. There are

$$\varphi\,(999\,999\,999\,988) = (2^2 - 2) \times 10 \times 124\,846 \times 182\,040$$
$$= 454\,539\,316\,800$$

primitive elements in $GF_{999\,999\,999\,989}$. Thus over 45% of the numbers less than $999\,999\,999\,989$ are primitive. Hence the average gap between primitive elements is just over 1.

5. The expected number of unbroken squares (see Problem 4) is

$$16\left(\frac{15}{16}\right)^{16} \approx 5.\,697\,186\,087$$

7. Use something like

```
randomize
set trials = 1000000
for i from 1 to trials do
    if rnd^2 + rnd^2 + rnd^2 < 1 then set n = n+1
end for
print 8*n/trials
end
```

With 1000000 trials, this algorithm produced 4.192168 compared with the exact value $4\pi/3 \approx 4.188\,790\,205$.

9. Use something like

```
randomize
set trials = 1000000
for i from 1 to trials do
    set n = abs(rnd - rnd)
end for
print n/trials
end
```

With 1000000 trials, the algorithm produced an approximate length of 0.3335 381.

The expected length can be represented by the integral

$$2 \int_0^1 x\,(1-x)\ dx = \frac{1}{3}$$

where x is the expected length and $2\,(1-x)$ is the probability density function on the unit interval. It can also be represented by the double integral

$$\int_0^1 \int_0^1 |x-y|\ dx\,dy = \frac{1}{3}$$

where x and y are the endpoints and $f\,(x,y) = 1$ is the probability density function on the unit square.

11. Results of an experiment with 1000000 trials repeated 10 times:

0.10005552	0.10020009	0.10016799	0.09977582
0.10080621	0.09980621	0.10029138	0.09948852
0.09980552	0.10004283		

The expected area is given by the integral

$$3 \int_0^1 x^2\,(1-x)^2\ dx = \frac{1}{10}$$

where x^2 is the area of a square of side x and $3\,(1-x)^2$ is the probability density function. Note that the vertex in the lower left corner of a square must be chosen from a square of edge $1 - x$.

Chapter 10   Error-Correcting Codes

Problems 10.1

1. If $x^3 + x + 1$ were reducible over GF_2, then it would have a root in $GF_2 = \{0, 1\}$ because the degree of $x^3 + x + 1$ is 3. However, $0^3 + 0 + 1 \bmod 2 = 1$

and $1^3 + 1 + 1 \bmod 2 = 1$, so $x^3 + x + 1$ has no roots in GF_2. Thus $x^3 + x + 1$ is irreducible over GF_2.

3. If θ is a root of $x^3 + x + 1$, then $m_1(x) = m_2(x) = x^3 + x + 1$, so that $g(x) = x^3 + x + 1$. Given a plaintext polynomial $a(x) = a_3 x^3 + a_2 x^2 + a_1 x + a_0$ with $a_i \in \{0, 1\}$, the codeword is given by

$$a(x) g(x) \bmod 2$$

The following is a list of all polynomials of degree ≤ 3 with their corresponding codewords.

$a(x)$	$a(x) g(x) \bmod 2$	$a(x)$	$a(x) g(x) \bmod 2$
0	0	x^3	$x^6 + x^4 + x^3$
1	$x^3 + x + 1$	$x^3 + 1$	$x^6 + x^4 + x + 1$
x	$x^4 + x^2 + x$	$x^3 + x$	$x^6 + x^3 + x^2 + x$
$x + 1$	$x^4 + x^3 + x^2 + 1$	$x^3 + x + 1$	$x^6 + x^2 + 1$
x^2	$x^5 + x^3 + x^2$	$x^3 + x^2$	$x^6 + x^5 + x^4 + x^2$
$x^2 + 1$	$x^5 + x^2 + x + 1$	$x^3 + x^2 + 1$	$x^6 + x^5 + x^4 + x^3$
			$+x^2 + x + 1$
$x^2 + x$	$x^5 + x^4 + x^3 + x$	$x^3 + x^2 + x$	$x^6 + x^5 + x$
$x^2 + x + 1$	$x^5 + x^4 + 1$	$x^3 + x^2 + x + 1$	$x^6 + x^5 + x^3 + 1$

Observe that the sum of any two distinct codewords is another codeword since

$$a(x) g(x) + b(x) g(x) \bmod 2 = (a(x) + b(x)) g(x) \bmod 2$$

and the degree of $a(x) + b(x) \leq 3$. Since all nonzero codewords have at least three nonzero terms, it follows that the minimum distance between codewords is 3.

5. If $\theta + 1$ is a root of an irreducible polynomial $m(x)$, then

$$(\theta + 1)^3 \bmod 3 = 1 + 3\theta + 3\theta^2 + \theta^3 \bmod 3 = 1 + \theta^3$$

must also be a root of $m(x)$. Since $\theta + 1$ must be a root of a polynomial of degree ≤ 2, it follows that the minimal polynomial of $\theta + 1$ must be

$$(x - (\theta + 1)) (x - (\theta^3 + 1)) \bmod a^2 + 1$$
$$= x^2 - 2x - \theta^3 x + 1 + \theta^3 - \theta x + \theta + \theta^4 \bmod \theta^2 + 1$$
$$= x^2 - 2x + 2 \bmod 3$$
$$= x^2 + x + 2$$

As a check, note that

$$\left((\theta + 1)^2 + (\theta + 1) + 2 \bmod \theta^2 + 1\right) \bmod 3 = 0$$

so $\theta + 1$ is indeed a root of $x^2 + x + 2$.

7. If $\theta^3 + \theta^2$ is a root of a polynomial over GF_2, then so are $\left(\theta^3 + \theta^2\right)^2$, $\left(\theta^3 + \theta^2\right)^4$, and $\left(\theta^3 + \theta^2\right)^8$. The minimal polynomial of $u = \theta^3 + \theta^2$ is

$$\left((x - u)\left(x - u^2\right)\left(x - u^4\right)\left(x - u^8\right) \bmod \theta^4 + \theta^3 + 1\right) \bmod 2 = x^4 + x + 1$$

9. If $u = \theta^4 + 2\theta$ is a root of a polynomial $m(x)$, then so are u^3, u^9, u^{27}, and u^{81}. The minimal polynomial is

$$\left((x - u)\left(x - u^3\right)\left(x - u^9\right)\left(x - u^{27}\right)\left(x - u^{81}\right) \bmod v\right) \bmod 3$$
$$= x^5 + 2x^4 + 2x^3 + 2$$

where $v = \theta^5 + \theta^3 + \theta + 1$.

11. The minimal polynomial is

$$\left((x - (3\theta + 2))\left(x - (3\theta + 2)^5\right) \bmod \theta^2 - 2\right) \bmod 5 = x^2 + x + 1$$

13. Substituting $-b$ for b in the equation $(a + b)^p = a^p + b^p$ gives $(a - b)^p = a^p + (-1)^p b^p$. If p is odd, then $(-1)^p = -1$, so this is just $a^p - b^p$. If p is even, then $p = 2$ so $c = -c$ for any c because $2c = 0$.

15. A polynomial code encodes the polynomial $p(x)$ by the codeword $g(x)p(x)$. This code encodes the polynomial $p(x)$ by the codeword $\left(1 + x + x^2\right)p\left(x^3\right)$, which does not have the right form. Suppose the plaintext word is

$$a_0 a_1 a_2 a_3 a_4$$

so the polynomial $p(x)$ is $a_0 + a_1 x + a_2 x^2 + a_3 x^3 + a_4 x^4$. Then $p\left(x^3\right)$ is $a_0 + a_1 x^3 + a_2 x^6 + a_3 x^9 + a_4 x^{12}$ and $\left(1 + x + x^2\right)p\left(x^3\right)$ is

$$a_0 + a_0 x + a_0 x^2 + a_1 x^3 + a_1 x^4 + a_1 x^5 + \cdots$$

Problems 10.2

1. Since long division produces

$$\frac{x^5 + x^4 + x^3 + 1}{x^3 + x + 1} = x^2 + x + \frac{1 - 2x^2 - x}{x^3 + x + 1}$$

where the remainder $1 - 2x^2 - x \bmod 2 = x + 1$ is not zero, we assume a single error. Thus $c(x) + x^k = x^5 + x^4 + x^3 + 1$, where $c(\theta) = 0$, so evaluation produces

$$\theta^k = \left(\theta^5 + \theta^4 + \theta^3 + 1 \bmod \theta^3 + \theta + 1\right) \bmod 2$$
$$= 1 - 2\theta^2 - \theta \bmod 2 = \theta + 1 = \theta^3$$

and hence $k = 3$. Thus $c(x) = x^5 + x^4 + 1$ and long division produces

$$\frac{x^5 + x^4 + 1}{x^3 + x + 1} = x^2 + x - 1 + \frac{2 - 2x^2}{x^3 + x + 1}$$

where the remainder $2 - 2x^2 \bmod 2 = 0$ and hence the plaintext is $x^2 + x - 1 \bmod 2 = x^2 + x + 1$.

3. We have

$$m_1(x) = x^2 + x + 2$$
$$m_2(x) = ((x - \theta^2)(x - \theta^6) \bmod \theta^2 + \theta + 2) \bmod 3 = x^2 + 1$$

and hence

$$g(x) = (x^2 + x + 2)(x^2 + 1) \bmod 3$$
$$= x^4 + 3x^2 + x^3 + x + 2 \bmod 3 = x^4 + x^3 + x + 2$$

5. The order of θ must divide 31, so $o(\theta) = 1$ or $o(\theta) = 31$. Since

$$(\theta \bmod \theta^5 + \theta^2 + 1) \bmod 2 - 0 \neq 1$$

it follows that $o(\theta) = 31$ and hence $q(x)$ is a primitive polynomial.

7. The plaintext is $x^{19} + x^{10} + x + 1$. (See disc for solution details.)

9. Note that

$$m_1(x) = (x - \theta)(x - \theta^3)(x - \theta^9)$$
$$= ((x - \theta)(x - \theta^3)(x - \theta^9) \bmod q(\theta)) \bmod 3$$
$$= x^3 + 2x + 1$$
$$m_2(x) = (x - \theta^2)(x - \theta^6)(x - \theta^{18})$$
$$= ((x - \theta^2)(x - \theta^6)(x - \theta^{18}) \bmod q(\theta)) \bmod 3$$
$$= x^3 + x^2 + x + 2$$

so

$$p(x) = m_1(x) m_2(x)$$
$$= (x^3 + 2x + 1)(x^3 + x^2 + x + 2) \bmod 3$$
$$= x^6 + x^5 + 2x^3 + 2x + 2$$

Plaintext polynomials in $GF_3[x]$ can be of degree at most $3^3 - 6 - 2 = 19$.

11. From problem 6, $m_1(x) = x^3 + 2x + 1$ and $m_2(x) = x^3 + x^2 + x + 2$. Since $m_3(x) = m_1(x)$ (why?), we need only find

$$
\begin{aligned}
m_4(x) &= (x - \theta^4)(x - \theta^{12})(x - \theta^{36}) \\
&= (x - \theta^4)(x - \theta^{12})(x - \theta^{10}\theta^{26}) \\
&= ((x - \theta^4)(x - \theta^{12})(x - \theta^{10}) \bmod q(\theta)) \bmod 3 \\
&= x^3 + x^2 + 2
\end{aligned}
$$

Hence

$$
\begin{aligned}
g(x) &= [m_1(x), m_2(x), m_3(x), m_4(x)] \\
&= m_1(x) m_2(x) m_4(x) \\
&= (x^3 + 2x + 1)(x^3 + x^2 + x + 2)(x^3 + x^2 + 2) \bmod 3 \\
&= x^9 + 2x^8 + x^7 + x^6 + x^5 + 2x^4 + 2x^3 + 2x^2 + x + 1
\end{aligned}
$$

Plaintext polynomials in $GF_3[x]$ can be of degree at most $3^3 - 9 - 2 = 16$.

Problems 10.3

1. Let θ be a root of $x^4 + x + 1$. Note that

$$
\begin{aligned}
(\theta^3 \bmod q(\theta)) \bmod 2 &= \theta^3 \\
(\theta^5 \bmod q(\theta)) \bmod 2 &= \theta + \theta^2 \\
(\theta^{15} \bmod q(\theta)) \bmod 2 &= 1
\end{aligned}
$$

and hence θ is primitive.

3. Note that $2^8 \bmod 17 = 1$ and hence 2 has order at most 8. Since $3^8 \bmod 17 = 16$, it follows that 3 has order 16 because the only prime dividing $17 - 1 = 16 = 2^4$ is 2.

5. Note that $2^{128} \bmod 257 = 1$ and hence 2 has order at most 128. Note that $3^{128} \bmod 257 = 256$. Since 2 is the only prime dividing $257 - 1 = 256 = 2^8$, it follows that 3 is primitive.

7. Let $q(x) = x^8 + x^4 + x^3 + x^2 + 1$ and let θ be a root in some extension field. We show that $q(x)$ is primitive by showing that $o(\theta) = 255 = 3 \times 5 \times 17$. This is established by the following:

$$
\begin{aligned}
\theta^{255} \bmod q(\theta) &= 1 \\
\theta^{255/3} \bmod q(\theta) &= \theta^7 + \theta^6 + \theta^4 + \theta^2 + \theta \neq 1 \\
\theta^{255/5} \bmod q(\theta) &= \theta^3 + \theta \neq 1 \\
\theta^{255/17} \bmod q(\theta) &= \theta^5 + \theta^2 + \theta \neq 1
\end{aligned}
$$

Chapter 11   Advanced Encryption Standard

Problems 11.1

1. Let $u_i \subset \{0,1\}$ and define $f(u_0, u_1, u_2, u_3) = (u_0 + u_1, u_0, u_1 + u_2, u_2 + u_3)$. Given an 8-bit word, let L denote the left 4 bits and R the right 4 bits. Define

$$F(L, R) = (R, L + f(R))$$

The first iterations of F applied to (11111111) is given by

$$
\begin{pmatrix}
0 & 0 & 0 & 0 & 1 & 0 & 0 & 0 \\
0 & 0 & 0 & 0 & 0 & 1 & 0 & 0 \\
0 & 0 & 0 & 0 & 0 & 0 & 1 & 0 \\
0 & 0 & 0 & 0 & 0 & 0 & 0 & 1 \\
1 & 0 & 0 & 0 & 1 & 1 & 0 & 0 \\
0 & 1 & 0 & 0 & 1 & 0 & 0 & 0 \\
0 & 0 & 1 & 0 & 0 & 1 & 1 & 0 \\
0 & 0 & 0 & 1 & 0 & 0 & 1 & 1
\end{pmatrix}
\begin{pmatrix}
1 \\ 1 \\ 1 \\ 1 \\ 1 \\ 1 \\ 1 \\ 1
\end{pmatrix}
\bmod 2 =
\begin{pmatrix}
1 \\ 1 \\ 1 \\ 1 \\ 1 \\ 0 \\ 1 \\ 1
\end{pmatrix}
$$

whereas the first iteration applied to (11101111) yields

$$
\begin{pmatrix}
0 & 0 & 0 & 0 & 1 & 0 & 0 & 0 \\
0 & 0 & 0 & 0 & 0 & 1 & 0 & 0 \\
0 & 0 & 0 & 0 & 0 & 0 & 1 & 0 \\
0 & 0 & 0 & 0 & 0 & 0 & 0 & 1 \\
1 & 0 & 0 & 0 & 1 & 1 & 0 & 0 \\
0 & 1 & 0 & 0 & 1 & 0 & 0 & 0 \\
0 & 0 & 1 & 0 & 0 & 1 & 1 & 0 \\
0 & 0 & 0 & 1 & 0 & 0 & 1 & 1
\end{pmatrix}
\begin{pmatrix}
1 \\ 1 \\ 1 \\ 0 \\ 1 \\ 1 \\ 1 \\ 1
\end{pmatrix}
\bmod 2 =
\begin{pmatrix}
1 \\ 1 \\ 1 \\ 1 \\ 1 \\ 0 \\ 1 \\ 0
\end{pmatrix}
$$

See the disc for the complete solution.

3. The inverse of the function $F(L, R) = (R, L + f(R))$ is

$$G(L, R) = (R + f(L), L)$$

since

$$G(F(L, R)) = G(R, L + f(R)) = (L + f(R) + F(R), R) = (L, R)$$

5. The Hamming distance between the two vectors $(0, 1, 1, 0, 0, 1, 0, 0)$ and $(1, 1, 0, 1, 0, 1, 1, 1)$ is 5.

Problems 11.2

1. The product is

$$\left(\left(\theta^6 + \theta^3 + \theta\right)\left(\theta^7 + \theta^2 + 1\right) \bmod \theta^8 + \theta^4 + \theta^3 + \theta + 1\right) \bmod 2 = \theta^6 + \theta + 1$$

3. One solution is $x^8 + x^4 + x^3 + x^2 + 1$.

5. The answer is $x = \theta^7 + \theta + 1$ and $y = \theta^7 + \theta^6 + \theta^5 + \theta^2 + \theta$. See the disc for details of the solution.

7. The inverse is

$$A^{-1} = \begin{pmatrix} \theta^7 + \theta^5 + \theta & \theta^7 + \theta^6 + \theta^5 + 1 \\ \theta^6 + \theta^4 + \theta^3 + \theta^2 + \theta + 1 & \theta^7 + \theta^6 + \theta^4 + \theta^3 \end{pmatrix}$$

See the disc for details of the solution.

Problems 11.3

1. Direct evaluation shows the product of these two matrices modulo 2 is the identity matrix. See the disc for details.

3. The inverse of $\theta^4 + \theta$ is

$$\left(\left(\theta^4 + \theta\right)^{254} \mod \theta^8 + \theta^4 + \theta^3 + \theta + 1 \right) \mod 2 = \theta^7 + \theta^5 + \theta^3 + \theta$$

$$= (10101010)$$

Apply the affine transformation

$$\begin{pmatrix} 1 & 0 & 0 & 0 & 1 & 1 & 1 & 1 \\ 1 & 1 & 0 & 0 & 0 & 1 & 1 & 1 \\ 1 & 1 & 1 & 0 & 0 & 0 & 1 & 1 \\ 1 & 1 & 1 & 1 & 0 & 0 & 0 & 1 \\ 1 & 1 & 1 & 1 & 1 & 0 & 0 & 0 \\ 0 & 1 & 1 & 1 & 1 & 1 & 0 & 0 \\ 0 & 0 & 1 & 1 & 1 & 1 & 1 & 0 \\ 0 & 0 & 0 & 1 & 1 & 1 & 1 & 1 \end{pmatrix} \begin{pmatrix} 0 \\ 1 \\ 0 \\ 1 \\ 0 \\ 1 \\ 0 \\ 1 \end{pmatrix} + \begin{pmatrix} 1 \\ 1 \\ 0 \\ 0 \\ 0 \\ 1 \\ 1 \\ 0 \end{pmatrix} \mod 2 = \begin{pmatrix} 1 \\ 0 \\ 0 \\ 1 \\ 0 \\ 0 \\ 1 \\ 1 \end{pmatrix}$$

Thus the byte (00010010) gets replaced by (11001001).

5. Apply the transformation

$$\begin{pmatrix} \theta & \theta+1 & 1 & 1 \\ 1 & \theta & \theta+1 & 1 \\ 1 & 1 & \theta & \theta+1 \\ \theta+1 & 1 & 1 & \theta \end{pmatrix} \begin{pmatrix} 1 & \theta & \theta^2 & \theta^3 \\ 1 & \theta^2 & \theta^4 & \theta^6 \\ 1 & \theta^3 & \theta^6 & \theta^9 \\ 1 & \theta^4 & \theta^8 & \theta^{12} \end{pmatrix} \mod \theta^8 + \theta^4 + \theta^3 + \theta + 1$$

$$= \begin{pmatrix} 1 & \theta^4 & \theta^6 + \theta^5 + \theta + 1 & \theta^6 + \theta^3 + \theta^2 + 1 \\ 1 & \theta & \theta^7 + \theta^6 + \theta^5 + \theta^4 + \theta^3 + \theta^2 + \theta + 1 & \theta^6 + \theta^5 + \theta^4 + \theta^3 + 1 \\ 1 & \theta^5 + \theta^2 + \theta & \theta^7 + \theta^5 + \theta^4 + \theta^3 + 1 & \theta^7 + \theta^6 + \theta \\ 1 & \theta^5 + \theta^3 + \theta & \theta^6 + \theta^5 + \theta^3 + \theta & \theta^5 + \theta + 1 \end{pmatrix}$$

7. Let β be a primitive element of GF_{p^n}. Then $u = \beta^k$ for some k. Thus

$$u^{p^n-1} = \left(\beta^k\right)^{p^n=1} = \left(\beta^{p^n-1}\right)^k = 1^k = 1$$

Chapter 12   Polynomial Algorithms and Fast Fourier Transforms

Problems 12.1

1. The content and primitive polynomials are

   (a) $2x^2 - \dfrac{1}{3}x = \dfrac{1}{3}\left(6x^2 - 1\right)$

   (b) $\dfrac{4}{15}x^3 - \dfrac{6}{7}x^2 + 10 = \dfrac{2}{105}\left(14x^3 - 45x^2 + 525\right)$

   (c) $\dfrac{1}{3}x + \dfrac{1}{2} = \dfrac{1}{6}\left(2x + 3\right)$

   (d) $14x^2 - 21x + 35 = 7\left(2x^2 - 3x + 5\right)$

3. The interpolating polynomial is

$$p\left(x\right) = y_0 \frac{x - x_1}{x_0 - x_1} + y_1 \frac{x - x_0}{x_1 - x_0}$$

$$= y_0 - y_0 + y_0 \frac{x - x_1}{x_0 - x_1} + y_1 \frac{x - x_0}{x_1 - x_0}$$

$$= y_0 + \frac{y_1 - y_0}{x_1} \frac{}{x_0}\left(x - x_0\right)$$

5. Let $p\left(x\right) = a + bx + cx^2 + dx^3$. The solution to the system of equations

$$p\left(2\right) = 7,\ p\left(4\right) = 10,\ p\left(7\right) = 23,\ p\left(11\right) = 5$$

   is

$$\left\{ d = -\frac{64}{315}, c = \frac{2021}{630}, b = -\frac{7597}{630}, a = \frac{896}{45} \right\}$$

7. The polynomial is

$$p\left(x\right) = \frac{41}{8}x^4 - \frac{5}{12}x^3 - \frac{165}{8}x^2 + \frac{35}{12}x + 8$$

   See the disc for details of the solution.

9. Substituting r_n for x yields

$$p(r_n) = \sum_{i=0}^{k} d_i \prod_{j \neq i} \frac{r_n - r_j}{r_i - r_j}$$

   If $i \neq n$, then for $j = n$ it follows that $\frac{r_j - r_j}{r_i - r_j} = 0$, so $\prod_{j \neq i} \frac{r_n - r_j}{r_i - r_j} = 0$. Thus

$$p\left(r_n\right) = d_n \prod_{j \neq i} \frac{r_n - r_j}{r_n - r_j} = d_n$$

Problems 12.2

1. If the polynomial $f(x) = x^4 + x^2 + 1$ factors, then one of the factors must have degree at most two. Three evaluation points are required to determine a polynomial of degree 2. Evaluate $f(x)$ at -1, 0, and 1 to get

$$f(-1) = 3, \; f(0) = 1, \; f(1) = 3$$

The possible function values for a polynomial of degree at most 2 that divides $f(x)$ are given by

$$D_0 = \{\pm 1, \pm 3\}, \; D_1 = \{\pm 1\}, \; D_2 = \{\pm 1, \pm 3\}$$

Since $f(x) \geq 1$ for all x, it suffices to choose from among positive values. The choices

$$d_0 = 1, \, d_1 = 1, \, d_2 = 1$$

yield the polynomial

$$(1)\,\frac{(x-0)\,(x-1)}{(-1-0)\,(-1-1)} + (1)\,\frac{(x+1)\,(x-1)}{(0+1)\,(0-1)} + (1)\,\frac{(x+1)\,(x-0)}{(1+1)\,(1-0)} = 1$$

which is a trivial factor. The choices

$$d_0 = 3, \, d_1 = 1, \, d_2 = 3$$

yield the polynomial

$$(3)\,\frac{(x-0)\,(x-1)}{(-1-0)\,(-1-1)} + (1)\,\frac{(x+1)\,(x-1)}{(0+1)\,(0-1)} + (3)\,\frac{(x+1)\,(x-0)}{(1+1)\,(1-0)} = 2x^2 + 1$$

and this is not a factor of $x^4 + x^2 + 1$ (since the leading term would necessarily divide 1). The choices

$$d_0 = 3, \, d_1 = 1, \, d_2 = 1$$

yield the polynomial

$$(3)\,\frac{(x-0)\,(x-1)}{(-1-0)\,(-1-1)} + (1)\,\frac{(x+1)\,(x-1)}{(0+1)\,(0-1)} + (1)\,\frac{(x+1)\,(x-0)}{(1+1)\,(1-0)} = x^2 - x + 1$$

and the choices

$$d_0 = 1, \, d_1 = 1, \, d_2 = 3$$

yield the polynomial

$$(1)\,\frac{(x-0)\,(x-1)}{(-1-0)\,(-1-1)} + (1)\,\frac{(x+1)\,(x-1)}{(0+1)\,(0-1)} + (3)\,\frac{(x+1)\,(x-0)}{(1+1)\,(1-0)} = x^2 + x + 1$$

A quick check shows that indeed

$$\left(x^2 - x + 1\right)\left(x^2 + x + 1\right) = x^4 + x^2 + 1$$

3. If $f(x) = 2x^6 + 2x^5 + 3x^4 + 3x^3 + 3x^2 + x + 1$ factors, then it must have a factor of degree at most 3. This requires four evaluation points. In order to minimize the number of guesses, we choose the evaluation points -3, -1, 0, and 1 to get

$$f(-3) = 1159 = 19 \times 61$$
$$f(-1) = 3$$
$$f(0) = 1$$
$$f(1) - 15 = 3 \times 5$$

Note that $f(-2) = 99 = 3^2 11$ and $f(2) = 279 = 3^2 31$ have more divisors to test. Hence

$$D_0 = \{\pm 1, \pm 19, \pm 61, \pm 1159\}$$
$$D_1 = \{\pm 1, \pm 3\}$$
$$D_2 = \{\pm 1\}$$
$$D_3 = \{\pm 1, \pm 3, \pm 5, \pm 15\}$$

The lucky guesses

$$d_0 = 19,\ d_1 = 3,\ d_2 = 1,\ d_2 = 3$$

yield the polynomial

$$(19)\frac{(x+1)(x-0)(x-1)}{(-3+1)(-3-0)(-3-1)} + (3)\frac{(x+3)(x-0)(x-1)}{(-1+3)(-1-0)(-1-1)}$$
$$+ (1)\frac{(x+3)(x+1)(x-1)}{(0+3)(0+1)(0-1)} + (3)\frac{(x+3)(x+1)(x-0)}{(1+3)(1+1)(1-0)} = 2x^2 + 1$$

and long division produces $\frac{f(x)}{2x^2+1} = x^4 + x^3 + x^2 + x + 1$, so

$$f(x) = (2x^2 + 1)(x^4 + x^3 + x^2 + x + 1)$$

5. If $x^4 + 2x + 2$ factors in $\mathbb{Z}[x]$, then it would also factor in $GF_3[x]$. Let θ be a root of $x^4 + 2x + 2$ in some Galois field containing GF_3. The order of θ is 80 since

$$\left(\theta^{80} \bmod \theta^4 + 2\theta + 2\right) \bmod 3 = 1$$
$$\left(\theta^{80/2} \bmod \theta^4 + 2\theta + 2\right) \bmod 3 = 2$$
$$\left(\theta^{80/5} \bmod \theta^4 + 2\theta + 2\right) \bmod 3 = \theta^3 + 2\theta + 2$$

Thus $x^4 + 2x + 2$ is irreducible in $GF_3[x]$ and hence irreducible in $\mathbb{Z}[x]$.

7. The only possible rational roots are ± 1 and ± 2. Since $f(-2) = 0$, it follows that $x + 2$ is a factor. Knowing that, it is easy to see that

$$x^4 + x^3 + x^2 + x + 1$$

is the other factor, irreducible by problem 4.

9. There are no linear factors because the only possible rational roots are ± 1, and $f(1) = 1$ and $f(-1) = 1$. If $f(x)$ factors, then it must have an irreducible factor of degree 2. Assume that $g(x) = ax^2 + bx + c$ is such a factor. Since $f(-1) = 1$, $f(0) = 1$, and $f(1) = 1$, it follows that $g(-1) \in \{-1, 1\}$, $g(0) \in \{-1, 1\}$, and $g(1) \in \{-1, 1\}$. This yields the 8 possible factors

$$
\begin{array}{lll}
g(x) = 1 & g(x) = 2x^2 - 1 & g(x) = -x^2 - x + 1 \\
g(x) = -x^2 + x + 1 & g(x) = x^2 + x - 1 & g(x) = -2x^2 + 1 \\
g(x) = x^2 - x - 1 & g(x) = -1 &
\end{array}
$$

Trial division by the six polynomials of degree 2 shows that none is a factor, and hence f must be irreducible.

Problems 12.3

1. We have

$$
q(1,1) = \frac{(x - x_0) y_1 - (x - x_1) y_0}{x_1 - x_0}
$$

$$
= y_0 - y_0 + \frac{y_1 x - y_1 x_0 - y_0 x + y_0 x_1}{x_1 - x_0}
$$

$$
= y_0 + \frac{y_1 - y_0}{x_1 - x_0} (x - x_0)
$$

3. Note that

$$
q(1,1) = \frac{(x - (-2))(-5) - (x - (-1))(3)}{-1 - (-2)} = -8x - 13
$$

$$
q(2,1) = \frac{(x - (-1))(1) - (x - 1)(-5)}{1 - (-1)} = 3x - 2
$$

$$
q(3,1) = \frac{(x - 1)(1) - (x - 2)(1)}{2 - (1)} = 1
$$

$$
q(2,2) = \frac{(x - (-2))(3x - 2) - (x - 1)(-8x - 13)}{1 - (-2)}
$$

$$
= \frac{11}{3} x^2 + 3x - \frac{17}{3}
$$

$$
q(3,2) = \frac{(x - (-1))(1) - (x - 2)(3x - 2)}{2 - (-1)}
$$

$$
= 3x - 1 - x^2
$$

$$
q(3,3) = \frac{(x - (-2))(3x - 1 - x^2) - (x - 2)\left(\frac{11}{3} x^2 + 3x - \frac{17}{3}\right)}{2 - (-2)}
$$

$$
= \frac{4}{3} x^2 + \frac{25}{6} x - \frac{7}{6} x^3 - \frac{10}{3}
$$

The graphs are as follows:

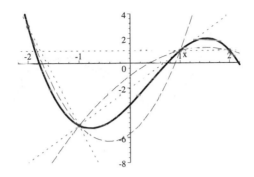

$$p\,(-2) = 3, \ p\,(-1) = -5, \ p\,(1) = 1, \ p\,(2) = 1$$

5. We have

$$q\,(1,1) = \frac{(x-0)\,3 - (x-1)\,2}{1-0} = x + 2$$

$$q\,(2,1) = \frac{(x-1)\,5 - (x-2)\,3}{2-1} = 2x + 1$$

$$p\,(x) = \frac{(x-0)\,(2x+1) - (x-2)\,(x+2)}{2-0} = 4x^2 + 4x + 2$$

Problems 12.4

1. Let $f\,(x) = a + bx + cx^2 + dx^3$ and $g\,(x) = A + Bx + Cx^2 + Dx^3$ and solve the systems

$$
\begin{array}{ccc}
f\,(0) = 0 & & g\,(0) = 5 \\
f\,(1) = 2 & \text{and} & g\,(1) = 2 \\
f\,(2) = 4 & & g\,(2) = 4 \\
f\,(3) = 1 & & g\,(3) = 1
\end{array}
$$

to get

$$\left\{ a = 0, d = -\frac{5}{6}, c = \frac{5}{2}, b = \frac{1}{3} \right\}$$

$$\left\{ A = 5, C = \frac{15}{2}, D = -\frac{5}{3}, B = -\frac{53}{6} \right\}$$

The graphs

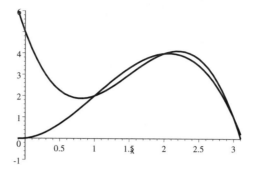

may be distinguished by their y-intercepts.

3. Let $p(1), p(2), \ldots, p(n-1)$ have fixed values and take p_0 to be randomly chosen. Neville's iterated interpolation algorithm generates a unique polynomial $q(x)$ with values

$$q(0) = p_0$$
$$q(1) = p(1)$$
$$\vdots$$
$$q(n-1) = p(n-1)$$

Thus the possible values of $p(0)$ have not been limited at all.

Problems 12.5

1. Let $f(x) = 3(x+1)(x-2)(x-5)$.

    (a) Since $f(x) = 3x^3 - 18x^2 + 9x + 30$, its coefficients are 3, -18, 9, and 30.

    (b) The function values at $x = 0, 1, 2, 3$ are $f(0) = 1$, $f(1) = 15$, $f(2) = 279$, $f(3) = 2299$.

    (c) The values at $x = i, i^2, i^3, i^4$ are $f(i) = -1$, $f(i^2) = 3$, $f(i^3) = -1$, $f(i^4) = 15$.

    (d) Its leading coefficient is 3. It is zero at $x = -1$, 2, and 5.

3. The table is

	Step 1		Step 2	Step 3
y_0	0	0	0	1
y_1	0	0	0	ω
y_2	0	0	0	ω^2
y_3	0	0	0	ω^3
y_4	1	1	1	-1
y_5	0	1	1	$-\omega$
y_6	0	0	1	$-\omega^2$
y_7	0	0	1	$-\omega^3$

This is correct because $-1 = \omega^4$, $-\omega = \omega^5$, $-\omega^2 = \omega^6$, and $-\omega^3 = \omega^7$.

5. We have

Permute

$$
\begin{array}{ll}
x_0 = 3 & 3 \\
x_1 = 6 & 5 \\
x_2 = 4 & 4 \\
x_3 = 0 & 1 \\
x_4 = 5 & 6 \\
x_5 = 0 & 0 \\
x_6 = 1 & 0 \\
x_7 = 8 & 8 \\
\end{array}
$$

Permutation phase

and

Combine		Combine		Combine	
3		8		13		10
5		15		10		9
4		5		3		12
1		3		3		1
6		6		14		16
0		6		8		11
0		8		15		11
8		9		4		5

Combination phase

See the disc for details of the arithmetic. As a check, direct calculation shows that

$$
\begin{array}{llll}
f(1) = 10 & f(2) = 9 & f\left(2^2\right) = 12 & f\left(2^3\right) = 1 \\
f\left(2^4\right) = 16 & f\left(2^5\right) = 11 & f\left(2^6\right) = 11 & f\left(2^7\right) = 5
\end{array}
$$

7. The ij-entry of the product is

$$\sum_{k=0}^{n-1} \left(\omega^{i-1}\right)^k \left(\zeta^{j-1}\right)^k = \sum_{k=0}^{n-1} \left(\omega^{i-1}\zeta^{j-1}\right)^k$$

$$= \sum_{k=0}^{n-1} \left(\omega^{i-1-j+1}\omega^{j-1}\zeta^{j-1}\right)^k$$

$$= \sum_{k=0}^{n-1} \left(\omega^{i-j}\right)^k$$

because $\zeta = \omega^{-1}$. If $i = j$, this is

$$\sum_{k=0}^{n-1} \left(\omega^0\right)^k = \sum_{k=0}^{n-1} 1^k = n.$$

From $\omega^n = 1$ we get

$$\left(\omega^{i-j}\right)^n = (\omega^n)^{i-j} = 1^{i-j} = 1.$$

If $i \neq j$, then $\omega^{i-j} \neq 1$ because ω is a *primitive* nth root of 1. But ω^{i-j} is a root of the polynomial $1 - x^n = (1 - x)\left(1 + x + x^2 + \cdots + x^{n-1}\right)$, so ω^{i-j} is a root of $1 + x + x^2 + \cdots + x^{n-1}$. Thus when $i \neq j$,

$$\sum_{k=0}^{n-1} \left(\omega^{i-j}\right)^k = 0$$

9. Recall that if two rows in a matrix are equal, then the determinant of the matrix is zero. Our idea is to show that the polynomial

$$f(x) = \det \begin{pmatrix} 1 & \alpha_0 & \alpha_0^2 & \cdots & \alpha_0^{n-1} \\ 1 & \alpha_1 & \alpha_1^2 & \cdots & \alpha_1^{n-1} \\ 1 & \alpha_2 & \alpha_2^2 & \cdots & \alpha_2^{n-1} \\ \vdots & \vdots & \vdots & \ddots & \vdots \\ 1 & x & x^2 & \cdots & x^{n-1} \end{pmatrix}$$

is equal to the polynomial

$$g(x) = \left(\prod_{n-1>i>j} (\alpha_i - \alpha_j)\right) \prod_{j=0}^{n-2} (x - \alpha_j)$$

Then $f(\alpha_{n-1}) = g(\alpha_{n-1})$, which is the desired result.

Suppose two of $\alpha_0, \alpha_1, \ldots, \alpha_{n-2}$ are equal. Then clearly g is the zero polynomial. But f is also the zero polynomial because two of the rows in the defining determinant are equal. So we may assume that $\alpha_0, \alpha_1, \ldots, \alpha_{n-2}$

are all different. Both f and g are of degree $n-1$. The leading coefficient of f is equal to the determinant of the matrix obtained by removing the last row and column from the matrix whose determinant is f. By induction, this determinant is equal to $\displaystyle\prod_{n-1>i>j}(\alpha_i - \alpha_j)$, which is the leading coefficient of g. Clearly $g(\alpha_j) = 0$ for $j = 0, 1, \ldots, n-2$, and $f(\alpha_j) = 0$ for $j = 0, 1, \ldots, n-2$ because $f(\alpha_j)$ is the determinant of a matrix with two equal rows. So f and g have the same leading coefficient and the same zeros. Thus they are equal.

11. The left-hand side is equal to

$$\theta\theta^4 - \theta^2\theta^2 + \theta^2 - \theta^4 + \theta^2 - \theta$$

which is $\theta^5 - 2\theta^4 + 2\theta^2 - \theta$. The right-hand side is equal to

$$
\begin{aligned}
\theta\,(\theta-1)^2\,(\theta^2-1) &= (\theta^3 - 2\theta^2 + \theta)\,(\theta^2 - 1)\\
&= \theta^5 - 2\theta^4 + \theta^3 - \theta^3 + 2\theta^2 - \theta\\
&= \theta^5 - 2\theta^4 + 2\theta^2 - \theta
\end{aligned}
$$

Notice that the equation holds for any θ whatsoever, and for any field. But we knew that.

Problems 12.6

1. To find such a prime, look at the numbers $2^5 k + 1$ for $k = 1, 2, 3, \ldots$ and hope you find a prime quickly. To go from k to $k+1$ you simply add 32 to the number you had before. So, starting with $k = 1$, which gives 33 (not a prime), successively adding 32 gives 65 (not a prime), then 97, bingo! Thus $97 = 2^5 3 + 1$ is a prime. That was easy. The next part is a little trickier. As $96 = 32 \cdot 3$, we get $(u^3)^{32} \bmod 97 = 1$ so we can look for cubes b such that $b^{16} \bmod 97 \neq 1$. Computing, $8^{16} \bmod 97 = 1$, $27^{16} \bmod 97 = 1$, $125^{16} \bmod 97 = 96$. Bingo! So $125 \equiv 28 \pmod{97}$ is an element of order 32.

3. Note that $3^{128} \bmod 257 = 256$ and $3^{256} \bmod 257 = 1$ implies that 3 has order 256.

5. Use a computer algebra system to verify that $p = 1 + 2^{30} \cdot 3^7 \cdot 5^8 \cdot 7^3 \cdot 11^2$.

$13 \cdot 17^2 \cdot 19 \cdot 23$ is a prime. Let $q = p - 1$ and observe that

$$29^{q/2} \bmod p = 62\,504\,431\,582\,015\,232\,409\,600\,000\,000$$
$$29^{q/3} \bmod p = 39\,510\,906\,027\,632\,123\,741\,196\,378\,894$$
$$29^{q/5} \bmod p = 20\,739\,905\,493\,567\,049\,891\,681\,480\,849$$
$$29^{q/7} \bmod p = 46\,525\,541\,236\,811\,572\,209\,996\,095\,299$$
$$29^{q/11} \bmod p = 25\,821\,169\,224\,421\,318\,975\,300\,997\,529$$
$$29^{q/13} \bmod p = 9966\,500\,034\,696\,371\,632\,146\,125\,001$$
$$29^{q/17} \bmod p = 27\,534\,573\,559\,640\,223\,122\,138\,911\,033$$
$$29^{q/19} \bmod p = 2797\,174\,485\,912\,474\,274\,185\,207\,652$$
$$29^{q/23} \bmod p = 57\,885\,642\,771\,173\,560\,912\,474\,071\,472$$

and hence 29 is a primitive element of GF_p. Let

$$n = q/2^{30} = 58\,211\,788\,145\,839\,453\,125$$

Then
$$29^n \bmod p = 58\,183\,380\,006\,354\,634\,332\,026\,032\,462$$
is an element of order 2^{30}. To verify this, note that

$$58\,183\,380\,006\,354\,634\,332\,026\,032\,462^{2^{30}} \bmod p = 1$$
$$58\,183\,380\,006\,354\,634\,332\,026\,032\,462^{2^{29}} \bmod p = -1$$

7. To show that the order of ω is $242 = 2 \times 11^2$, it is sufficient to evaluate

$$\omega^{242} \bmod \omega^5 + 2\omega + 1 = 1$$
$$\omega^{242/2} \bmod \omega^5 + 2\omega + 1 = 2$$
$$\omega^{242/11} \bmod \omega^5 + 2\omega + 1 = 2\omega^3 + 2\omega^2 + \omega + 1$$

We have $\omega^{-1} = \omega^{241} \bmod \omega^5 + 2\omega + 1 = 2\omega^4 + 1$. As a check, note that $\omega\left(2\omega^4 + 1\right) \bmod \omega^5 + 2\omega + 1 = 1$. The minimal polynomial of ω^{-1} is

$$\left(x - \left(2\omega^4 + 1\right)\right)\left(x - \left(2\omega^4 + 1\right)^3\right)\left(x - \left(2\omega^4 + 1\right)^9\right)\left(x - \left(2\omega^4 + 1\right)^{27}\right) \cdot$$
$$\left(x - \left(2\omega^4 + 1\right)^{81}\right) \bmod \omega^5 - \omega + 1 = x^5 + 2x^4 + 1$$

9. The polynomial f has $a + 1$ coefficients and the polynomial g has $b + 1$ coefficients. Each of the coefficients of f must be multiplied by each of the coefficients of g, for a total of $(a + 1)(b + 1)$ multiplications.

Now we have to add these $(a + 1)(b + 1)$ products to form $a + b + 1$ sums $(\deg fg + 1)$. To add up m numbers and get n sums, you need $m - n$ additions. For example, to add up 7 numbers to get 1 sum, you need 6 additions. To add up 8 numbers to get 2 sums you need 6 additions: for example, $1 + 2 + 3$ and $4 + 5 + 6 + 7 + 8$. So to add up $(a+1)(b+1)$ numbers to get $a + b + 1$ sums, you need $(a+1)(b+1) - (a+b+1) = ab$ additions.

Bibliography

[1] Adámek, Jiří. *Foundations of Coding*, John Wiley & Sons, Inc., 1991

[2] Akritas, Alkiviadis G. *Elements of Computer Algebra with Applications*, Wiley Interscience, New York, 1989

[3] Albert, A. A. *Fundamental Concepts of Higher Algebra*, University of Chicago Press, Chicago, 1956

[4] Ax, J. "The elementary theory of finite fields," *Annals of Math.* 88 (239–271), 1968

[5] Barr, Thomas H. *Invitation to Cryptology*, Prentice Hall, Upper Saddle River, NJ, 2002

[6] Berlekamp, E. R. *Algebraic Coding Theory*, McGraw-Hill, New York, 1968

[7] Beutelspacher, A. *Cryptology*, The Mathematical Association of America, Washington, DC, 1994

[8] Blahut, R. E. *Theory and Practice of Error Control Codes*, Addison-Wesley, Reading, MA, 1983

[9] Bose, R. C. "On the application of the properties of Galois fields to the construction of hyper-Graeco-Latin squares," *Sankhya* 3 (323–328), 1938

[10] Ellis, Graham. *Rings and Fields*, Oxford, 1992

[11] Galois, E. "Sur la Théorie des Nombres," *Bull. Sci. Math. de M. Ferussac* 13 (428–435), 1830

[12] Garrett, Paul. *Making, Breaking Codes: An Introduction to Cryptology*, Prentice Hall, Upper Saddle River, NJ, 2001

[13] Garrett, Paul. *The Mathematics of Coding Theory*, Prentice Hall, Upper Saddle River, NJ, 2004

[14] Hamming, R. W. *Coding and Information Theory*, Prentice Hall, Englewood Cliffs, NJ, 1986

[15] Harmon, C. K. and R. Adams. *Reading Between the Lines*, Helmers Publishing, Peterborough, NH, 1989

[16] Heisler, J. "A characterization of finite fields," *Amer. Math Monthly* 74 (1211), 1967

[17] Herstein, I. N. *Topics in Algebra*, 2nd ed., Xerox College Publ., Lexington, MA 1975

[18] Jacobson, N. *Lectures in Abstract Algebra*, D. Van Nostrand, Princeton, NJ, 1964

[19] Lewand, Robert E. *Cryptological Mathematics*, The Mathematical Association of America, 2000

[20] Lidl, R. and H. Niederreiter. *Introduction to Finite Fields and Their Applications*, Cambridge University Press, Cambridge, 1986

[21] Mao, Wenbo. *Modern Cryptography*, Prentice Hall PTR, Upper Saddle River, NJ, 2004

[22] Ore, O. "Contributions to the theory of finite fields," *Trans. Amer. Math. Soc.* 36 (243–274), 1934

[23] Pless, V. *Introduction to the Theory of Error Correcting Codes*, Wiley, New York, 1989

[24] Rhee, M. Y. *Cryptography and Secure Communications*, McGraw-Hill, Singapore, 1994

[25] Rosen, K. H. *Elementary Number Theory and Its Applications*, Addison-Wesley, Reading, MA, 1984

[26] Schneier, B. *Applied Cryptography*, Wiley, New York, 1994

[27] Trappe, Wade and Lawrence C. Washington. *Introduction to Cryptography with Coding Theory*, Prentice Hall, Upper Saddle River, NJ 2002

Notation

Absolute value: $\quad |x| = \sqrt{x^2} = \begin{cases} x & \text{if} \quad x \geq 0 \\ -x & \text{if} \quad x < 0 \end{cases}$

Addition modulo m: $\quad a \oplus b = (a+b) \bmod m$

Base 2 logarithm: $\quad \log_2$

Base 2 representation: $\quad (b_n b_{n-1} \ldots b_2 b_1 b_0)_2 = \sum_{i=0}^{n} b_i 2^i$

Base b representation: $\quad (a_n a_{n-1} \ldots a_2 a_1 a_0)_b = \sum_{i=0}^{n} a_i b^i$

Binomial coefficient: $\quad \dbinom{n}{k}$

Bitwise and: $\quad x \wedge y$

Bitwise complement: $\quad \bar{x}$

Bitwise or: $\quad x \vee y$

Bitwise xor: $\quad x \oplus y$

Ceiling function: $\quad \lceil x \rceil$

Chinese remainder function: $\quad CRA(\mathbf{a}, \mathbf{m}, r)$

Congruence class modulo p:

$\qquad [a] = \{\ldots, a-3p, a-2p, a-p, a, a+p, a+2p, a+3p, \ldots\}$

Congruent modulo m: $\quad a \equiv b \pmod{m}$

Congruent modulo $m(x)$: $\quad a(x) \equiv b(x) \pmod{m(x)}$

Cyclic subgroup generated by a: $\quad \langle a \rangle$

Degree of polynomial $p(x)$: $\quad \deg p(x)$

Divides: $\quad n \mid m$

Euler phi function: $\quad \varphi(n)$

Factorial: $\quad n!$

Field of complex numbers: $\quad \mathbb{C}$

Floor function: $\quad \lfloor x \rfloor$

Galois field of order p: $\quad GF_p$

Galois field of order p^n: $\quad GF_{p^n}$

Greatest common divisor of integers: $\quad \gcd(a, b)$

Hamming distance: $\quad d(\mathbf{u}, \mathbf{v})$

397

Integers: $\quad \mathbb{Z} = \{\ldots, -3, -2, -1, 0, 1, 2, 3, \ldots\}$

Integers a and b are relatively prime: $\quad a \perp b$

Integers modulo n: $\quad \mathbb{Z}_n = \{0, 1, 2, \ldots, n - 1\}$

Inverse modulo m: $\quad a^{-1} \bmod m$

Least common multiple of integers: $\quad \text{lcm}(a, b)$

Least common multiple or polynomials: $\quad \text{lcm}(a(x), b(x))$

Logarithm base 2 $\quad \log_2$

Matrix inverse modulo m: $\quad \begin{pmatrix} a & b \\ c & d \end{pmatrix}^{-1} \bmod m$

Minimal polynomial of a field element α: $\quad m_\alpha(x)$

Mod function: $\quad n \bmod m = n - \lfloor n/m \rfloor m$

Multiplication modulo m: $\quad a \otimes b = ab \bmod m$

Multiplicative group of GF_{p^n}: $\quad (GF_{p^n})^*$

Next prime larger than m: $\quad \text{nextprime}(m)$

Nonnegative integers: $\quad \mathbb{N} = \{0, 1, 2, 3, \ldots\}$

Order of group G: $\quad o(G)$

Permutation as images: $\quad \sigma = \begin{pmatrix} 1 & 2 & 3 & 4 & 5 & 6 & 7 & 8 & 9 \\ 5 & 8 & 2 & 6 & 9 & 4 & 1 & 3 & 7 \end{pmatrix}$

Permutation as disjoint cycles: $\quad \sigma = (1597)(283)(46)$

Permutation σ acting on x: $\quad x^\sigma$

Polynomial remainder: $\quad r(x) = a(x) \bmod m(x)$

Polynomials over the field F: $\quad F[x]$

Polynomials over the Galois field of order p: $\quad GF_p[x]$

Polynomials with integer coefficients: $\quad \mathbb{Z}[x]$

Positive integers: $\quad \mathbb{P} = \{1, 2, 3, 4, \ldots\}$

Product: $\quad \prod a_n$

Rational numbers: $\quad \mathbb{Q}$

Real numbers: $\quad \mathbb{R}$

Secure hash algorithm: $\quad \text{SHA}(M)$

Shift left: $\quad x << n$

Shift right: $\quad x >> n$

Sum: $\quad \sum a_n$

Algorithms

Figures

Tables

403

Index